AIR POLLUTION EFFECTS ON BIODIVERSITY

AIR POLLUTION EFFECTS ON BIODIVERSITY

Edited by

Jerry R. Barker
ManTech Environmental Technology, Inc.
U.S. EPA Environmental Research Laboratory—Corvallis

and

David T. Tingey
U.S. Environmental Protection Agency
Environmental Research Laboratory—Corvallis

VAN NOSTRAND REINHOLD
New York

Library of Congress Catalog Card Number 91-47171
ISBN 0-442-00748-5

This book is based on a workshop that evaluated the effects of air pollution on biodiversity. The workshop was sponsored by the U.S. Environmental Protection Agency Environmental Research Laboratory—Corvallis, the U.S. Fish and Wildlife Service National Fisheries Research Center—Leetown, and the Electric Power Research Institute. Although the individual chapters have been peer reviewed in accordance with EPA's policy and approved for publication, they do not necessarily reflect the views and policies of the sponsors, and no official endorsement should be inferred. The opinions expressed within the individual chapters reflect the views of the authors, and mention of trade names or commercial products does not constitute endorsement by the sponsors or recommendation for use.

Printed in the United States of America.

Van Nostrand Reinhold
115 Fifth Avenue
New York, New York 10003

Chapman and Hall
2-6 Boundary Row
London, SE1 8HN, England

Thomas Nelson Australia
102 Dodds Street
South Melbourne 3205
Victoria, Australia

Nelson Canada
1120 Birchmount Road
Scarborough, Ontario MIK 5G4, Canada

16 15 14 13 12 11 10 9 8 7 6 5 4 3 2 1

Library of Congress Cataloging-in-Publication Data

Air pollution effects on biodiversity/edited by Jerry R. Barker and David T. Tingey.
p. cm.
Includes index.
ISBN 0-442-00748-5
1. Air—Pollution—Environmental aspects—Congresses. 2. Biological diversity conservation—Congresses. I. Barker, Jerry R. II. Tingey, David T.
QH545.A3A38 1992
574.5'222dc20

91-47171
CIP

Contents

Contributors

Dr. Jerry R. Barker, Editor
ManTech Environmental Technology, Inc.
US EPA Environmental Research Laboratory
Corvallis, Oregon, USA

Dr. David T. Tingey, Editor
US Environmental Protection Agency
Environmental Research Laboratory
Corvallis, Oregon, USA

Dr. Thomas V. Armentano
South Florida Research Center
Everglades National Park
Homestead, Florida, USA

Dr. James P. Bennett
US National Park Service
Cooperative Research Unit
University of Wisconsin
Madison, Wisconsin, USA

Dr. Jeremy J. Colls
Department of Physiology and Environmental Sciences
Faculty of Agricultural and Food Sciences
University of Nottingham
Sutton Bonington, England

Dr. Roger M. Cox
Forestry Canada—Maritimes Region
Fredericton, New Brunswick, Canada

Dr. David Fowler
Institute of Terrestrial Ecology
Bush Estate, Penicuik
Midlothian, Scotland

Dr. Douglas G. Fox
US Department of Agriculture
Forest Service
Rocky Mountain Forest and Range Experiment Station
Fort Collins, Colorado, USA

Ms. Sandra Henderson
ManTech Environmental Technology, Inc.
US EPA Environmental Research Laboratory
Corvallis, Oregon, USA

Dr. Orie L. Loucks
Department of Zoology
Miami University
Oxford, Ohio, USA

Dr. Peter W. Lucas
Institute of Environmental and Biological Sciences
University of Lancaster
Lancaster, England

Prof. Terry A. Mansfield
Institute of Environmental and Biological Sciences
University of Lancaster
Lancaster, England

Dr. Robert McKelvey
University of Montana
US EPA Environmental Research Laboratory
Corvallis, Oregon, USA

Dr. William H. Moir
US Department of Agriculture
Forest Service
Rocky Mountain Forest and Range Experiment Station
Fort Collins, Colorado, USA

Dr. Robert C. Musselman
US Department of Agriculture
Forest Service
Rocky Mountain Forest and Range Experiment Station
Fort Collins, Colorado, USA

Dr. Jeffrey C. Nekola
Curriculum in Ecology
University of North Carolina
Chapel Hill, North Carolina, USA

Dr. James R. Newman
KBN Engineering and Applied
Sciences, Inc.
Gainesville, Florida, USA

RNDr. Ing. Eliska Novakova, Dr.Sc.
Institute of Applied Ecology
Kostelec n. Cer. lesy
Czechoslovakia

Dr. Louis F. Pitelka
Biological Studies Program
Electric Power Research Institute
Palo Alto, California, USA

Dr. Paul G. Risser
University of New Mexico
Albuquerque, New Mexico, USA

Dr. R. Kent Schreiber
US Fish and Wildlife Service
National Fisheries Research Center
Kearneysville, West Virginia, USA

Dr. Charles G. Shaw III
US Department of Agriculture
Forest Service
Rocky Mountain Forest and Range
Experiment Station
Fort Collins, Colorado, USA

Dr. William H. Smith
School of Forestry and
Environmental Studies
Yale University
New Haven, Connecticut, USA

Dr. George E. Taylor, Jr.
Biological Sciences Center
Desert Research Institute
Reno, Nevada, USA

Prof. Michael H. Unsworth
Department of Physiology and
Environmental Sciences
Faculty of Agricultural and Food Sciences
University of Nottingham
Sutton Bonington, England

Dr. Peter S. White
Department of Biology and North Carolina
Botanical Garden
University of North Carolina
Chapel Hill, North Carolina, USA

Dr. Jenny Wolfenden
Institute of Environmental and
Biological Sciences
University of Lancaster
Lancaster, England

Dr. Philip A. Wookey
Institute of Terrestrial Ecology
Grange-over-Sands
Cumbtia, England

Acknowledgments

The editors are indebted to the distinguished contributors of this volume for sharing of their knowledge and expertise. The efforts of Technical Resources, Inc., for providing technical support throughout the project are also greatly appreciated. The following scientists provided peer reviews of one or more of the chapters:

Chris Anderson
Tom Armentano
Spencer Barrett
J.N.B. Bell
James Bennett
Jan Bonga
Ron Bradow
Anthony Bradshaw
Bruce Burry
Jeremy Colls
Roger Cox
J.H.B. Garner
Chad Gubala
Sandra Henderson
Ross Kiester
Judy Loo-Dinkins
Tonnie Maniero
Terry Mansfield
Rich Meganck
Paul Miller
Ron Neilson
James Newman
Kevin Percy
David Peterson
Don Phillips
Lou Pitelka
John Pye
Kent Schreiber
Lena Skärby
George Taylor
Kathy Tonnessen
Rick Vong
Jim Weber
Bill Winner

Preface

The continuing loss of plant, animal, and microbial species around the world is a concern to scientists and conservationists. Indeed, biodiversity has emerged as a worldwide policy issue with organizations such as the United Nations Environmental Programme, the United Nations Educational, Scientific and Cultural Organization, the United States Committee for Man and the Biosphere Program, the International Union of Biological Sciences, the U.S. Council on Environmental Quality, and the Society of Conservation Biology taking the lead. The concern is that the species extinction rate is currently much higher now than at any time in the past. The loss of plant, animal, and microbial species is due to anthropogenic stress resulting from a growing human population and unwise natural resource management practices. Habitat fragmentation and destruction, plant and animal overexploitation, competition with exotic species, and the discharge of toxic chemicals into the air, water, and soil are the main anthropogenic activities causing the decline in worldwide biodiversity.

The effects of habitat fragmentation and destruction, species overexploitation, and exotic species on ecosystems are well documented in the scientific literature. However, the effects of air pollution on biota and ecological processes are not very well understood. Emphasis in past research has been on evaluating the effects of the dominant pollutants (e.g., ozone, sulfur dioxide, nitrogen oxides) on crop production and to some extent tree growth and development. This research has provided some information in evaluating the impacts of air pollution on natural vegetation. In comparison, the database for evaluating the effects of air pollution on other biota (e.g., animals, fungi) is relatively small. Recent research on the role of air pollution in forest-tree decline shows that toxic airborne contaminants may directly or indirectly affect ecological processes, resulting in decreased growth and perhaps mortality. A documented example is the changes that occurred within the forests of the San Bernardino Mountains of southern California in response to ozone. Vegetal dominance shifted from the ozone-sensitive ponderosa and Jeffrey pines to the ozone-tolerant shrub and oak species. The demise of the pine species resulted from such impacts as foliar injury and premature needle fall, increased susceptibility to root rot infections and pine beetle attacks, and changes in nutrient cycling and carbon allocation. Undoubtedly, the shift in vegetal composition resulted in changes in wildlife habitat quality, increased soil erosion, and effects on human welfare in recreational and commercial value.

To address the issue of air pollution impacts on biodiversity, the U.S. Environmental Protection Agency Environmental Research Laboratory—Corvallis, the U.S. Fish and Wildlife Service National Fisheries Research Center—Leetown, and the Electric Power Research Institute convened a workshop to evaluate current knowledge, identify information gaps, provide direction to research, and assess policy issues. In order to obtain the most current and authoritative information possible, air pollution and biodiversity experts were invited to author the chapters of the book and participate in a workshop. At the workshop, the chapters were presented and discussed for scientific merit and relationship with the other chapters. The authors used the workshop discussions to prepare the final manuscripts. Thus, the book represents the current thinking of the workshop participants and authors on the effects of air pollution on biodiversity.

The book has been organized into four parts that best represent the subject matter of the chapters:

- Introduction
- Overview of Air Pollution Exposure and Effects
- Consequences of Air Pollution Exposure on Biodiversity
- Policy Issues and Research Needs

The editors hope that this volume will provide a framework for a comprehensive evaluation of air pollution impacts on biodiversity and provide direction for research and policy making. An understanding of these complex issues is essential for wise air quality and biodiversity management.

Jerry R. Barker
David T. Tingey

Part I

Introduction

1

The Effects of Air Pollution on Biodiversity: A Synopsis

Jerry R. Barker and David T. Tingey

INTRODUCTION

Thousands of chemicals are commonly used throughout the world for industrial, agricultural, and domestic purposes with many new ones being produced yearly (Maugh 1991; Schroeder and Lane 1988). The majority of these chemicals, many of which are toxic or radiatively active, eventually enter into the atmosphere and may pose a risk to the well-being of plants, animals, and microorganisms. The consequences of air pollution to biota and the resulting impacts on biodiversity are not clearly known; only fragmentary information is available. The purpose of this book is to evaluate what is known, identify information gaps, explore policy issues, and provide direction for research.

AIR POLLUTION

Air pollution is the presence in the atmosphere of one or more contaminants in quantities and durations such as to be injurious to humans, animals, plants, property, or to impair the enjoyment of life and property. This broad definition covers an array of contaminants including dust, smoke, odors, fumes, mists, and gases. Historically, the primary concern with airborne pollutants has been human health effects in industrial-urban areas. However, ecological research strongly suggests that air pollutants, many of which are distributed worldwide, can also adversely affect biota and ecosystems (Moser, Barker, and Tingey 1991; Schreiber and Newman 1988; Woodwell 1970) which in turn may impact human health and welfare through contaminated food and water and the loss of ecosystem services and products.

A good example of air pollution effects on biodiversity is the change in forest vegetation of the San Bernardino Mountains of southern California (Miller, Taylor, and Wilson 1982). Prolonged exposure of the vegetation to photochemical oxidants resulted in a shift in vegetation dominance from the ozone-sensitive pines to the ozone-tolerant oaks and deciduous shrubs. The main causes for the decline of the pine trees were decreased photosynthetic capacity due to foliar injury and premature needle fall, suppressed radial growth of the stems, and reduced nutrient retention in the needles. Indirect effects on the pine trees included increased susceptibility to insect attacks and fire. Other processes that were affected in the forest ecosystem were changes in water flow, loss of nutrients, changes in carbon allocation, and altered patterns of spatial and temporal diversity. All of these changes resulted in reduced commodities and amenities of anthropogenic value such as commercial forestry and recreation. Continued research and monitoring by Miller et al. (1989) has documented that the improved air quality of the last few years corresponds with increased pine tree growth and vigor.

Realizing the importance of air pollution effects on biota and ecosystems, Tingey, Hogsett, and Henderson (1990) argued for greater emphasis on the consideration of both known and potential ecological impacts when developing ambient air quality standards. Airborne chemicals are emitted into the atmosphere from an array of point, area, and mobile sources such as chemical, metal, plastic, and paper/pulp industries; energy processing plants; motor vehicles and aircraft; municipal waste incinerators; agricultural practices such as pesticide usage and field burning; and from household heating and cooking (Moser, Barkey, and Tingey 1992). The emission of these contaminants into the atmosphere may occur directly through deliberate or inadvertent releases or indirectly through volatilization following deliberate or accidental discharge into water or soil.

Once the chemicals enter into the atmosphere, they are subjected to physical, chemical, or photochemical processes that determine their ultimate environmental exposure and fate (Schroeder and Lane 1988). The compounds are mixed and transported, with all but the most chemically stable (e.g., chlorofluorohydrocarbons) being transformed into new products. These reactions may result in the formation of compounds such as ozone or peroxacetyl nitrate (PAN) that are more toxic than their precursors. The prevailing meteorological conditions and the physicochemical properties of the contaminants will dictate the atmospheric residence times and deposition velocities to the ecosystem receptors—biota, soil, and water.

The potential ecological impact of air pollution is determined by the contaminant's environmental partitioning, exposure pattern, toxicity, and species sensitivity (Weinstein and Birk 1989). For example, trace metals tend to accumulate in humus and organic matter and may reduce plant growth and vigor through the disruption of nutrient uptake by the roots and decrease organic matter decomposition (Urban 1991). Gaseous pollutants partition to the atmosphere with the potential to disrupt plant-leaf biochemical processes after absorption through the stomata or cuticle, with subsequent impacts on plant growth and reproduction (Foster 1991). Transfer of toxic chemicals among ecosystem compartments may occur through processes such as the volatilization from water, soil, or vegetation surfaces, the leaching of soils, and the decay of plant and animal tissues.

BIODIVERSITY

Diversity is characteristic of biological systems at all levels of organization (Noss 1990; Solbrig 1991a, b). Natural variation occurs among genes, species, populations, communities, ecosystems, and landscapes. Diversity originates at the genetic level, and is extended to the higher levels of organization through populations. New diversity within a population arises from genetic processes such as gene mutation or recombination and immigration of individuals. Natural selection and emigration are responsible for removal of diversity from populations. Thus, diversity at any one time is a function of both genetic and natural selection processes (Solbrig 1991b).

The purposes of biological diversity are not completely understood (Solbrig 1991a, b). Diversity is believed to provide genes, populations, communities, and ecosystems with the ability to adapt and respond to environmental uncertainty. Biological systems are constantly changing in response to environmental stimuli. One consequence of biological fluctuation is the inadvertent loss of genes and species, i.e., natural background extinction. Many of the environmental changes are transient or cyclic and to some extent predictable, such as recurring fire in many forest and grassland ecosystems. Other changes are directional and have long-lasting effects such as climate change. All change, regardless of its severity and whether it is cyclic or directional, introduces a degree of randomness into biological systems. For example, prolonged drought may favor those genes within plant and animal populations that confer drought tolerance to individuals at the expense of nontolerant genes.

There is growing concern among scientists, natural-resource managers, and conservationists regarding the loss of species, habitats, ecosystems, and the genetic diversity of agricultural crops (Wilson 1988, 1989). Scientific investigation suggests that biodiversity may be declining throughout the world in response to anthropogenic stressors such as habitat modification or the release of toxic chemicals into air, water, or soil (Barker et al. 1991; Reid and Miller 1989). The present species extinction rate is estimated to be higher than at any previous time including the massive extinctions of the Ordovician, Devonian, Permian, Triassic, and Cretaceous periods (Wilson 1989). The issue of species loss has received the greatest attention. However, the loss of genes and habitats are also important (Plucknett 1987; Ehrlick and Mooney 1983) and intimately linked with species extinction. The deforestation of tropical forests may eliminate between 5 and 15 percent of the world's species by the year 2020, a loss of 15,000 to 90,000 species per year (Reid and Miller 1989). Also, the loss of native plant species or unique gene complexes may reduce the options for plant breeders to develop new crop varieties to sustain a growing human population (Plucknett 1987). Associated with all types of biodiversity losses are reduced options for biota and humans to an ever-changing environment including disruption of ecological processes, services, and products (Abramovitz 1991; Barker et al. 1991).

Through the millennia of time, species have acquired attributes through genetic mutation and natural selection that allow adaptation to environmental change (Solbrig 1991b). Even so, rapid or drastic environmental changes resulting from catastrophic events such as volcanic eruptions have resulted in species extinctions during past geological periods because the biota were not adapted to the new environment (Signor 1990). In many ways, anthropogenic activities such as deforestation are similar to natural catastrophic events because habitats are modified or destroyed and the biota are exposed to different environmental conditions in which they may not be adapted. The end result is local species extinction. For example, the draining of a wetland over several months leaves the biota faced with much drier conditions. Those species that are not adapted to the new environment will perish. Natural successional change from a wetland to a grassland would have occurred over decades with a replacement of biota adapted to the drier conditions.

The most important threats to biodiversity are habitat fragmentation and destruction, species overexploitation, competition with exotic species, and toxic chemicals released into air, water, and soil (Barker et al. 1991; Reid and Miller 1989). Habitat fragmentation and destruction are currently viewed as the leading threat to biodiversity and, therefore, justifiably require immediate attention (Wilson 1988, 1989). However, the release of toxic chemicals into air, water, and soil could pose an immediate but little understood threat to biodiversity (Reid and Miller 1989; Solbrig 1991a, b). The impact of toxic chemicals on biota will most likely increase with continued industrial growth, especially in developing countries (Speth 1988).

Air pollution has the potential to adversely impact biodiversity on local, regional, and global scales. Local plume effects resulting from emissions of sulfur dioxide, hydrogen fluoride, trace metals, and other toxics from pulp and paper mills, ore smelters, chemical manufacturers, power plants, etc., can reduce vegetation cover, biodiversity, and ecosystem integrity downwind from point sources (Gordon and Gorham 1963). Airborne pollutants may also impact biota on a regional scale through exposure to ozone, industrial chemicals, acid deposition, and/or excess nutrient deposition and indirectly through the interactions with natural stressors such as drought, nutrient deficiencies, and insect infestations (Smith 1990). Worldwide biodiversity is at risk from the global distribution of some toxic chemicals (e.g., DDT, dioxins, polychlorinated biphenyls, mercury) and greenhouse/stratospheric ozone depletion gases (e.g., carbon dioxide, methane, chlorofluorocarbons), which may induce global warming and exposure to elevated levels of ultraviolet-B radiation (Warrick and Jäger 1986; Worrest and Caldwell 1986).

The purpose of this book is to evaluate the effects of air pollution on biodiversity. Air pollution may impact biodiversity if it: (1) alters genetic diversity within populations; (2) reduces the reproductive potential of biota; (3) reduces crop or natural vegetation production; and (4) impairs the structure and function of ecosystems. Air pollution effects on biodiversity are difficult to document. Unlike habitat destruction, which results in a pronounced and rapid environmental change, the effects of air pollution on biota many times are subtle and elusive because of their interactions with natural stressors. Years may be required before ecological change/damage within ecosystems resulting from continuous or episodic exposure to toxic airborne contaminants or the effects of global climate change become evident.

ACCEPTABLE LEVEL OF RISK

The effects of air pollution on biodiversity can be scientifically assessed through the risk assessment process (Cohrssen and Covello 1989). The key to the assessment is an estimation of the probability, extent, and magnitude that the hazard may pose to the environment. The risk that air pollution will impact biodiversity is a function of exposure pattern, chemical toxicity, and species sensitivity. Scientific research often validates that airborne contaminants can adversely affect plants, animals, or microorganisms, and establishes the probability that damage (e.g., reduced or abnormal growth, reduced reproduction, mutation) will occur.

A complicating and important factor in protecting the environment is defining the level of adversity that is acceptable to society (Jacobson 1981). This, however, is not the role of the risk assessment. The level of acceptable adversity can only be defined by society. Adversity can be viewed in an air pollution-biodiversity context as the willingness of society to accept the loss of a certain number of species, the reduced growth and reproduction of vegetation, or the reduction of habitat quality.

The definition and level of adversity can only be achieved once societal values are defined (Jacobson 1981). People perceive adverse effects differently depending on their interests, perceptions, and values (Randall 1988). What is of value to one group may be of lesser value to another or not be of value at all to a third. Therefore, value is the crux in determining the limits of acceptable adverse effects. Negotiations among the interest groups become necessary to achieve the policy that will be the most beneficial to society.

The risk assessment process helps find a balance among competing or varying values by providing scientific information to the negotiating interest groups.

The interplay among science, society, and economics is crucial to formulating an acceptable level of adversity for air quality management (Tingey, Hogsett, and Henderson 1990). Scientific data are critical to the solution because research establishes the validity of cause and effect. Without the supporting data, the issue is not based on scientific validity but rather is a concern based on public perception. The role of science is to identify and explain the potential adverse effects of air pollution to society. Policy makers (the representatives of society) then use scientifically defensible information to establish air quality standards that will limit the consequences of the effects. The public in turn influences the policy makers through the established political process and the willingness to pay for an acceptable level of air quality.

For example, science has shown that air pollution can reduce crop quality and yield, resulting in a loss of billions of dollars annually (Adams, Glyer, and McCarl 1988). Air quality standards mandated by law and achieved through risk reduction such as pollution control technology can improve air quality, resulting in improved crop quality and yield. The level of acceptable adversity (i.e., effects on crop production) then is defined by society based on public perception.

The final decision to protect biodiversity from the effects of air pollution rests with society (Jacobson 1981; Randall 1988; Tingey, Hogsett, and Henderson 1990). However, society is composed of people with differing values, traditions, educational levels, and ethnic backgrounds, which at times may conflict. Certain groups of people are willing to protect biodiversity from the effects of air pollution, while others are more concerned with immediate economic growth at the expense of conserving natural resources. Therefore, public participation is critical to defining an acceptable level of air pollution effects on biodiversity. Science has a critical role not only to establish cause and effect through research, but to educate society about its findings. Only when biodiversity is viewed by society to be of value will action be taken to reduce or prevent adverse effects by air pollution.

FRAMEWORK

The intent of this effort is to evaluate in a detailed and critical manner the literature dealing with air pollution effects on biodiversity. The information and databases necessary to carry out risk assessments and cost-benefit analyses are not available. This volume will begin to fill this void by identifying information gaps, evaluating policy issues, and providing direction to needed research.

The chapters of this book were prepared by biodiversity and air pollution scientists who were invited to participate in a workshop (5-7 February 1991, at Otter Rock, Oregon, USA) sponsored by the U.S. Environmental Protection Agency Environmental Research Laboratory—Corvallis, U.S. Fish and Wildlife Service National Fisheries Research Center—Leetown, and Electric Power Research Institute to discuss the effects of air pollution on biodiversity. The authors presented their chapters to the workshop participants for evaluation of scientific merit and relevance to the book. The authors then used the workshop critiques to finalize the chapters. The revised chapters were peer reviewed by three scientific experts in the particular subject matter. Thus, the book reflects the current thinking of the chapter authors and the other workshop participants.

The book is divided into four parts that best represent the contents of the chapters: (1) Introduction; (2) Overview of Air Pollution Exposure and Effects; (3) Consequences of Air Pollution Exposure on Biodiversity; and (4) Policy Issues and Research Needs. The authors' major premise is that a sound understanding of air pollution exposure and effects on biodiversity is essential for wise management of these valuable natural resources.

REFERENCES

1. Abramovitz, J.N. 1991. Biodiversity: Inheritance from the past, investment in the future. *Environ. Sci. & Technol.* 25:1817-18.
2. Adams, R.A., J.D. Glyer, and B.A. McCarl. 1988. The NCLAN assessment: Approach, findings, and implications. In *Assessment of crop loss from air pollutants*, ed. W.H. Heck, O.C. Taylor, and D.T. Tingey. London: Elsevier Applied Science.
3. Barker, J.R., S. Henderson, R.F. Noss, and D.T. Tingey. 1991. Biodiversity and human impacts. In *Encyclopedia of earth system science*, ed. W.A. Nierenberg. San Diego: Academic Press, Inc.
4. Cohrssen, J.J., and V.T. Covello. 1989. *Risk analysis: A guide to principles and methods for analyzing health and environmental risks.* Springfield, VA: National Technical Information Service.
5. Ehrlick, P.R., and H.A. Mooney. 1983. Extinction, substitution, and ecosystem services. *Bioscience* 33:251-52.
6. Foster, J.R. 1991. Effects of organic chemicals in the atmosphere on terrestrial plants. In *Ecological exposure and effects of airborne toxic chemicals: An overview*, ed. T.J. Moser, J.R. Barker, and D.T. Tingey. US EPA Environmental Research Laboratory—Corvallis: EPA Report No. 600/3-91/001.
7. Gordon, R.J., and E. Gorham. 1963. Ecological aspects of air pollution from an iron-sintering plant at Wawa, Ontario. *Can. J. Bot.* 41:1063-78.
8. Jacobson, J.S. 1981. Acid rain and environmental policy. *J. Air Pollut. Control Assoc.* 31:1071-73.
9. Maugh, T.H. II 1991. Chemicals: How many are there? *Science* 199:162.
10. Miller, P.R., O.C. Taylor, and R.G. Wilson. 1982. *Oxidant air pollution effects on a western coniferous forest ecosystem.* US EPA Report No. 600/D-82-276.
11. Miller, P.R., J.R. McBride, S.L. Schilling, and A.P. Gomez. 1989. Trend of ozone damage to conifer forests between 1974 and 1988 in the San Bernardino Mountains of southern California. In *Effects of air pollution on western forests*, ed. R.K. Olson and A.S. Lefohn. Pittsburgh: Air and Waste Management Association.
12. Moser, T.J., J.R. Barker, and D.T. Tingey, eds. 1991. *Ecological exposure and effects of airborne toxic chemicals: An overview.* US EPA Environmental Research Laboratory—Corvallis: EPA Report No. 600/3-91/001.
13. Moser, T.J., J.R. Barker, and D.T. Tingey. 1992. Atmospheric transport, deposition, and potential effects on terrestrial ecosystems. In *The science of global change,* ed. D.A. Dunnette and R.J. O'Brien. ACS Symposium Series 483, pp. 134-48. Washington, DC: American Chemical Society.
14. Noss, R.F. 1990. Indicators for monitoring biodiversity: A hierarchical model. *Conserv. Biol.* 4: 355-64.
15. Plucknett, D.L. 1987. *Gene banks and the world's food.* Princeton: Princeton Univ. Press.
16. Randall, A. 1988. What mainstream economists have to say about the value of biodiversity. In *Biodiversity*, ed. E.O. Wilson and F.M. Peter. Washington, DC: National Academy Press.

17. Reid, W.V., and K.R. Miller. 1989. *Keeping options alive: The scientific basis for conserving biodiversity*. Washington, DC: World Resources Institute.
18. Schreiber, R.K., and J.R. Newman. 1988. Acid precipitation effects on forest habitats: Implications for wildlife. *Conserv. Biol.* 2:249-59.
19. Schroeder, W.H., and D.A. Lane. 1988. The fate of toxic airborne pollutants. *Environ. Sci. & Technol.* 22:240-46.
20. Signor, P.W. 1990. The geological history of diversity. *Annu. Rev. Ecol. System.* 21:509-39.
21. Smith, W.H. 1990. *Air pollution and forests*. New York: Springer-Verlag.
22. Solbrig, O.T. 1991a. *Biodiversity*. MAB Digest 9. Paris: United Nations Educational, Scientific, and Cultural Organization.
23. Solbrig, O.T. 1991b. The origin and function of biodiversity. *Environment* 33:17-38.
24. Speth, J.G. 1988. Environmental pollution. In *Earth '88: Changing geographic perspectives*, ed. H.J. De Blij. Washington, DC: National Geographic Society.
25. Tingey, D.T., W.E. Hogsett, and S. Henderson. 1990. Definition of adverse effects for the purpose of establishing secondary national ambient air quality standards. *J. Environ. Qual.* 19:635-39.
26. Urban, N.R. 1991. Effects of atmospheric pollutants on peatlands. In *Ecological exposure and effects of airborne toxic chemicals: An overview*, ed. T.J. Moser, J.R. Barker, and D.T. Tingey. US EPA Environmental Research Laboratory—Corvallis: EPA Report No. 600/3-91/001.
27. Warrick, B., and J. Jäger, eds. 1986. *The Greenhouse Effect, climate change, and ecosystems*. New York: John Wiley & Sons.
28. Weinstein, D.A., and E.M. Birk. 1989. The effects of chemicals on the structure of terrestrial ecosystems: Mechanisms and patterns of change. In *Ecotoxicology: Problems and approaches*, ed. S.A. Levin, M.A. Harwell, J.R. Kelly, and K.D. Kimball. New York: Springer-Verlag.
29. Wilson, E.O. 1988. The current status of biological diversity. In *Biodiversity*, ed. E.O. Wilson and F.M. Peter. Washington, DC: National Academy Press.
30. Wilson, E.O. 1989. Threats to biodiversity. *Sci. Am.* 261:108-17.
31. Woodwell, G.M. 1970. Effects of pollution on the structure and physiology of ecosystems. *Science* 168:429-33.
32. Worrest, R.C., and M.M. Caldwell, eds. 1986. *Stratospheric ozone reduction, solar ultraviolet radiation and plant life*. Berlin: Springer-Verlag.

2

Biological Diversity in an Ecological Context

Peter S. White and Jeffrey C. Nekola

INTRODUCTION

The most common definitions of biological diversity focus on state variables, such as genes, species, and communities, but processes, such as gene flow, survivorship, competition, and energy flow, ultimately determine the nature of these state variables and are critical to the survival of biological diversity itself (Noss 1990). The relationship between biological diversity, ecological process, and human activities is now a critical concern for scientists and policy makers (Lubchenco et al. 1991).

In this chapter, we will discuss the history of the biological diversity issue, elaborate more fully on a definition, and briefly describe the value of and threats to diversity. We will then address the general linkage between ecological processes and biological diversity. Because detecting change in diversity is critical to understanding human effects, we will turn to a discussion of major issues in the measurement of biological diversity.

THE BIOLOGICAL DIVERSITY ISSUE

A Historical Perspective

Although conservationists have been concerned with the survival of the individual species for more than 100 years, the goal of conserving "biological diversity" represents a shifted conservation focus that has occurred during the past two decades. The past emphasis has been on three separate goals. We will illustrate these with examples from the United States, although the three categories are universal elements of conservation philosophies.

The first of these conservation goals was the protection of pristine natural areas, with the implicit assumption that this would result in the survival of the species found there. In the United States, this philosophy had its roots in the mid-1800's through the writings of individuals like John Muir and Henry David Thoreau. The grounds for preservation in these early years were usually aesthetic, rather than biological, and a scientific understanding of such issues as minimum viable populations and ecosystem dynamics was lacking (Nash 1976). Today, conservationists continue to place emphasis on protecting large wilderness areas, but cite scientific, as well as aesthetic, criteria.

A second emphasis in conservation has been on the survival of particular species groups perceived to be valuable (e.g., wildlife, fisheries, and forest trees, dating from the mid-1800's in the United States), vulnerable (e.g., endangered species, dating from the mid-1950's), or aesthetically pleasing (e.g., wildflowers and the charismatic megavertebrates, dating from the early 1900's). Legislation in the United States to protect fisheries and wildlife dates from the late 1800's and early 1900's, but the culmination of the species-oriented approach at the Federal level was the passage of the Endangered Species Act in 1973. Interest in endangered species also heightened the awareness of the importance of genetic diversity and minimum viable population sizes (Frankel and Soule 1981; Soule and Simberloff 1986; Burgman, Akcakaya, and Loew 1988). Conservation of endangered species has also introduced the idea that ex situ management, intensive management in artificial systems, and direct human intervention in natural populations were legitimate conservation means to insure endangered species survival (Templeton and Read 1983; Falk and McMahan 1988; Jordan, Peters, and Allen 1988).

Protection of endangered species has mostly focused on larger animals and higher plants, with less attention to more obscure groups, such as soil insects and fungi, which may be critical to ecological function. Proponents have argued that protection of large, wide-ranging animals would insure, through habitat protection, the survival of these more obscure species, a suggestion that has not been critically investigated. Some have argued that species with key ecosystem roles should be the highest priority for conservation efforts (e.g., Terborgh 1988).

A third conservation emphasis has been on the use of resources in a way that could be continued with no long-term decline in the productivity, thus producing a "sustained yield" of renewable resources (Nash 1976). In the United States, this form of conservation was formulated by individuals who supported resource use, but saw the exploitive and destructive "mining" of natural resources of the late 1800's as eroding the resource base. Conservation in this sense has often resulted in some level of regulation for use of wildlife, fisheries, and forests.

Current interest in biological diversity is, however, more than the sum of interest in pristine natural areas, special groups of species, and sustained productivity of natural resources. Since 1980, awareness of the biological diversity issue has been heightened by increased threats to tropical ecosystems. More than 50 percent of all terrestrial species are found in the tropics, with the rate of habitat loss being estimated at greater than 1 percent per year (Myers 1988). It has been estimated that this habitat loss is causing 1,500-10,000 species to become extinct per year (Wilson 1987; Myers 1988). The rate of species loss is currently much faster than the rate at which new species are discovered and described (Wilson 1987; Ehrlich and Wilson 1991).

The current era of air pollution and global climate change has further brought into question historical conservation emphases. The number of remaining pristine areas is becoming ever fewer, and even those that are fully protected from direct human disturbance will see major changes if global warming occurs (Peters 1988). A host of other direct and indirect human impacts also permeate wilderness areas (White and Bratton 1980). In addition, we have learned that some biological diversity can be protected in the midst of human activity and that biological diversity can be reintroduced to damaged lands in ecological restoration (Jordan, Peters, and Allen 1988). Further, a narrow focus on legally endangered species, including the use of such methods as ex situ conservation and the protection of small areas of critical habitat for rare species, would leave much of biological diversity unprotected until individual species approached extinction.

In summary, contemporary developments have served to focus attention on biological diversity as a property to be conserved, whether in pristine natural areas, intensively managed and artificial populations, or areas of human resource use. This new emphasis has served to diversify acceptable conservation tactics (e.g., the "eight paths" of Soule 1991), has raised awareness about the tremendous biological diversity of our planet, and has underscored our poor understanding of both the amount of diversity present and the ecological processes that support or depend upon this diversity. At present, estimates of the total number of species range over an order of magnitude (4-30 million), of which only 1.4 million have been formally described by the scientific community (Ehrlich and Wilson 1991; Erwin 1988). Biologically rich (e.g., the tropics) or remote (e.g., the ocean floor) habitats are poorly known taxonomically (Ehrlich and Wilson 1991; Ray 1988). Ecologically important but obscure groups such as soil insects, fungi, and bacteria are also poorly known.

A Working Definition of Biological Diversity

The phrase "biological diversity" obviously implies enumeration of the variety in living things. However, an account of biological diversity must be more than an enumeration of the living things themselves; it must include the structures and processes that maintain this diversity (Franklin 1988; Noss 1990), as well as levels of organization above and below the species level. Noss (1990) discussed three attributes of biological diversity: composition (the number of "things"—e.g., alleles, species, or ecosystems); structure (the physical arrangement of the "things"—e.g., biomass distribution within a forest or ecosystem arrangement on a landscape); and function (the natural processes). The variety of living things would not exist without a host of ecological processes, such as natural disturbance regimes, nutrient transformations, symbioses, and food webs. Because the species and individuals are obvious, while the natural processes are not, the phrase "biological diversity," if superficially applied, has the danger of focusing attention away from the underlying processes and dynamics.

The structure of ecosystems and landscapes is critical to the maintenance of biological diversity (Franklin 1988). Two forests may be dominated by the same tree species but have very different structures, with such features as standing dead trees, fallen logs, and soil leaf litter playing a major role in defining animal habitat, nutrient cycling, and the interface with stream ecosystems within riparian ecosystems. The soil compartment of ecosystems, influenced by geology, mineral substrates, the deposition of organic matter, and the subterranean activities of plants and animals, is a key aspect of ecosystem structure and function and one that is potentially influenced by pollutant deposition. On a landscape scale, two areas may possess similar ecosystems and yet be very different in the physical arrangement of those ecosystems, with this physical arrangement influencing ecosystem processes such as animal migration and the spread of disturbance (Gardner et al. 1989). Noss and Harris (1986) have described the effects of landscape structure on ecological function and retention of biological diversity.

Although the species level is often the most obvious element of biological diversity, genetic and ecosystem diversity are also critical. It has been argued that genetic diversity is the foundation of all other aspects of biodiversity; ultimately, it supplies the abilities that produce functioning ecosystems. Small population sizes can result in reduced genetic diversity, lowered fitness, and increased extinction risk (e.g., Frankel and Soule 1981; Schonewald-Cox et al. 1983), although genetic stochasticity may be outweighed as short-term risks by demographic and environmental stochasticity in many populations (Menges 1986). Further, not all populations of a species are equivalent; the genetic variation among populations can be the basis of both future evolution and response to environmental change. Human impact results in substantial loss in the number of original populations and, presumably, genetic diversity, within a species (Peet, Glenn-Lewin, and Wolf 1983). Loss of alleles can remove variation that is critical in times of change in the physical or biological environment. All of these perspectives suggest that biological impoverishment can proceed on the genetic level in the absence of immediate species extinctions.

Natural variation in ecosystems across a landscape or region is also important. Ignoring for the moment cases of spatial constraint (see below), the distribution of this variation is ultimately controlled by environment, including the dynamics of disturbance regimes. Thus, a wilderness landscape of varying topography and habitats contains ecosystems for a range of environments and, thus, provides for response to environmental change.

Finally, the most efficient way to protect species is through protection of the ecosystems in which they occur.

An undue emphasis on the number of species does not take into account the distinctiveness of biological lineages nor the fact that some groups are rapidly evolving and others are more static. Vane-Wright, Humphries, and Williams (1991) and Erwin (1991) advocate conservation strategies based on information about lineages and evolutionary rate, rather than the equal consideration of all species. A lineage that is monotypic depends on the survival of its one species; a lineage with thousands of species does not depend as greatly on each of its species. Thus, some have argued for the conservation of unique lineages and taxa of higher rank (i.e., making sure each genus, family, order, or taxon of higher rank is conserved). The proposition that the amount of diversity varies with taxonomic level is illustrated by comparisons between marine and terrestrial systems: the oceans possess a greater diversity of higher level animal taxa (e.g., there are more distinct lineages at the phylum level) than do tropical rainforests, where diversity is concentrated at the species level (Ray 1988). However, a greater number of species at lower ranks (within a genus or family) may mean that the group is rapidly evolving and has high evolutionary potential, while monotypic categories may represent "living fossils" with great uniqueness but little potential (Erwin 1991). Arguments based on the uniqueness of a lineage and the conservation of taxa of higher rank and evolutionary potential sometimes conflict, but all support the idea that the number of species need not be the sole conservation goal.

In light of this discussion, we define biological diversity in the broadest sense as the variety of life and life processes at all organizational levels, with usual emphasis on: (1) genetic diversity within species, other taxa, or populations and/or the sum of genetic diversity within a community or geographic area; (2) species or other taxon diversity within a community or geographic area; and (3) community or ecosystem diversity across a landscape or larger region.

Our working definition emphasizes state variables (i.e., "things," such as genes, species, communities, biomass, composition, and structure), but processes (e.g., gene flow, survivorship, reproduction, migration, competition, energy flow, and mineral cycling) provide the critical bases for and are themselves derived from these state variables. Although states and processes are both part of biological diversity, state variables, such as genetic diversity, species richness, and community pattern, are often the easiest elements of diversity to measure, particularly in short-term studies that essentially take a "snap-shot" of the state of nature. State variables are also usually the focus of concern about biological impoverishment. Variety in genes, species, and ecosystems is threatened; this variety, particularly for genes and species, is the unique product of past evolution and is impossible to recover once lost. Further, as we will discuss later, the exact correspondence between state variables and ecological function remains an area for continued research. This, again, suggests that documenting change in biological diversity centers on the state variables.

The Value of Biodiversity

Utilitarian arguments are those that base the conservation of biological diversity on current or future human benefit. These benefits may be material or nonmaterial, and economic or noneconomic. Material benefits include ones with direct economic value (e.g., genetic resources for crop plants, tourism and nature conservation, and sources of medicines) and ones without economic benefit (e.g., biological diversity as a reservoir of genetic resources for unknown future needs). Some economic benefits are hard to quantify; e.g., the economic value of clean air is difficult to define because it is not "available" in the marketplace. Environmental quality, ecosystem function, and ecological "services" depend on biological variety in the sense that roles such as organic production, decomposition, and chemical transformation depend on biological variety. Nonmaterial benefits, which include the psychological and spiritual benefits of natural beauty and wilderness, may have economic importance (e.g., tourism).

In terms of economic, material benefits, biological diversity provides potentially useful medicines, foods, chemical products, and fuels. Natural compounds have been the template for the synthesis of many medicines; even if we ignore this past use of biological diversity, 25 percent of U.S. prescription drugs (for a value of approximately $4.5 billion per year) have at least one ingredient that is extracted directly from higher plants (Farnsworth 1988). Only a small percentage of the world's plant species have been tested for usefulness and are being used at present (Plotkin 1988). Although at least 75,000 higher plants have edible parts (at least 25 percent of all higher plant species), only 7,000 have had recorded use, and the bulk of the human diet today comes from only 20 species (Plotkin 1988; Veitmeyer 1986).

Although utilitarian arguments are often the most politically compelling, there is a danger in basing conservation solely on these grounds because this can wrongly equate the importance of biological diversity with short-term economic benefits, whereas human survival and quality of life also are utilitarian values, albeit ones that require a longer term view. Alternatives to utilitarian arguments are such ideas as species rights (Norton 1988) and the land ethic (Leopold 1949).

Threats to Biological Diversity

Threats to biological diversity are ultimately driven by the growth of human population and by the per capita amount of impact. There are five major categories of threats:

1. Direct species loss (e.g., large predators and herbivores, vulnerable because of low population density and dependence on large areas of undisturbed land or because of unusual behaviors).
2. Habitat loss and fragmentation (Harris 1984), with the loss often being nonrandom—the most productive sites are preferentially used.

3. Exotic species invasions; i.e., the purposeful and accidental release of species from natural barriers (Elton 1958).
4. Change in natural processes; i.e., human control of natural dynamic processes, such as fire, hydrology, and coastal processes (White and Bratton 1980; White 1987).
5. Air and water pollution (alteration of the physical and chemical environment).

Because of the magnitude of these threats, particularly habitat loss in the tropics, contemporary extinction rates are thought to be 2-3 orders of magnitude higher than natural background rates (Raup 1988; see also discussion in Jablonski 1991). Species can be quickly lost due to human influence (10^{0-2} years) but are generally slow to evolve (10^{3-5} years). In addition, each species is unique; once lost, that exact configuration of genetic structure is lost forever. This introduces a large risk and an asymmetry into the process: losses cannot be easily reversed. Further, local losses in diversity can either be global or not (that is, the lost species or genotypes may persist in other places); on the other hand, local increases in diversity due to human activities (migration of new species or genotypes due to climatic warming) are almost never global increases in diversity.

The threats listed above combine in nature. In particular, the habitat loss and fragmentation influence make more severe the loss of biological diversity caused by the other four categories of threats. If the time course of environmental change is fast, sensitive species will be lost before tolerant species invade.

BIOLOGICAL DIVERSITY IN AN ECOLOGICAL CONTEXT

Species Richness and Ecological Function

Impacts to biological diversity and ecosystems can consist of any of the following combinations: impacts to diversity may result in a change in ecological function and/or a loss of ecological integrity; impacts to ecosystem function may result in changes to diversity; and changes to diversity, whether or not these affect ecological function, may be viewed as impacts in their own right. These cases overlap when ecological function changes and diversity is lost, but there are important differences in emphasis. The first view places emphasis on the role of diversity in such attributes as ecosystem health and stability, the second view places emphasis on the role of ecosystem function in supporting diversity, and the last simply sees any loss of diversity as a problem in and of itself.

Functioning ecosystems depend on the presence of a variety of primary producers, decomposers, other heterotrophs, and symbionts. It frequently has been conjectured that variety is necessary within these functional groups, as it conveys a robustness or stability to ecosystems in the face of environmental change. This argument also predicts that monocultures will be more vulnerable to change, with external forces quickly producing instability. Such ideas have a long history in ecology (Odum 1969; Ehrlich and Ehrlich 1981).

While these statements must be true at some level, the exact correspondence between biological diversity and ecological function requires continued research. Part of the problem lies in the definition and quantitative measure of such terms as ecological function, health, stability, balance, and integrity (e.g., Norse 1990). While ecosystems must have organisms that carry out basic processes—carbon fixation, mineralization, and nutrient transformations—no one definition of terms like ecosystem "health" exists (Noss 1990).

Karr (1990) has defined biological integrity as "the capability of supporting and maintaining a balanced, integrated, adaptive community of organisms having a species composition and functional organization comparable to that of a natural habitat of the region" and further states that "a biological system . . . can be considered healthy when its inherent potential is realized, its condition [relatively] stable, its capacity for self-repair when perturbed is preserved, and minimal external support for management is needed." The phrases "comparable to that of a natural habitat" and "inherent potential" show that these definitions are based on comparisons to an original or undisturbed condition, rather than on an underlying theory that directly predicts diversity from integrity or vice versa.

Beyond the definition and measurement of ecological function, an additional problem lies in defining the expected relationship between diversity and function. Does the importance of biological diversity to ecosystem function require the demonstration of a one-to-one predictive relationship between diversity and integrity? At what time and space scales should this question be asked? In short-term observations, some species and genotypes may appear redundant or extraneous to immediate ecological function, while these same elements might be critical components of diversity over longer time periods. A final problem is that genetic, species, and ecosystem levels of biological diversity are not necessarily coupled in their responses to human-caused changes. Some may even be negatively correlated. For example, a decrease in species diversity within a community may lead to an increase in genetic diversity within the surviving species if lowered competition favors the survival of additional genotypes.

We summarize these points by saying that while ecological function and diversity are linked and while diversity is required for long-term ecological and evolutionary flexibility, a precise relationship predicting changes in integrity from changes in diversity or vice versa has not yet been shown; this subject remains an exciting area for future research. In the absence of a clearer understanding, an alternative is to follow Karr (1990) by viewing human-caused changes in diversity and ecological function as departures from the original condition.

Contingency, Environment, and Biological Diversity

The amount of biological diversity (e.g., the number of species or the amount of genetic diversity) existing in any area is partially contingent on the past course of evolution. Given the quickness of extinction compared to the origin and spread of new species, the present amount of species and genetic diversity in a given setting is probably not in equilibrium with the present environment. The importance of history may hinder the development of generalities in investigations of both the ecological role of biological diversity and the impacts of pollution because ecosystems, even in similar environmental settings, may have had different amounts of original biological diversity. For example, in the North Temperate zone, diversity (both at the species and genetic level) in many systems may still be recovering from the last advance of glacial ice some 20,000 years ago (Davis 1981). It has been suggested that air pollution impacts on red spruce (*Picea rubens*) in the Appalachians were more severe than on other species because of low genetic diversity in red spruce, a condition that may have resulted from the restriction of this species to a small Pleistocene refugium (Johnson et al., in press).

Below we will develop two contrasting concepts for describing how species richness is distributed spatially, one based on environment-species relations and the other on spatial and temporal constraint.

The Niche Difference Model

Species have different physiological tolerances to the physical and biological environment and these determine overall distribution and local abundance. Models of this type fall into two subcategories: equilibrium competition-based models and nonequilibrium models (Pickett 1980). In equilibrium models, time is sufficient for species occurrences to reflect niche differences and competitive relations. In nonequilibrium models, time is insufficient for competitive sorting of species, and a component of local diversity (i.e., within communities) is due to the persistence of noncompetitive species or genotypes, and, thus, expansion of species-realized niche space (Peet, Glenn-Lewin, and Wolf 1983).

Among competing organisms, the conjectured basis for niche differences is the idea of evolutionary tradeoffs, which result in species differing in tolerance, competitive ability, and life history strategy (MacArthur and Wilson 1967; Grime 1979; Huston 1979; Tilman 1988). In addition to competition, species occurrence is determined by other biological interactions such as predation and mutualism. Biological interactions also include the influence of community structure on species distributions, e.g., bird species diversity as a function of vegetation structure and prey items (MacArthur 1957), and on the physical environment.

The turnover of species through succession reflects niche differences among these species. Natural processes in ecosystems such as fire, flooding, windfall, and avalanche create successional patches within older vegetation, allowing more species to persist than would if the landscape were either completely undisturbed or entirely within a disturbance patch (White 1979; Denslow 1985). The intermediate disturbance hypothesis predicts that a regime of "intermediate" disturbance will maximize species richness (Connell 1978; Huston 1979). A positive feedback, where the probability of disturbance increases with successional time as a function of the changing structure and age of the patch, can produce a patch dynamic equilibrium in which the locus of disturbance shifts through time but the average area in various patch age classes remains constant (White and Pickett 1985). Shugart (1984) has used simulation models to suggest that the relationship of disturbance patch size to landscape area determines whether a patch dynamic equilibrium occurs or not; the importance of this in conservation and persistence of species is clear (Pickett and Thompson 1978; Romme and Knight 1982).

Huston (1979), Grime (1979), and Peet and Christensen (1988) have presented models of community-level plant species richness. In general, these models suggest a unimodal response of richness to both fertility or "rate of replacement" (Huston 1979), and disturbance, or "stress," (Grime 1979) gradients.

Peet and Christensen (1988) suggest that the descending part of the species richness curve results from a transition from symmetrical (individuals capture resources in proportion to their size) to asymmetrical competition (larger individuals capture resources at a higher rate than is predicted from their size and, thus, increasingly dominate resource use). In forests, they suggest that competition for light during succession follows the full progression from symmetrical to asymmetrical competition (and hence should produce a decline in richness late in succession), but that the use of soil resources is symmetrical throughout (thus, richness increases from the poorest to the richest sites). They also stress the different behavior of various plant guilds and note that published reports of the trajectory of richness through secondary succession to forest showed increases, decreases, and stability. In contrast to forests, species richness in forests almost always declines from the moderately fertile to the most fertile sites (called the "paradox of enrichment" by Peet, Glenn-Lewin, and Wolf 1983). On the most fertile sites, dominant species are able to competitively exclude other species. Stress, grazing, fire, mowing, or other disturbances in such systems may act in a similar way to keystone predators—that is, they reduce the dominant species and result in increased richness on small scales (Peet, Glenn-Lewin, and Wolf 1983). Presumably, prolonged exposure to these disturbances would result in evolutionary adaptation in the species present; thus, systems that are exposed to novel disturbances would not respond in the same way as systems long exposed to disturbances.

The niche difference model suggests that population and species differences, combined with patterns in the environment, including gradients in physical factors and natural disturbance regimes, explain the original amount and distribution of biological diversity. In general, we would expect that diversity (i.e., genetic diversity among populations within a species, species diversity, and ecosystem variation) would increase with the amount of spatial variation in physical factors, with biological interactions (e.g., those produced by coevolution), with moderate to high resource abundance (where resources are more abundant, a few species dominate their use, according to some models), and with moderate disturbances that prevent a few species from dominating but do not erode site productivity. As an extreme, communities early in primary succession have low productivity, low resources, simple structures, and low diversity. Human activities, including air pollution, sometimes result in parallel changes (Smith 1990). However, the pattern of niche division in more complex communities, and hence the amount of species richness within given environments, has not been yielded to a general model, and attempts to closely correlate diversity with ecosystem processes like productivity or fertility have not yet produced clear patterns.

Models of Spatial and Temporal Constraint

Species and gene occurrences also can be affected by distance and spatial configuration. The theory of island biogeography models species number as a consequence of the processes of immigration and extinction (MacArthur and Wilson 1967). In the simplest situations, immigration rate is determined by distance and extinction rate by island area. Developed to explain species richness patterns of oceanic islands, the theory has also been applied to habitat patches on land. Where the theory holds, spatial configuration, in addition to environment, plays a role in the distribution of species. With the environment held constant, the smaller and more isolated the habitat patch, the lower the expected species richness.

Island biogeography has furnished one of the few predictive models concerning species richness and human effects (Peet, Glenn-Lewin, and Wolf 1983), namely that habitat fragmentation and isolation will result in species loss, called species relaxation, over time (Terborgh 1974; Diamond 1975; Soule, Wilcox, and Holtby 1979; Harris 1984). Once remnants of a formerly contiguous habitat become isolated and reduced in size, they will receive fewer immigrants from the surrounding landscape to balance local extirpations. In the extreme, no source of immigrants remains. Species loss is concentrated in "extinction-prone" species (Terborgh 1974). In addition, edge effects will increase with the increase in the perimeter-to-area ratio, producing environmental change within remnant habitat patches. A substantial literature has developed on design strategies for minimizing the predicted species loss through such features as corridors and networks (e.g., Noss 1983; Noss and Harris 1986). One of the most important considerations with human-caused stresses, such as air pollution, is that they affect natural areas that very often have been fragmented by human activity (Peet, Glenn-Lewin, and Wolf 1983).

Several other mechanisms of spatial constraint also have been proposed. For example, mass effect describes the situation in which species are present on a site because of a high population density nearby and continual immigration (Shmida and Wilson 1985; Shmida and Ellner 1984). At greater time and space scales, the vicariant evolution of several descendant species from a single ancestor species and the convergent evolution of functional equivalents from unrelated ancestors develop through isolation of gene pools and hence are cases of spatial constraint. The apparent commonness of these situations in evolutionary history suggests the importance of spatial constraints at the continental and global scales as mechanisms for increased biological diversity. Preston (1962) suggested that the global diversity of land birds was four times higher than it would be if all of the land were present in a single continental land mass. Local richness may be increased by immigration and subsequent co-occurrence of vicariant and trophically equivalent species, although niche displacement and reproductive isolation may be required for stability of this diversity.

Spatial constraint implies temporal constraint. For example, the probability of dispersal can be described as a decay function of distance. As long as probabilities are above zero, the limit represented by distance and physical barriers can be overcome by long time spans. Given enough time and the lack of absolute barriers, ranges can expand to the full physiological capabilities of the species.

Auerbach and Shmida (1987) have linked models of niche difference and spatial and temporal constraint. As a function of scale, they rank the determinants of species richness as niche relations (important from 10^{-1} to 10^3 m^2), habitat heterogeneity (important from 10^0 to 10^9 m^2), mass effects (important from 10^1 to 10^7 m^2), and trophic equivalence (important at $>10^9$ m^2). The first two of these fall under niche difference models and the second two under models of spatial and temporal constraint.

MEASURING BIOLOGICAL DIVERSITY

Measurement and monitoring of biological diversity is made challenging because the state variables (composition and structure) are much easier to measure and understand than the function variables (ecological processes), although the latter are critical to the existence of the former. Essentially, we must often take "snap-shot pictures" of the state variables at several points in time and then use the data to describe change, infer cause, and determine what processes should be studied in detail. However, the organizing concept for any research in this area must be the relationship between state variables and function.

While a catalogue of methods for measuring and monitoring biological diversity is beyond the scope of this chapter, several key issues must be addressed because the chances for spurious results are great. Because all measures of the state variables of biological diversity produce different results at different scales, we will end with a discussion of the issue of scale dependence. The emphasis will be on the species level because that has been the most generally studied. However, the general comments apply to genetic and ecosystem levels as well.

Continuous Variation

Change in some aspects of biological diversity within a study area is continuous. For example, the unambiguous classification of variety in communities and ecosystems may be impossible. In these cases, classification is an inappropriate technique for describing biological diversity and methods for describing continuous variation, such as ordination, should be employed.

Richness and the Distribution of Abundances

Some kinds of biological diversity are best represented through assessments of the number of "kinds" present (see discussion in Peet 1974). Classifications require that the variation within a class is less than the variation between classes. The classes present may be alleles, species, or ecosystem types. Even when classification is appropriate, it may be practically impossible to enumerate all classes and a representative sample may be required (e.g., the alleles of only some genes, rather than the structure of whole genomes, or the enumeration of only higher plants, rather than all species of all kingdoms). Diversity at any of these levels of organization includes two components: the number of classes and the distribution of abundances across those classes.

Given appropriate classification, the simplest and most widely used measure of diversity is the number of kinds or classes present, a property called richness (e.g., species richness). The number of classes present is, in addition to being a consequence of the ecological setting of the investigation, a function of the size of the sample. This can be stated either as the number of individuals examined or the area inventoried.

In addition to richness or the number of classes present, a second component of diversity consists of the distribution of abundances (e.g., biomass or the number of individuals) across the classes. For example, a community of 10 species in which dominance (e.g., the number of individuals or biomass) is highly concentrated in one species has a lower "apparent" diversity because samples of that community are likely to contain the dominant species only or the dominant and only a few additional species (Peet 1974). On the other hand, a community with the same species richness (10 species) in which abundance is evenly distributed among the species will have a higher "apparent" diversity because samples are more likely to contain more of the species. This may have ecological consequences in the sense that species interactions are likely to increase as evenness increases. The distribution of abundances has also been taken as a reflection of niche division or the way in which the species divide the functional roles in the community (MacArthur 1957; Whittaker 1975).

A simple graphical way of describing both richness and the distribution of abundances is the dominance-diversity curve (Whittaker 1975; Wilson 1991; Figure 2-1). Species are arranged from left to right from most abundant to least abundant. The length of the sequence is richness; the shape of the curve shows how abundances are distributed. Again there is a sample size effect: the more abundant species will be easy to capture, even in relatively small samples, whereas the rare species will usually be found only in larger samples.

Several indices have been used to represent the length and shape of the dominance-diversity curve. The absolute representation of the distribution of abundances among entities is called evenness (the shape of the dominance-diversity curve); when this quantity is expressed relative to a standard model (e.g., the maximum possible evenness for the sample), the measure is called equitability. Some diversity measures combine richness and evenness in a single measure; Peet (1974) proposed the term heterogeneity for these measures. The several indices of heterogeneity, equitability, and evenness have different sensitivities to sample size, richness, evenness, and kind of temporal trend (Peet 1974; Boyle et al. 1991).

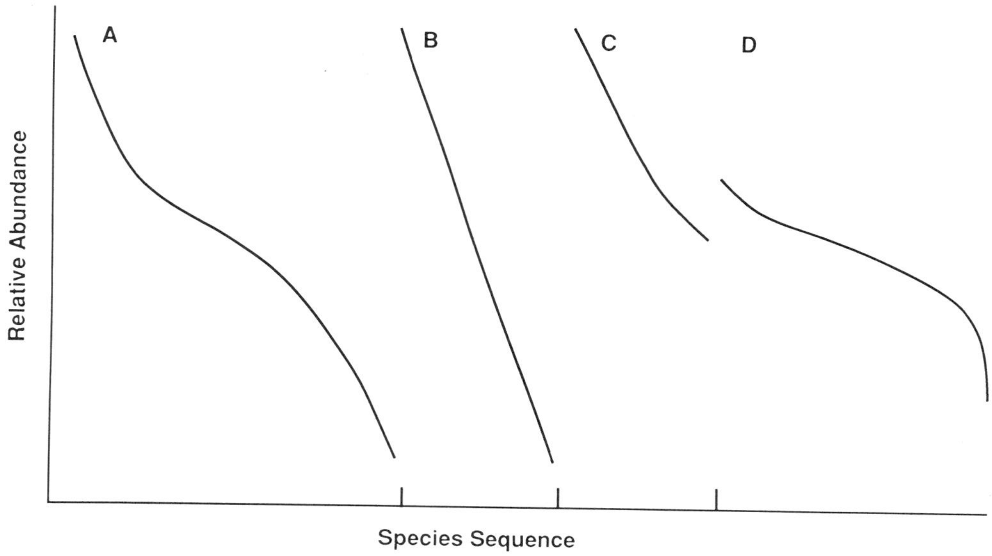

FIGURE 2-1. Dominance diversity curves. A. Log-normal distribution of abundance. B. Geometric distribution of abundance, implying strong dominance. C. Log-normal distribution of abundances after reduction of rare species and increases in dominance. D. Log-normal distribution after reductions in dominant species and increases in evenness.

A change in species diversity could include both the length and the shape of the dominance-diversity curve (Figure 2-1). For example, air pollution might eliminate particularly vulnerable species, thus truncating the sequence of species. However, the sequence could also be lengthened, as in the immigration of southern species into an area with climatic warming. Species tolerant of air pollution could also become more dominant, thus steepening the shape of the curve. Alternatively, if air pollution acts on the most dominant species more than others, the dominance-diversity curve would become flatter. Moderate stress that affects dominant species only may, in fact, increase local richness, even though this effect represents only a local, not global, source of richness.

Scale Dependence

Observed biological diversity is, in part, a function of sample size and the spatial and temporal scale of sample. For example, species richness is a curvilinear function of the area sampled (e.g., Preston 1962). Although the gain of species number often slows as the accumulated area of the sample increases, we cannot a priori assume that an asymptote exists in any particular circumstances. All richness values are contingent on sample size; a corollary is that sample size must be specified in order for any richness value to be useful.

Scale includes two components: grain and extent (Wiens 1989; Figure 2-2). Grain is the size of the unit of observation (e.g., quadrat size). Extent is the distance over which the observations (e.g., individual quadrats) are distributed. The two components of scale can independently influence the levels of diversity measured. Beyond grain and extent, sample number (e.g., the number of quadrats), and the way a given sample is arranged within the

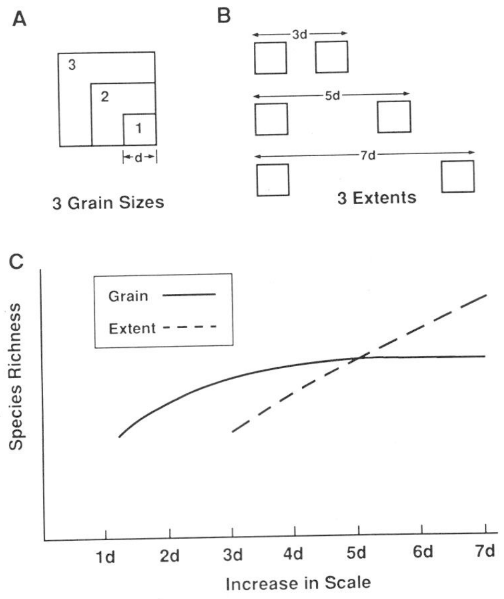

FIGURE 2-2. Grain, extent, and species richness. A. Three grain sizes (d=a unit of distance). B. Three spatial extents. C. Species accumulation as a function of increasing grain and extent. In this hypothetical case, more different kinds of environments are sampled with an increase in extent than an increase in grain; thus, species number (even at lower total area sampled) increases faster with increase in extent than increase in grain.

study area (e.g., random, stratified random, or regular) also influence the observed level of diversity.

The number of species encountered increases with the grain of the sample, steeply at first, and then with a diminished rate. If two areas (or the same area at two times) are sampled at different grain sizes, the differences in the number of species will be partly an artifact of the sample (Figure 2-3).

Thus, the shape of the relationship between richness and grain size is probably more useful than a single richness value at an arbitrary grain size. Richness will usually increase with an increase in the extent of the sample. Samples of greater extent are likely to encounter more variation in environment or history than the same sample (same grain size and same number of grains) concentrated in a smaller area.

In addition to the importance of spatial grain and extent, scale dependence also occurs with samples that have a temporal dimension. The greater the continuous duration (grain size in time) of the sample and the greater the time over which individual samples are distributed (extent in time), the greater the number of species that will be encountered.

Detected richness will also increase with the number of samples (e.g., the total amount of area or total amount of time sampled at a specific grain size and extent) and will increase faster for samples stratified by change in environment (e.g., positioned along a spatial or temporal gradient in a way that will minimize similarity among samples).

Changes in species richness and in the distribution of abundances among species underscore the potential problems of scale dependence. Species richness could behave differently on different scales. Obviously, local decreases in species richness may not be global losses. Chronic stress (e.g., grazing) increases species richness at small, but not

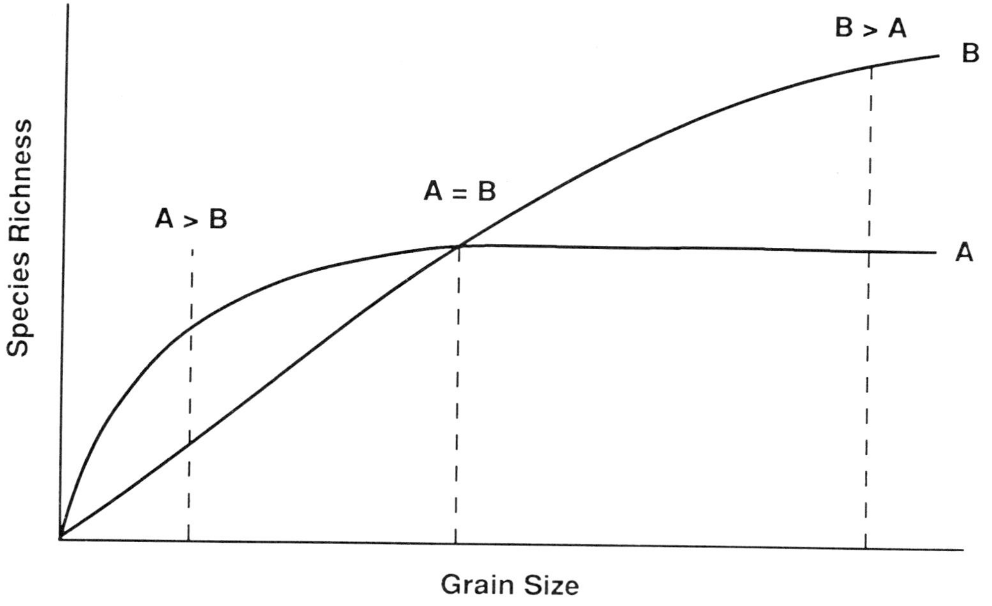

FIGURE 2-3. Species-area curves for two sites, showing the potential importance of grain size and scale dependence in detecting differences.

large grain sizes (Peet, Glenn-Lewin, and Wolf 1983; Van der Maarel 1988). In the absence of this stress, individual plant biomass is higher and dominance may be concentrated in a few species; thus, species richness may decrease in small grains with decrease in stress or disturbance rate. The alpha, beta, gamma, and delta components of diversity (Whittaker 1975; Cody 1986) are related to the phenomenon of scale dependence. Alpha diversity is the diversity within a community; in the extreme this is sometimes called "point diversity" (Cody 1986). Beta diversity is the turnover of species along an environmental gradient; a higher rate of turnover means that more species are present. Gamma diversity is the diversity present within a landscape in which many communities and gradients are often present.

Cody (1986) has used the term delta diversity for the geographic turnover of species, including those caused by climatic gradients and spatial and temporal constraints (e.g., the trophic equivalency of Auerbach and Shmida 1987). At small grain sizes, alpha diversity dominates; as grain size increases, beta and gamma diversity, and finally delta diversity, contribute to the species richness observed. Increase in spatial extent at a fixed grain size and sample number will more quickly detect beta, gamma, and delta diversity compared to an increase in grain size at a fixed spatial extent.

Although this discussion has centered on species richness, there is likely to be a similar scale dependence in the amount of genetic and ecosystem diversity observed as well. Scale dependence means that there is a great danger that sampling designs will introduce artifacts into comparisons of diversity between human-impacted systems and pristine systems or between systems before and after exposure to stress. Quality control and assurance protocols will be essential in assessments of biological diversity and its change. This is important because studies of biological diversity will almost always be conducted with samples due to the impossibility of inventorying large areas.

SUMMARY

Contemporary efforts to conserve biological diversity represent a shifted conservation focus. The emphasis on biological diversity includes but is more than the conservation of pristine ecosystems, particular species groups, or sustained use of resources. The protection of biological diversity now requires a range of conservation tactics, including protection of natural areas, better management of resource-use areas, and the manipulation of artificial ex situ populations.

Biological diversity implies the enumeration of living "things," but an undue emphasis on "things" can divert attention from the underlying processes that are critical to the survival of biological diversity. The attributes of biological diversity include diversity in composition, structure, and function. These attributes must be investigated at four levels: genetic diversity, species diversity, community or ecosystem diversity, and landscape.

Measuring change in biological diversity is made complex by the phenomenon of scale dependence; reported changes will vary with the scale of observation.

The amount of species diversity present has generally been explained by patterns in the physical environment, including the frequency of disturbances, and by spatial and temporal constraint. Although a general and dependent relationship has been found between diversity and ecological function, the precise formulation of this relationship remains an important area for research. More information is needed on the predictability of biological diversity from environment and history, on the dependence of ecological function on diversity, and on the historic and spatial contingencies present.

REFERENCES

1. Auerbach, M., and A. Shmida. 1987. Spatial scale and the determinants of plant species richness. *Tree* 2:238-42.
2. Boyle, T.P., G.M. Smillie, J.C. Anderson, and D.R. Beeson. 1991. A sensitivity analysis of niche diversity and seven similarity indices. *J. Water Pollut. Control Fed.* 62:749-62.
3. Burgman, M.A., H. Resit Akcakaya, and S.S. Loew. 1988. The use of extinction models for species conservation. *Biol. Conserv.* 43:9-25.
4. Cody, M.L. 1986. Diversity, rarity, and conservation in Mediterranean climate regions. In *Conservation biology*, ed. M.E. Soule, pp. 122-52. Sunderland, MA: Sinauer Assoc.
5. Connell, J.H. 1978. Diversity in tropical rain forests and coral reefs. *Science* 199:1302-10.
6. Davis, M.B. 1981. Quaternary history and the stability of forest communities. In *Forest succession*, ed. D.C. West, H.H. Shugart, and D.B. Botkin, pp. 132-53. New York: Springer-Verlag.
7. Denslow, J.S. 1985. Disturbance-mediated coexistence of species. In *The ecology of natural disturbance and patch dynamics*, ed. S.T.A. Pickett and P.S. White, pp. 307-23. New York: Academic Press.
8. Diamond, J.M. 1975. The island dilemma: Lessons of modern biogeographic studies for the design of nature reserves. *Biol. Conserv.* 7:129-46.
9. Ehrlich, P.R., and A.H. Ehrlich. 1981. *Extinction: Causes and consequences of the disappearance of species*. New York: Random House.
10. Ehrlich, P.R., and E.O. Wilson. 1991. Biodiversity studies: Science and policy. *Science* 253:758-62.
11. Elton, C.S. 1958. *The ecology of invasions by plants and animals*. New York: Chapman and Hall.
12. Erwin, T.L. 1988. The tropical forest canopy, the heart of biotic diversity. In *Biodiversity*, ed. E.O. Wilson and F.M. Peter, pp. 123-29. Washington, DC: National Academy Press.

13. Erwin, T.L. 1991. An evolutionary basis for conservation strategies. *Science* 253:750-53.
14. Falk, D.A., and L.R. McMahan. 1988. Endangered plant conservation: Managing for diversity. *Natural Areas* J. 8:91-99.
15. Farnsworth, N.R. 1988. Screening plants for new medicines. In *Biodiversity*, ed. E.O. Wilson and F.M. Peter, pp. 83-96. Washington, DC: National Academy Press.
16. Frankel, O.H., and M.E. Soule. 1981. *Conservation and evolution*. Cambridge: Cambridge Univ. Press.
17. Franklin, J.F. 1988. Structural and functional diversity in temperate forests. In *Biodiversity*, ed. E.O. Wilson and F.M. Peter, pp. 166-75. Washington, DC: National Academy Press.
18. Gardner, R.H., R.V. O'Neill, M.G. Turner, and V.H. Dale. 1989. Quantifying scale-dependent effects of animal movement with simple percolation models. *Landscape Ecol*. 3:217-28.
19. Grime, J.F. 1979. *Plant strategies and vegetation processes*. New York: John Wiley & Sons.
20. Harris, L.D. 1984. *The fragmented forest*. Chicago: Univ. of Chicago Press.
21. Huston, M. 1979. A general hypothesis of species diversity. *Am. Nat*. 95:137-46.
22. Jablonski, D. 1991. Extinctions: A paleontological perspective. *Science* 253:754-57.
23. Johnson, A.H., S.B. McLaughlin, M.B. Adams, E.R. Cook, D.H. DeHayes, C. Eagar, I.J. Fernandez, D.W. Johnson, R.J. Kohut, V.A. Mohnen, N.S. Nicholas, D.R. Peart, G.A. Schier, and P.S. White. In press. Synthesis and conclusions from epidemiological and mechanistic studies of red spruce decline. In *The ecology and decline of red spruce in the eastern United States*, ed. M.B. Adams and C. Eagar. New York: Springer-Verlag.
24. Jordan, W.R. III, R.L. Peters II, and E.B. Allen. 1988. Ecological restoration as a strategy for conserving biological diversity. *Environ. Manage*. 12:55-72.
25. Karr, J.R. 1990. Biological integrity and the goal of environmental legislation: Lessons for conservation biology. *Conserv. Biol*. 4:244-50.
26. Leopold, A. 1949. *A Sand County almanac and sketches here and there*. Oxford: Oxford Univ. Press.
27. Lubchenco, J., A.M. Olson, L.B. Brubaker, S.R. Carpenter, M.M. Holland, S.P. Hubbell, S.A. Levin, J.A. MacMahon, P.A. Matson, J.M. Melillo, H.A. Mooney, C.H. Peterson, H.R. Pulliam, L.A. Real, P.J. Regal, and P.G. Risser. 1991. The sustainable biosphere initiative: An ecological research agenda. *Ecology* 72:371-412.
28. MacArthur, R.H. 1957. *On the relative abundance of bird species. Proc. Nat. Acad. Sci. USA* 43:293-95.
29. MacArthur, R.H., and E.O. Wilson. 1967. *The theory of island biogeography*. Princeton: Princeton Univ. Press.
30. Menges, E.S. 1986. Predicting the future of rare plant populations: Demographic monitoring and modeling. *Natural Areas J*. 6:12-25.
31. Myers, N. 1988. Tropical forests and their species. In *Biodiversity*, ed. E.O. Wilson and F.M. Peter, pp. 28-35. Washington, DC: National Academy Press.
32. Nash, R., ed. 1976. The American conservation movement. In *Forums in history: Readings in the history of conservation*. Reading, MA: Addison-Wesley Pub. Co.
33. Norse, E. 1990. *Threats to biological diversity in the United States*. PM-223X. US EPA, Office of Planning and Evaluation.
34. Norton, B. 1988. Commodity, amenity, and morality: The limits of quantification in valuing biodiversity. In *Biodiversity*, ed. E.O. Wilson and F.M. Peter, pp. 200-205. Washington, DC: National Academy Press.
35. Noss, R.F. 1983. A regional landscape approach to maintain diversity. *BioScience* 33:700-706.
36. Noss, R.F. 1990. Indicators for monitoring biodiversity: A hierarchical model. *Conserv. Biol*. 4:355-64.

37. Noss, R.F., and L.D. Harris. 1986. Nodes, networks, and MUMs: Preserving diversity at all scales. *Environ. Manage.* 10:299-309.
38. Odum, E.P. 1969. The strategy of ecosystem development. *Science* 164:262-70.
39. Peet, R.K. 1974. The measurement of species diversity. *Ann. Rev. Systematics Ecol.* 5:285-307.
40. Peet, R.K., and N.L. Christensen. 1988. Changes in species diversity during secondary forest succession on the North Carolina Piedmont. In *Diversity and pattern in plant communities*, ed. H.J. During, M.J.A. Werger, and J.H. Willems, pp. 235-45. The Hague: SPB Academic Publishing.
41. Peet, R.K., D.C. Glenn-Lewin, and J.W. Wolf. 1983. Prediction of man's impact on plant species diversity: A challenge for vegetation science. In *Man's impact on vegetation*, ed. W. Holzner, M.J.A. Werger, and I. Ikusima, pp. 41-54. The Hague: Dr. W. Junk Publishers.
42. Peters, R.L. II. 1988. The effect of global climatic change on natural communities. In *Biodiversity*, ed. E.O. Wilson and F.M. Peter, pp. 450-61. Washington, DC: National Academy Press.
43 Pickett, S.T.A. 1980. Non-equilibrium coexistence of plants. *Bull. Torrey Bot. Club* 107:238-48.
44. Pickett, S.T.A., and J.N. Thompson. 1978. Patch dynamics and the size of nature reserves. *Biol. Conserv.* 13:27-37.
45. Plotkin, M.J. 1988. The outlook for new agricultural and industrial products from the tropics. In *Biodiversity*, ed. E.O. Wilson and F.M. Peter, pp. 106-16. Washington, DC: National Academy Press.
46. Preston, F.W. 1962. The canonical distribution of commonness and rarity: Part I. *Ecology* 43:185-215.
47. Raup, D.M. 1988. Diversity crises in the geological past. In *Biodiversity*, ed. E.O. Wilson and F.M. Peter, pp. 51-57. Washington, DC: National Academy Press.
48. Ray, G.C. 1988. Ecological diversity in coastal zones and oceans. In *Biodiversity*, ed. E.O. Wilson and F.M. Peter, pp. 36-50. Washington, DC: National Academy Press.
49. Romme, W.H., and D.H. Knight. 1982. Landscape diversity: The concept applied to Yellowstone Park. *BioScience* 32:664-70.
50. Schonewald-Cox, C.M., S.M. Chambers, B. MacBryde, and W. Lawrence Thomas, eds. 1983. *Genetics and conservation*. Menlo Park, CA: Benjamin/Cummings Publishing.
51. Shmida, A., and S. Ellner. 1984. Coexistence of plant species with similar niches. *Vegetatio* 58:29-55.
52. Shmida, A., and M.V. Wilson. 1985. Biological determinants of species diversity. *J. Biogeogr.* 12:1-20.
53. Shugart, H. 1984. *A theory of forest dynamics*. New York: Springer-Verlag.
54. Smith, W.H. 1990. *Air pollution and forests*. New York: Springer-Verlag.
55. Soule, M.E. 1991. Conservation: Tactics for a constant crisis. *Science* 253:744-50.
56. Soule, M.E., and D. Simberloff. 1986. What do genetics and ecology tell us about the design of nature reserves? *Biol. Conserv.* 35:19-40.
57. Soule, M.E., B.A. Wilcox, and C. Holtby. 1979. Benign neglect: A model of faunal collapse in the game reserves of east Africa. *Biol. Conserv.* 15:259-72.
58. Templeton, A.R., and B. Read. 1983. The elimination of inbreeding depression in a captive population of Speke's gazelle. In *Genetics and conservation*, ed. C.M. Schonewald-Cox, S.M. Chambers, B. MacBryde, and W. Lawrence Thomas, pp. 241-61. Menlo Park, CA: Benjamin/Cummings Publishing.
59. Terborgh, J. 1974. Preservation of natural diversity: The problem of extinction-prone species. *BioScience* 24:715-22.

60. Terborgh, J. 1988. The big things that run the world—a sequel to E.O. Wilson. *Conserv. Biol.* 2:402-3.
61. Tilman, D. 1988. *Plant strategies and the dynamics and structure of plant communities.* Princeton: Princeton Univ. Press.
62. Van der Maarel, E. 1988. Species diversity in plant communities in relation to structure and dynamics. In *Diversity and pattern in plant communities*, ed. H.J. During, M.J.A. Werger, and J.H. Willems, pp. 1-14. The Hague: SPB Academic Publishing.
63. Vane-Wright, R.I., C.J. Humphries, and P.H. Williams. 1991. What to protect?—systematics and the agony of choice. *Biol. Conserv.* 55:235-54.
64. Veitmeyer, N. 1986. Lesser known plants of potential use in agriculture and forestry. *Science* 232:1379-84.
65. White, P.S. 1979. Pattern, process, and natural disturbance in vegetation. *Bot. Rev.* 45:229-99.
66. White, P.S. 1987. Natural disturbance, patch dynamics, and landscape pattern in natural areas. *Natural Areas J.* 7:14-22.
67. White, P.S., and S.P. Bratton. 1980. After preservation: The philosophical and practical problems of change. *Biol. Conserv.* 18:241-55.
68. White, P.S., and S.T.A. Pickett. 1985. Natural disturbance and patch dynamics: An introduction. In *The ecology of natural disturbance and patch dynamics*, ed. S.T.A. Pickett and P.S. White, pp. 3-16. New York: Academic Press.
69. Whittaker, R.H. 1975. *Communities and ecosystems.* New York: MacMillan Publishing Co.
70. Wiens, J.A. 1989. Spatial scaling in ecology. *Funct. Ecol.* 3:385-97.
71. Wilson, E.O. 1987. Biological diversity as a scientific and ethical issue. In Papers Read at a Joint Meeting of the Royal Society and the American Philosophical Society, Volume 1, pp. 29-48. Philadelphia: American Philosophical Society.
72. Wilson, J.B. 1991. Methods for fitting dominance-diversity curves. *J. Veg. Sci.* 2:35-46.

Part II

Overview of Air Pollution Exposure and Effects

3

Air Pollution Transport, Deposition, and Exposure to Ecosystems

David Fowler

INTRODUCTION

This review chapter outlines the major processes influencing the transport, transformation, and deposition of the major pollutant gases and their oxidation products SO_4^{2-} NO_3^- in aerosols and cloud droplets.

The gases considered include those from fossil fuel combustion, sulfur dioxide (SO_2), nitric oxide (NO), nitrogen dioxide (NO_2), and to a lesser extent hydrogen chloride (HCl); the major gaseous products of photochemical smog, ozone (O_3), and peroxy-acetyl nitrate (PAN); and the most important regional pollutant resulting from agricultural activity, ammonia (NH_3).

The subject of air pollution effects on biodiversity draws together several different fields of interest in air pollutants. In particular, recent interest in the effects of deposited nitrogen (as NH_3, NH_4^+, and NO_3^-) on the species composition of seminatural ecosystems has developed quite independently of studies of the phytotoxic effects of the gaseous pollutants SO_2 and O_3 on plants. Similarly, interest in the phytotoxicity of "acidic" pollutants deposited as droplets on high-elevation forests may also be contrasted with interest in effects of the primary gaseous pollutants SO_2 and NO_2 on crop physiology and yield. There are, however, common processes effecting the transfer and uptake of some of these pollutants.

The ecological effects at any location result from a combination of the biological, physical, and chemical properties of the ecosystem, their response to pollutants, and the magnitude of pollutant "dose." The primary pollutants may cause crop loss close to sources through the dry deposition of SO_2 and NO_2. Or, following long-range transport ($\geq$ 1,000 km) and oxidation of the pollutants to SO_4^{2-} and NO_3^-, respectively, and deposition as cloud and rain droplets, the same emitted pollutants may cause a decline in the health of the high-altitude canopies of red spruce (*Picea rubens*) or Norway spruce (*P. abies*).

The scientific field is too large to cover comprehensively in the length of this chapter; therefore, it is necessary to focus on aspects of the subject most relevant to biodiversity. For this reason, a more detailed description of ammonia exchange between the atmosphere and vegetation than would commonly be found in texts considering atmospheric pollutants is provided. This emphasis reflects the recent research showing that inputs of atmospheric nitrogen as ammonia by heathland plant communities has a rapid and marked effect on the species composition and abundance. The influence of topography on pollutant exposure and fog and cloud droplet deposition are also highlighted, as these also have been shown to play an important part in "forcing" changes in biodiversity.

DISPERSION AND ATMOSPHERIC TRANSPORT

The major processes involved in the dispersion, transport, and removal of pollutants from the atmosphere are illustrated in Figure 3-1. These processes may arbitrarily be divided into three broad areas: dispersion, transformation, and deposition. The emphasis in this review is on the interaction of the pollutants with terrestrial vegetation in order to provide the necessary description of inputs that are considered in detail in later chapters. For this reason, the first two aspects of atmospheric behavior of pollutants are considered very briefly and the reader is referred to more detailed descriptions published elsewhere (Sandroni 1987). The remaining area, that of removal processes, is considered in more detail.

The gases are injected into a region of the atmosphere in which momentum, heat, moisture, and a variety of other gases are exchanged between the atmosphere and terrestrial (or aquatic) surfaces, known as the boundary or mixing layer. Over land the boundary layer varies in depth with time of day, from a maximum during early afternoon in the range from 500 m to 2,000 m, to a minimum during the night as small as a few tens of meters, depending on windspeed, surface roughness (vegetation height), and the vertical temperature structure of the air. The volume of air into which the pollutant is injected therefore varies with time and meteorological conditions.

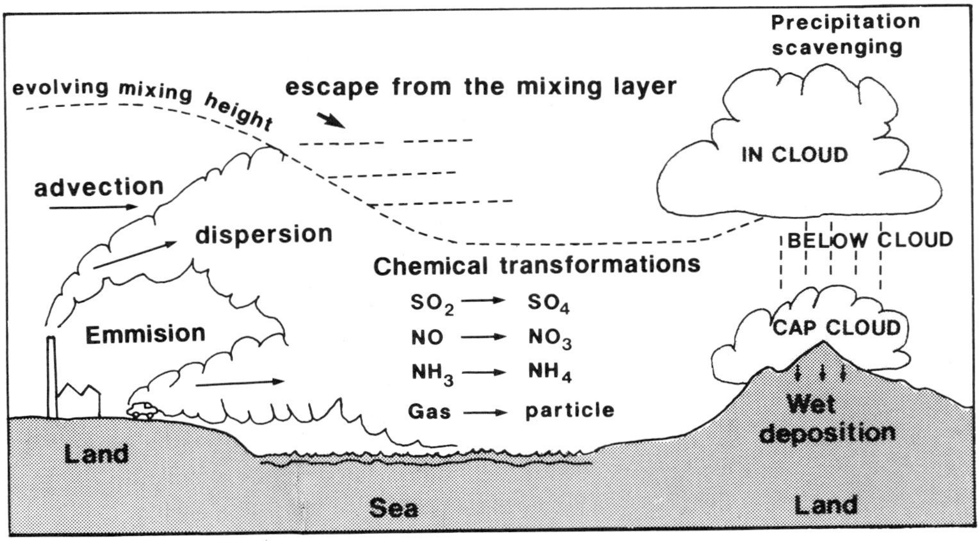

FIGURE 3-1. Processes involved in the transport and deposition of atmospheric pollutants.

In the presence of a low-level temperature inversion, as on a cold still night, any pollutants released at ground level will tend to be trapped close to the ground. Higher level pollutant emissions, from large power plants for example, are injected into layers of the atmosphere that are temporarily isolated from the ground by stratification of the air near the ground. Thus, the concentration of any pollutant gas close to the ground varies with atmospheric mixing and the proximity to sources. Following emission, the gases are subject to mixing processes in the boundary layer driven by turbulence generated by frictional drag of the surface on the airflow, and by convection.

The mixing dilutes initial concentrations and the dispersion and diffusion processes, which are reasonably well understood, provide the basis for much of the modelling of the short-range (0.1 to 10 km) dispersion of pollutant gases (Stull 1988). It has therefore become possible to build industrial plants with emission stacks sufficiently tall that maximum concentrations of pollutants in the plume when it reaches ground level are below the levels at which acute injury to vegetation or fauna occur. The damage to vegetation close to the sources of smelter emissions, for example at Sudbury, Ontario (Freedman and Hutchinson 1980), which were the result of very large emissions at or very close to ground level, is no longer common in the proximity of industrial plants in Western Europe or North America. There remain examples of such practices in Eastern Europe and in parts of Asia. With the development of taller stacks, the mean transport distance of SO_2 has gradually increased, simply by reducing the quantity that is dry deposited close to sources.

CHEMICAL TRANSFORMATION

The atmospheric transformations of SO_2, NO, and NO_2 lead to their oxidation to SO_4^{2-} and NO_3^-, respectively. In the case of SO_2 oxidation, the products are invariably present in particle or droplet form. The NO_3 may be present in particles or droplets as NO_3^- or as HNO_3 in the gas phase. These phase changes have important consequences for their removal from the atmosphere and are considered later.

The oxidation of NO_2 and SO_2 may occur in the gas phase by homogeneous processes, the most important oxidant being the OH radical for both gases. For SO_2, the reaction may be written

$$OH + SO_2 \rightarrow HSO_3^-$$

$$HSO_3^- + O_2 \rightarrow HO_2 + SO_3.$$

The maximum rate of SO_2 conversion on a clear summer day is about 4% hour^{-1} at 50°N, with typical average daily rates of 1% hour^{-1}.

The rate coefficient for the reaction between NO_2 and the hydroxyl radical is almost an order of magnitude faster than the equivalent reaction with SO_2,

$$NO_2 + OH \rightarrow HNO_3$$

and is the major gas-phase oxidation pathway. At night, a further gas phase oxidation pathway for NO_2 is available with O_3 as the oxidant, through the nitrate radical

$$NO_2 + O_3 \rightarrow NO_3 + O_2.$$

In daylight, the NO_3 radical is photolysed very rapidly back to NO_2 but at night the NO_3 radical reacts with a further NO_2 molecule to form N_2O_5.

$$NO_3 + NO_2 \rightarrow N_2O_5$$

which then reacts with water, probably on particles or in cloud droplets, to form HNO_3

$$N_2O_5 + H_2O \rightarrow 2\,HNO_3$$

The gas phase atmospheric chemistry of the sulfur and nitrogen oxides is therefore coupled closely with the chemistry of the principal oxidant species (Hough 1988), the detail of which is outside the scope of this review.

The other major pathway for atmospheric transformation lies in the heterogeneous mechanisms, in which the gases are oxidized within or on the surfaces of particles or cloud or fog droplets. The favored heterogeneous reaction medium for SO_2 is that of droplets in stratiform cloud (Hough 1988). The SO_2 dissolves at typical cloudwater pH's mainly to the bisulphite ion.

$$SO_2 \rightarrow SO_2\,H_2O \rightarrow HSO_3^- + H^+$$

The oxidation of bisulphite yields the sulphate ion and a further hydrogen ion.

$$HSO_3^- + \underline{\text{oxidant}} \rightarrow SO_4^{2-} + H^+$$

Oxidation by atmospheric oxygen is very slow but may be catalyzed by metal ions such as Mn^{2+} or by amorphous carbon, a ubiquitous component of atmospheric aerosols over industrial countries. The most important oxidants for the droplet phase oxidation, however, are hydrogen peroxide and ozone. Hydrogen peroxide is very soluble and in solution reacts readily with the bisulphite ion at a rate given by

$$\frac{d}{dt}[HSO_3^-] = k\,[H^+]\,[H_2O_2]\,[HSO_3^-].$$

Where all concentrations are in solution, k is the rate coefficient that is clearly dependent on the acidity of the cloud droplet. At cloud droplet acidities below pH 5.0, the hydrogen peroxide reaction is the major droplet phase pathway, provided that concentrations of H_2O_2 are adequate. This is the case in remote areas, but over industrial regions the concentrations of SO_2 may considerably exceed those of H_2O_2 and in these conditions the supply of H_2O_2 becomes the rate-limiting factor.

The other major oxidant for heterogeneous SO_2 oxidation is O_3. O_3 is much less soluble than H_2O_2 (1.14×10^{-2} moles l^{-1} atmos^{-1} against 7.1×10^4 moles l^{-1} atmos^{-1} at 298K). Therefore, although in solution O_3 reacts more rapidly than H_2O_2 with SO_2 at acidities typical of cloud water over Western Europe and Eastern North America, the H_2O_2 oxidation pathway is initially the most rapid under most atmospheric conditions. The rate of reaction is given by

$$\frac{d}{dt}HSO_3^- = (K_2 + {}^{k3}/[H^+])\,[O_3]\,[HSO_3^-].$$

The combination of this rate expression with the dependence of the sulfur dioxide solubility on $1/[H^+]$ leads to a marked pH dependence of this oxidation pathway. The presence of atmospheric NH_3, which enters solution rapidly and neutralizes acidity produced through the solution and oxidation of SO_2, promotes the ozone reaction pathway.

In practice, at cloud droplet pH exceeding pH 5.5, the O_3 oxidation pathway is likely to dominate, while at pH's below 5 the H_2O_2 oxidation pathway prevails.

No droplet phase mechanisms with significant rates of NO_2 oxidation to NO_3^- have yet been demonstrated in laboratory or field conditions. However, the $N_2O_5 + H_2O$ reaction may take place either on aerosols or in cloud droplets.

There is a rather poor understanding of the mechanisms leading to oxidation of NO_2 to NO_3 in winter and hence to production of NO_3^- in cloud and rain in the mid-latitude Northern Hemisphere. There remains the possibility that an important heterogeneous oxidation mechanism has yet to be discovered.

The NH_3 emitted from agricultural land following application of animal waste or fertilizer (Buijsman 1987), or from crops during senescence (Sutton 1990) is readily incorporated in acidic atmospheric aerosols and in cloud droplets. Erisman, de Leeuw, and van Aalst (1987) have estimated that the atmospheric conversion rate for NH_3 to NH_4 is $6 \times 10^{-4}\ s^{-1}$ during the day and $4 \times 10^{-4}\ s^{-1}$ at night, yielding remarkably short atmospheric lifetime for NH_3, in the range of 0.5 to 1 hour.

The four major ions NH_4^+, SO_4^{2-}, NO_3^-, and H^+ are the dominant ions in aerosols, cloudwater, and precipitation over substantial areas of eastern North America and Europe. The concentrations of all four ions correlate well with each other (Marsh 1978) and lead to a range of compounds in the solid phase of particulate matter sampled such as $(NH_4)_2SO_4$, NH_4HSO_4, NH_4NO_3, $(NH_4)_2SO_4{\cdot}2NH_4NO_3$ (Hough 1988).

The remaining components that are important for the effects on vegetation are the photochemical oxidants O_3 and PAN. Ozone is produced naturally in the stratosphere and some is transported to the lower atmosphere by large-scale air movements. In addition, an uncertain, but substantial fraction of the ozone found in the lower troposphere is produced by chemical reactions involving man-made emissions.

In the absence of interfering atmospheric species, nitric oxide, the major nitrogen oxide emitted from combustion sources, reacts with ozone to form nitrogen dioxide.

$$NO + O_3 \rightarrow NO_2 + O_2$$

In the presence of sunlight, NO_2 is photolysed in air to reform NO and O_3.

$$NO_2 + hv \rightarrow NO + O$$

$$O_2 + O \rightarrow O_3$$

These reactions are fast, and a photostationary state is rapidly established such that:

$$[NO][O_3] = \frac{k_2[NO_2]}{k_1}$$

These processes cannot produce more ozone than that originally present, but in the presence of any reactive hydrocarbon (represented by RH, where R is a reactive hydrocarbon radical), the following reactions take place.

$$RH + OH \xrightarrow{O_2} RO_2 + H_2O$$

$$RO_2 + NO \xrightarrow{O_2} \text{Carbonyl} + NO_2 + HO_2$$

$$HO_2 + NO \rightarrow NO_2 + OH$$

$$\text{Overall} \quad RH + 2NO + 2O_2 \rightarrow \text{Carbonyl} + 2NO_2 + H_2O$$

The NO_2, which is produced without consumption of O_3, disturbs the stationary state. Some NO_2 then decomposes to NO and O_3 to restore the stationary state, thus providing a mechanism for net O_3 formation.

Hydroxyl radicals, OH, are formed by the photolysis of ozone in the presence of water vapor

$$O_3 \rightarrow O_2 + O^1D$$

$$O^1D + H_2O \rightarrow 2OH$$

and by the photolysis of carbonyl compounds in the presence of NO, e.g.,

$$HCHO \rightarrow 2HO_2 + CO$$

$$HO_2 + NO \rightarrow NO_2 + OH.$$

Hydrogen peroxide is produced by the recombination of hydroperoxyl radicals.

$$HO_2 + HO_2 \rightarrow H_2O_2 + O_2$$

These chemical processes lead to a transformation of the primary gaseous pollutants found in the proximity of sources (0-100 km) to secondary particulate pollutants following 1 to 3 days transport, typically over 500-2,000 km. In this way, the "acidic" pollutant inputs to ecosystems ≥ 1,000 km distant from major source areas are in the form of wet deposition (rain, snow, etc.) or deposited as particles or cloud droplets. In contrast, the pollutant input in the "near field," that is, 0-100 km (downwind) from source areas, is dominated by dry deposition. The precise distance at which the ratio of dry deposition/wet deposition becomes less than unity depends on local meteorological and land use properties and the pollutants under consideration.

Wet Deposition

The processes by which pollutants are transferred to the ground in precipitation (rain, snow, hail, etc.) are collectively termed wet deposition. A range of microphysical processes by which particles may be incorporated by cloud or raindrops have been described by Pruppacher and Klett (1980) and applied to the removal of pollutants by Garland (1978). Most of the mechanisms—diffusiophoresis, thermophoresis, diffusion, and impaction—make only a minor contribution to the observed concentrations of SO_4^{2-}, NO_3^-, NH_4^+ in precipitation (Figure 3-2).

The principal mechanism by which atmospheric particles become incorporated in precipitation is through nucleation scavenging, the particles forming the nuclei on which cloud droplets are formed. In this process, the condensation nuclei in an ascending parcel of air grow by accumulating water vapor once activated close to cloud base until the supersaturation of air close to the droplet ceases. The growth rates of individual droplets are determined by the rates of water vapor diffusion to the droplet and heat transfer from the droplet to the air.

The aerosols containing SO_4^{2-}, NO_3^-, and NH_4^+ form particularly effective condensation nuclei as a consequence of their hygroscopic nature. The contribution of nucleation scavenging to the observed SO_4^{2-} and NO_3^- in rain has been estimated by Garland (1978) at 55% and 85%, respectively.

The other mechanism for incorporating sulfur and nitrogen compounds from air into cloud and precipitation is through the solution and oxidation pathway and through impaction of cloud or rain drops with the gaseous and particulate compounds. For sulfur dioxide, as outlined earlier, two important heterogeneous oxidation pathways exist, contributing the majority of the remaining sulfur in rain.

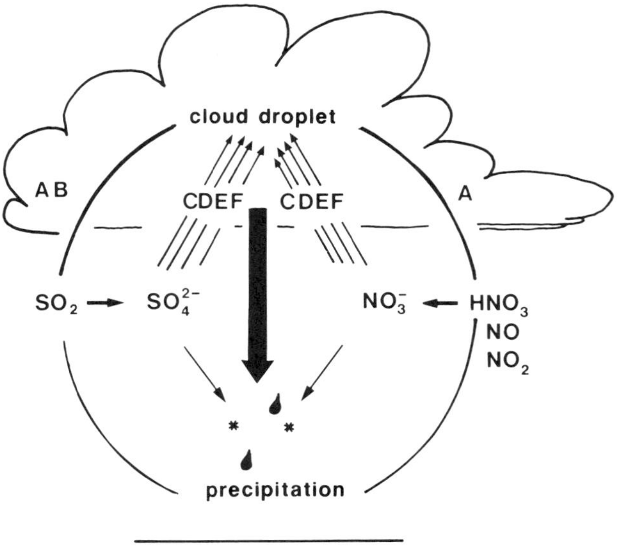

FIGURE 3-2. Pathways and mechanisms for the transfer of sulfur- and nitrogen-containing compounds to the ground in precipitation. A: dissolution; B: oxidation; C: diffusiophoresis; D: Brownian diffusion; E: impaction; F: cloud condensation nuclei pathway. Bold arrows indicate the major routes for wet removal of S- and N-containing compounds from the atmosphere.

The wet deposition of pollutants is extensively monitored in North America and Europe, and a global network of "background" stations has been developed (Galloway and Gandry 1984). Such networks provide satisfactory regional estimates for low elevation and uniform terrain.

For mountains and particularly in regions with large orographic enhancement of precipitation, there are difficulties with measurements as a consequence of the very large spatial variability, with consequent statistical errors. Practical problems also present limitations because no collectors have yet been designed that are suitable for long-term monitoring of wet deposition at hilltop locations.

The inputs by wet deposition at hilltop sites, therefore, have been studied in intensive field experiments in which the mechanism of precipitation scavenging has been investigated (Choularton et al. 1988; Fowler et al. 1988). Such work has shown that vegetation at hilltop locations is subject to larger inputs than valley locations as a consequence of the two processes. First, orographic cloud droplets are intercepted by vegetation, and, second, the wet deposition is enhanced by seeder-feeder scavenging. Such orographic cloud droplets show concentrations of the major ions between a factor of 2 and 8 greater than concentrations in rain at the same site (Fowler et al. 1988). While the number of sites where these studies have been performed is small (Unsworth and Fowler 1988), they do cover the regions where the effects of acidic deposition on vegetation are believed to be occurring. Such locations include high-elevation red spruce forests in the southern Appalachian Mountains, hilltop locations in Northern Britain, and high-altitude sites in the Black Forest.

In the second process, the orographic cloud droplets containing large ion concentrations are scavenged by falling rain from higher level clouds by the seeder-feeder process, leading to larger concentrations in rain at higher elevation (Figure 3-3). The underlying physical

TYPICAL CONCENTRATIONS ($\mu eq\ l^{-1}$)

	Cap cloud	Rain 200m	Rain 800m
SO_4^{2-}	100-2000	42	60
NO_3^-	30-2000	45	65
H^+	10-1000	25	40

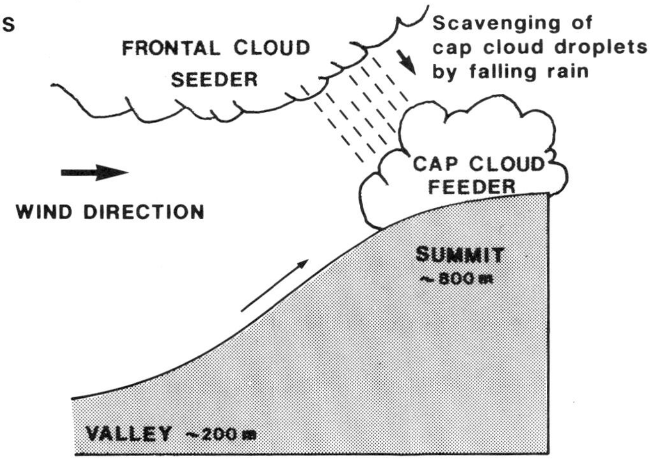

FIGURE 3-3. The seeder-feeder mechanism for enhancing precipitation amount and the concentrations of major ions in precipitation on hills.

processes have been simulated in modelling studies by Choularton et al. (1988) so that the regional wet deposition patterns deduced from low elevation may now be modified to simulate the seeder-feeder effects induced by orography. The important features of hill or mountain sites for vegetation/pollutant interactions are the following:

1. Exposure to large concentrations, up to 2,000 μeq l^{-1} SO_4^{2-}, NO_3^-, or NH_4^+ and pH2.0.
2. Increased inputs of all major ions in wet deposition up to altitudes of 1.5 km. At much higher altitudes (> 2,000 m, in the Alps, for example), the dominance of convective precipitation and the much cleaner air at such altitudes lead to lower concentrations.
3. Additional stress resulting from high windspeed/poor soils and low temperatures. High-elevation sites that are exposed in general and windy provide a closer coupling of foliage with the atmosphere (except in the case of very low-growing plants). As such, they are subject to potentially larger inputs of gaseous and particulate pollutants.

For ozone, the windy nature of such sites prevents the development of a nocturnal temperature inversion and vegetation remains "well coupled" to the atmosphere and in many cases to the free troposphere. Such effects are well illustrated by comparative measurements of ozone on high mountains and in valleys in Bavaria (Isaken 1987), but are equally clear in the contrast between a site at 847 m asl and one at approximately 200 m asl in northern Britain (Figure 3-4).

Particles

For particles smaller than 1.0 μm in diameter, transfer in the free atmosphere is effected by turbulent diffusion and depends on windspeed, surface roughness, and temperature stratification of the lowest layers of the atmosphere. However, transfer through the viscous

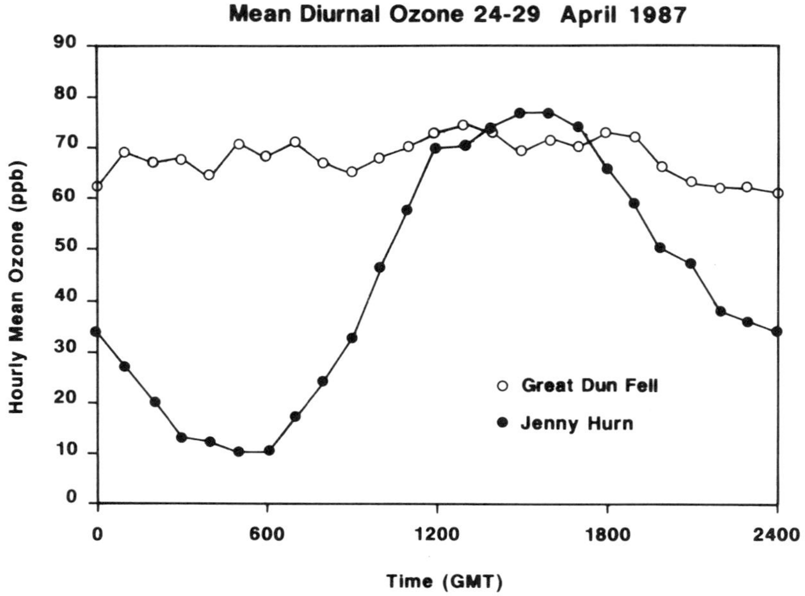

FIGURE 3-4. The mean diurnal variation in ozone concentration at an upland site (Great Dun Fell, 847 m above sea level) and a lowland site in England during the period 24 to 29 April 1987.

sublayer close to the surface presents a considerable restriction to the transfer of 0.1-1.0 μm particles. For much smaller particles, Brownian diffusion effects the transfer; for particles much larger than 5 μm, impaction provides the means of penetrating this viscous sublayer. The problem is similar whether particle transfer is to the ground or to cloud droplets. Consequences of the inefficient dry removal mechanism are a gradual increase in the particle concentration during calm, dry, anticyclonic weather and a reduction in visibility (Whelpdale 1978). The rates of deposition of particles in the size range from 0.1-1.0 μm to grass are small in laboratory studies (deposition velocity $V_g < 0.1$ mm s^{-1}; Chamberlain 1975). Field measurements show more variable results for particles containing S and N. Garland (1978) reports consistently small values ($V_g < 0.5$ mm s^{-1}); Hicks et al. (1983) report more variable and generally larger values (V_g of 1-8 mm s^{-1}). The possibility that a small concentration of large (over 5 μm) particles is present has been suggested (Garland 1978) but has not been confirmed by measurements. Until agreement is reached on the deposition velocities for these particles, inputs by this mechanism must remain uncertain.

The deposition of aerosols in the range from 0.1 to 2 μm on conifer forests has been studied recently by Wiman (1988). He showed from modelling the capture of pollutants at forest edges that deposition velocities to the edge trees may be large (> 3 mm s^{-1}) and of a similar order to the dry deposition velocities for SO_2 on similar canopies (Hallgren et al. 1982). There are, however, few data appropriate to test the predictions of such models, and this currently limits the estimates of particle inputs to the countryside.

Gases

The gases fall into two groups:

1. Those that exhibit no surface resistance: HNO_3, HCl, and for most unmanaged vegetation, NH_3.
2. Those gases whose exchange with terrestrial surfaces is under plant physiological control largely through stomatal resistance: O_3, NO_2, and to some extent SO_2.

HNO_3 and HCl

These gases are deposited on all natural surfaces (with the exception of snow at subzero temperatures) at rates that are determined only by the atmospheric resistances $r_{am} + r_{bg}$ (Figure 3-5). The surface in these conditions is referred to as a perfect sink, and the surface or canopy resistance is zero. The field studies to date are few (Huebert 1983), but all show large deposition velocities and negligible canopy resistances. Measurements of ambient HNO_3 concentrations are few. The available data for northern Europe show concentrations in the range from 0.1 to 1.0 ppbV, with typical summer values of 0.5 ppbV. Fluxes of HNO_3 onto short vegetation with such concentrations would contribute 2 to 4 kg N ha^{-1} annually, but for forest canopies the inputs may approach 8 kg N ha^{-1} and would represent a significant input. In continental Europe and in the polluted regions of North America, nitric acid inputs to vegetation may be much larger as a consequence of the larger ambient concentrations.

The uptake mechanism is therefore simple and the lack of any complicating canopy resistances make estimation of input fluxes quite straightforward. All that is required is an estimate of r_{am} and r_{bg} and air concentration of HNO_3, at the same reference height.

$$\text{Flux } HNO_3 = HNO_3 \text{ concentration}/(r_{am} + r_{bg})$$

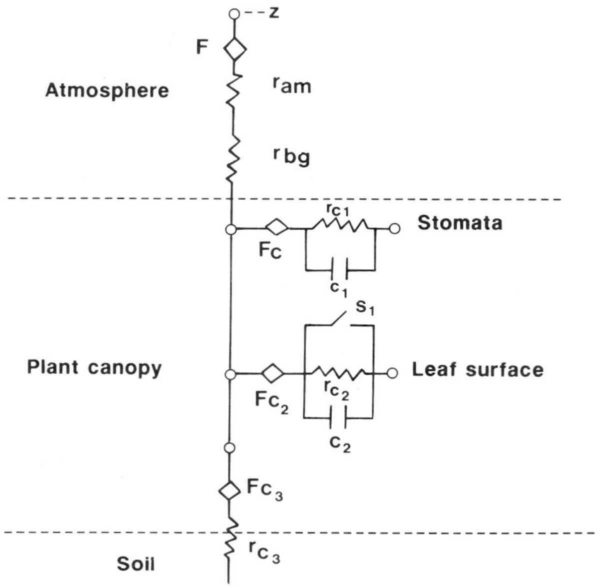

FIGURE 3-5. A simple resistance network to simulate fluxes (denoted F) and resistances (r) in the transport of pollutant gases between soil, vegetation, and the atmosphere. Symbols: r_{am} - atmospheric resistance, bg - viscous sublayer resistance, r_{c1} - stomatal resistance, c2 leaf surface resistance r_{c3} soil resistance, c1 and c2-capacitance of stomata and leaf surface, and s1-a switch that short circuits the r_{c2} term due to surface water.

The most important limitation for estimates of regional inputs of HNO_3 is the lack of monitored HNO_3 concentrations. Additionally, complications arise with the boundary layer resistance r_{bg} over rough surfaces, such as forests, as a consequence of the different empirical relationships available to calculate r_{bg}. Over forests and other complex rough surfaces (buildings and urban areas in general) these lead to uncertainties in HNO_3 deposition of at least 50%. Few measurements of HCl deposition have been reported, but those by Dollard, Davies, and Lundstrom (1987) show no surface resistance. The deposition velocities reported are therefore similar to those for nitric acid, differing only as a consequence of small differences in r_{bg}, resulting from their different molecular diffusities. Again, the extrapolation to tall canopies is subject to uncertainty in r_{bg}.

Stomatal Uptake of O_3 and NO_2

Field measurements of O_3 deposition over vegetation show that the majority of the daytime flux is absorbed via stomata with minimum canopy resistances between 50 and 100 sm^{-1} for crop plants (Wesely et al. 1982; Leuning et al. 1979). The nocturnal deposition rates represent nonstomatal uptake and are typically from 0.5 to 1.0 mm s^{-1}, which represent the leaf surface component of canopy resistance r_{c2} in the range from 1,000 to 2,000 sm^{-1}. In these conditions, rates of deposition are insensitive to r_{am} and r_{bg}. It is probable that rates of external leaf surface uptake are not constant and increase as ambient ozone concentrations increase following the morning breakdown in the nocturnal temperature inversion, just as all new surfaces (even PTFE) exposed to ozone show a rapid initial uptake. However, most field measurements show that the bulk of the flux is absorbed via stomata. The diurnal cycle in deposition resulting from this uptake is a feature of O_3 deposition on vegetation.

The relative importance of leaf surface and stomatal uptake for O_3 is illustrated in Figure 3-6 from laboratory studies with *Zea mays* from Unsworth (1981). The stomatal uptake represents almost 90% of the total flux, and the deposition rates to leaf surfaces in these conditions are only = 0.5 mm s^{-1}. The stomatal uptake pathway provides a convenient means of calculating deposition rates, assuming that there is no internal (mesophyll) resistance r_i. Field studies of ozone deposition have yielded no evidence of a significant mesophyll resistance (Unsworth 1981); however, laboratory studies have shown that ozone may induce stomatal closure although the mechanism and site of action in the stomatal complex remain unknown (Mansfield and Freer-Smith 1984).

Simple resistance models may be used to estimate ozone uptake in the field (Hicks and Matt 1988). In principle, the Penman-Monteith equation (Monteith 1973) also may be applied to provide hourly estimates of bulk canopy resistance to water vapor loss (r_v) from which the bulk resistance to stomatal ozone uptake may be readily obtained. There appears to be no difficulty estimating O_3 fluxes to vegetation using these simple approaches. Ozone is not conserved and following uptake and reaction, in the absence of physiological effects, its uptake is of no further consequence for terrestrial ecosystems, unlike compounds of sulfur and nitrogen.

Nitrogen Dioxide

Early measurements of NO_2 uptake showed that NO_2 was readily absorbed by vegetation (Hill 1971). Such measurements have been repeated by others (Law and Mansfield 1982); laboratory studies show an almost linear increase in NO_2 flux as NO_2 concentration increases from 20 to 100 ppbV (i.e., deposition velocity V_g is constant throughout this concentration range). However, more recent measurements, at smaller concentrations from

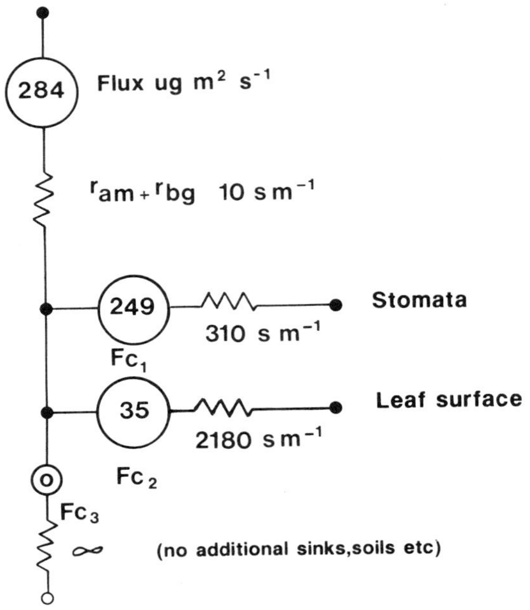

FIGURE 3-6. Fluxes (in circles) and resistances in the deposition of O_3 to *Zeu mays.* The nomenclature follows that described in Figure 3-5.

Source: Unsworth 1981.

1 to 20 ppb, NO_2 deposition rates have been shown to be more complex. Using a cuvette technique, Johansson (1987) showed that uptake of NO_2 by Scots pine was negligible in the range from 0.1 to 2 ppbV NO_2, and 1 mms^{-1} at 1.2 ppb NO_2. However, at 40 ppbV, NO_2 uptake was proportional to stomatal conductance, and throughout this concentration range there was no needle surface uptake. Similarly, Fowler, Cape, and Unsworth (1989) reported small rates of NO_2 deposition over *Eriophorum*-dominated moorland, with mean deposition velocities close to 2 mms^{-1} at an NO_2 concentration of 0.5 ppbV. More recent field measurements over grassland show larger rates of NO_2 deposition than those over pine shoots in the concentration range from 3 to 12 ppbV. The measurements made using micrometeorological methods showed that stomata were the major sink for NO_2, with deposition velocities in the range from 3 to 9 mms^{-1} and no evidence of a compensation point (Hargreaves et al. 1990). The modification of vertical profiles of NO, NO_2, and O_3 resulting from reaction between NO, and O_3, soil emission of NO, and photosynthesis and deposition on NO_2 led to corrections to gradient-derived fluxes of NO_2 of typically 40%.

These recent measurements show that NO_2 uptake by vegetation in the concentration range found in rural areas of Europe and North America may vary from 1 to 3 mms^{-1} in the concentration range from 1 to 5 ppb and up to 6 or 8 mms^{-1} at concentrations of $\geq$ 20 ppb. The dynamic response of the fluxes to the changing concentrations that are a feature of pollutant gases remains uncertain. However, the difficulties experienced in recent NO_2 flux measurement experiments may well be due in part to the enhancement of the nonstationarity problem, as vegetation responds to the natural variations in ambient concentration by changing the sign of the net flux as a new equilibrium in the absorption process is established. Such processes may be treated as the charging and discharging of a capacitor between the sites of uptake and the atmosphere. The role of plant intercellular fluids as a "capacitor" in the exchange of NO_2 with the atmosphere is discussed by Fowler, Cape, and Unsworth (1989), who argue that for ambient NO_2 concentrations, changes in air concentration make negligible contributions to the measured changes in field condition flux and, therefore, leaf surface effects predominate. Further, Lendzian and Kerstiens (1988) have reported much larger permeability of plant cuticles to NO_2 than to O_3, SO_2, or CO_2. Their results imply both reaction with cuticular components and "storage" of adsorbed NO_2 within the cuticle which are large enough to explain measured NO_2 fluxes both towards and away from the surface in the range v_g 1-3 mms^{-1}.

Ammonia

Ammonia is the most abundant alkaline component of the atmosphere. In western Europe, agricultural activities are the primary source of ammonia (Buijsman 1987), which contribute 98% of total emissions. The waste from livestock and particularly that from urine contribute approximately 80% of the total.

In the atmosphere, ammonia plays an important role in neutralizing strong acids such as HNO_3 or H_2SO_4, and may promote the heterogeneous oxidation of SO_2 in cloud with O_3 by raising the droplet pH (Hough 1988). In areas with high densities of farm animals (particularly in the Netherlands, the border counties of England and Wales, and in parts of Denmark), ambient NH_3 concentrations may reach 10 to 20 $\mu g\ m^{-3}$ while aerosol NH4+ concentrations are commonly in the range from 1 to 5 $\mu g\ NH_4\ m^{-3}$ and show less spatial variability. In remote regions (e.g., northern Scotland and northern Sweden), concentrations of ammonia are much smaller, typically 0.1 - 1.0 $\mu g\ m^{-3}$ (Sutton, Moncrieff, and Fowler 1991), and the bulk of the atmospheric NH_x is present as NH_4^+ in these regions.

The exchange of ammonia between terrestrial surfaces and the atmosphere is complex because the emission and deposition fluxes are both commonly observed. For acidic heathland and peat wetlands, the fluxes are invariably directed towards the ground (Duyzer et al. 1987; Sutton, Moncrieff, and Fowler 1991), whereas over agricultural grassland, fertilized using animal manure, fluxes are directed away from the surface. For arable cropland, fluxes may be of either sign depending on atmospheric conditions and the stage in the cropping cycle (Sutton 1990). Furthermore, the nitrogen metabolism of crop plants has been shown by Farquhar et al. (1980) to produce ammonia; as a result, there is a compensation point. At ambient ammonia concentrations exceeding the compensation point, ammonia is deposited and vice versa. To describe the current understanding of ammonia exchange, it is necessary to consider natural and managed vegetation separately.

Ammonia Deposition to Heathland, Moorland, and Unmanaged Grassland

Moorland and heathland surfaces are dominated by plant communities that have developed on poor soils with a limited supply of the major nutrients. The wet ombrotrophic blanket mires are an extreme example and rely entirely on the atmosphere for their nutrient supply, which in the absence of atmospheric pollutants is very limited. The recent changes in species composition in the Dutch heathlands has been shown to be linked to the supply of large quantities of nitrogen from the atmosphere as NH_4^+ and NH_3 (Heil and Diemont 1988). The effects of deposited fixed nitrogen on ombrotrophic wetland plants have also been identified (Lee et al. 1987). The interest in ammonia deposition, therefore, has centered on these sensitive ecosystems for which detailed measurements of ammonia deposition have recently been made (Duyzer et al. 1987; Sutton, Moncrieff, and Fowler 1991).

The measurements by Duyzer et al. (1987) over Dutch heathland show a consistent pattern of deposition. The rates of deposition are too large to be the consequence of stomatal uptake alone and the small surface resistances in dry conditions are reduced to zero when the surface is wet. These surfaces are almost perfect sinks for gaseous ammonia and with ambient concentrations in the range 3 to 10 ppbV, the annual input of nitrogen by dry deposition is in the range from 30 to 100 kg N ha^{-1}. The measurements by Sutton (1990) of ammonia deposition on moorland agree well with those of Duyzer et al. (1987), but being wetter surfaces than the dry Dutch heathland, the *Eriophorum*-dominated moorlands showed no significant surface resistance to ammonia deposition over extended periods, even when the surface appeared to be dry. Figure 3-7 shows a sequence of measured ammonia deposition rates by Sutton (1990) in which deposition velocities varied between 30 mms^{-1} and 60 mms^{-1} throughout the period of measurements, but no surface resistances were significantly different from zero. Similar results were obtained by Sutton, Moncrieff, and Fowler (1991) over unfertilized grassland with rates of deposition limited only by atmospheric resistances ($r_c \approx$ zero).

Forests

Interest in the role of atmospheric nitrogen inputs to forests has been stimulated by hypotheses linking recent forest decline with nitrogen inputs (Nihlgard 1985; van Breeman 1988). The potential rates of ammonia deposition are large and the deposition velocities, like those of HNO_3, may exceed 10 cm s^{-1}. However, such deposition rates have proved difficult to measure using micrometeorological methods. Sutton (1991) reported data in which no significant surface resistance was detected and deposition velocities in the range 70-120 mms^{-1} were measured. The associated uncertainties, however, were very large and

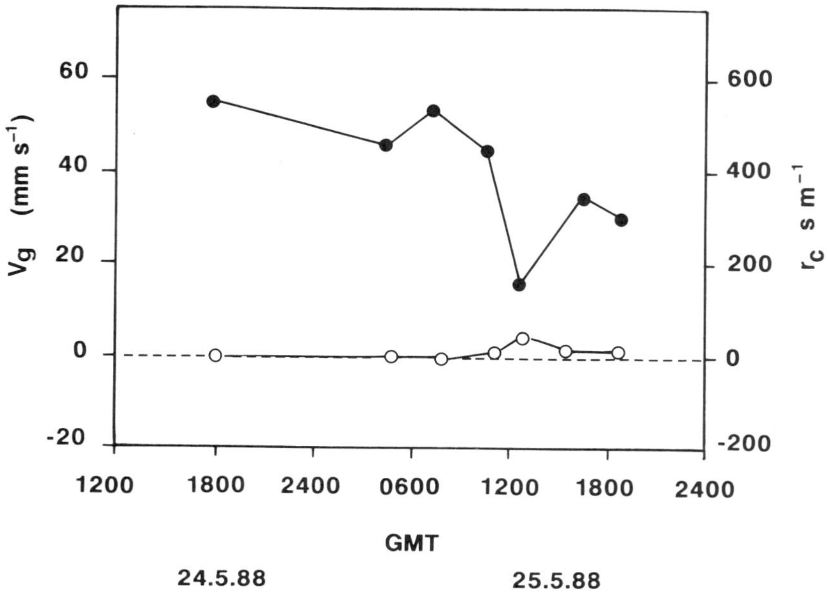

FIGURE 3-7. The rates of deposition of ammonia on moorland, V_g (solid circles) and canopy resistances, r_c (open circles) to ammonia uptake at the same site (Fala Moor) during windy late spring conditions.

expressed as a coefficient of variation approach ± 100% as a consequence of the very small ambient concentrations and small vertical gradients. Recently, however, more detailed measurements of ammonia deposition on a Douglas fir forest in the Netherlands confirm these very large deposition rates with negligibly small canopy resistances (Duyzer, Verhagen, and Westrate 1990).

Cropland

Measurements over agricultural crops have centered on the problem of volatilization losses of ammonia following application of urea and on the nitrogen budget of crop canopies. The upward fluxes from soil surfaces in these conditions may be large (up to 1.6 μg N m^{-2} s^{-1}), two orders of magnitude larger than deposition fluxes onto moorland. These fluxes may be reduced by crop uptake so that the flux to the free atmosphere above the crop canopy is almost an order of magnitude smaller (Denmead, Simpson, and Freney 1974). Measurements over grass, clover, and maize all show large emission fluxes following fertilizer application (Denmead, Simpson, and Freney 1974) or after cutting for hay, but in all cases deposition as well as emission fluxes were reported, depending on soil, vegetation, and atmospheric conditions.

The bidirectional nature of the ammonia flux has been investigated in laboratory conditions by Farquhar et al. (1980), in which a compensation point determines the equilibrium atmospheric NH_3 concentration; below this point emission occurs and above this point deposition occurs. This may be written

$$F_{NH_3} = (r_{am} + r_{bg} + r_c)^{-1} (1 - C_s/C_a)C_a$$

where the resistances are denoted with subscripts am for aerodynamic, bg for laminar sublayer, and c for canopy. C_s is the NH_3 concentration in soil (or plant) and C_a the concentration in air. The intermittent nature of emission fluxes and their dependence on soil and plant conditions make the estimation of annual net budgets a speculative process in the absence of mechanistic models to simulate the underlying processes. In general, the work on volatilization (Denmead, Simpson, and Freney 1974) has been used to deduce the loss of applied fertilizer rather than to deduce net annual NH_3 exchange over cropland. Sutton (1990) suggests that in UK conditions, the net exchange over cereals is likely to be very small (1-4 kg N ha^{-1}) as a consequence of small deposition fluxes over long periods and large emission fluxes for short periods following fertilizer application and during senescence.

Sulfur Dioxide

As the first pollutant identified as a major contributor to long-range transport of atmospheric acidity, sulfur dioxide and its oxidation products in the atmosphere have been the subject of three decades of study (see e.g., Husar 1980). The mechanism of dry deposition has been examined in laboratory experiments (Payrissat and Beilke 1978; Taylor and Tingey 1982) and in field studies (Garland 1978; Fowler and Unsworth 1979). These studies showed that rates of SO_2 deposition onto short vegetation were determined largely by stomatal uptake. A small rate of deposition onto external surfaces (at typically 1 to 2 mms^{-1}) was also identified. Superimposed on the leaf surface uptake, the large stomatal fluxes generate a diurnal cycle in deposition velocity from a daytime maximum of 15 mms^{-1} to nocturnal deposition velocities in dry conditions of 1 to 2 mms^{-1} (Fowler and Unsworth 1979). More detailed laboratory studies of the mechanisms of SO_2 uptake by plants (Taylor and Tingey 1982) show that stomatal resistance is in some cases augmented by a residual internal resistance. Similarly, field measurements of SO_2 deposition on a Scots pine forest provide further evidence of internal resistance (Fowler and Cape 1983). The net uptake of SO_2 uptake by forests is further complicated by the emission of reduced sulfur species (Hallgren et al. 1982), a process which has been investigated in other species more recently by Rennenberg (1984).

The deposition velocities from field and laboratory studies have been used extensively in long-range transport models to estimate the inputs of sulfur by dry deposition of regional scales. Such parameterization, however, fails to make full use of the available understanding of processes. The stomatal uptake may be simulated as described in the modelling section to follow. The effect of different vegetation across the landscape on the atmospheric resistances in the deposition pathway may also be simulated from measured or derived variables. The use of a fixed deposition velocity to estimate dry deposition to the countryside is therefore difficult to justify on scientific grounds.

Wet Surfaces

For a soluble, reactive gas such as SO_2, the presence of surface water provides additional sites of uptake. Measured rates of SO_2 deposition onto a dew-wetted wheat crop while dew was accumulating were close to the upper limit ($(r_{am} + r_{bg})^{-1}$) (Fowler and Unsworth 1979). However, during the period when dew was not accumulating or was evaporating, a significant surface resistance was detected, which reached 200 s m^{-1} shortly before the dew had completely evaporated. A rain-wetted Scots pine canopy may also exhibit a significant surface resistance, this time of ca 300 sm^{-1} (Fowler and Cape 1983). It appears from these measurements that the S^{IV} in water layers may frequently be in equilibrium with ambient

SO_2 concentrations and that slow oxidation of the dissolved S^{IV} species to S^{VI} or the pH of the water film was restricting further uptake (Liss 1971).

Co-deposition of NH_3 and SO_2

The uptake of SO_2 by surface water layers on vegetation is, in principle, promoted by the presence of ambient NH_3, as this on uptake by the water neutralizes acidity generated by the solution and oxidation of SO_2. The potential for NH_3 uptake by cloud droplets has been widely recognized as a mechanism to enhance the uptake and heterogeneous oxidation of SO_2 (Hough 1988). In the Netherlands, where the large ambient NH_3 concentrations are similar to SO_2 concentrations, there is strong circumstantial evidence that enhanced rates of deposition of the two gases on vegetation occur as a consequence of "co-deposition" (Heil and Diemont 1988).

Measurements in wind tunnel experiments by Adema (1986) show that the presence of NH_3 enhances SO_2 deposition. In field conditions using the ambient concentrations of these two pollutants, it is not so easy to demonstrate that co-deposition is occurring. However, measurements from a recent field study of trace gas exchange provide data that are entirely consistent with the phenomenon of "co-deposition."

The ammonia fluxes in this study were directed away from the surface, a grazed pasture, while SO_2 was being deposited. The concurrent measurements of latent and sensible heat flux enabled estimates of the canopy resistance to water loss, and bulk stomatal resistance to SO_2 uptake, to be made. These measurements showed that the total canopy resistance to SO_2 deposition was smaller than bulk stomatal resistance by almost a factor of 4 and implied that the bulk of the SO_2 was being deposited on external surfaces of the vegetation. The fluxes and their partitioning between leaf surface and stomatal uptake are provided in diagrammatic form in Figure 3-8.

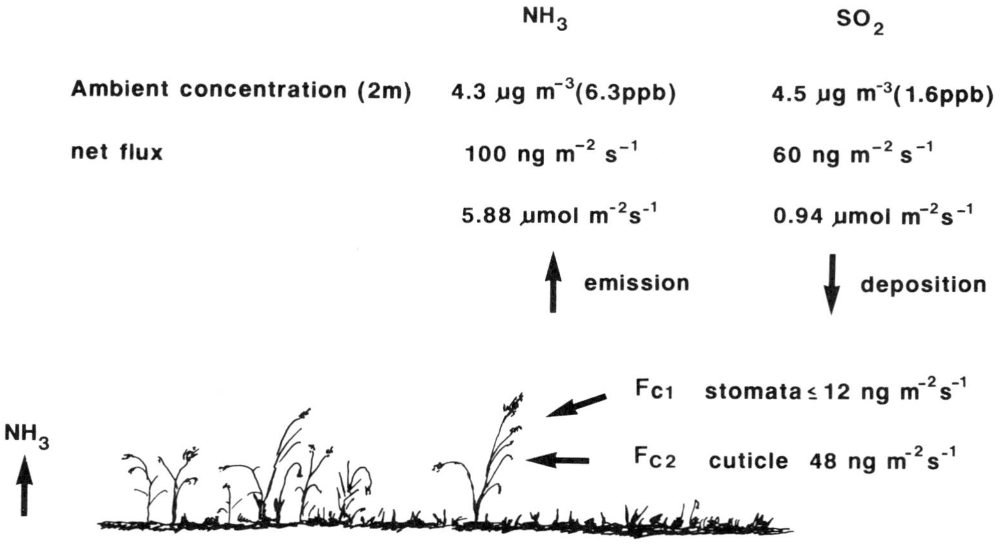

FIGURE 3-8. Fluxes and concentrations of ammonia and sulfur dioxide over the Halvergate marshes showing evidence of the large rates of leaf surface SO_2 uptake. The stomatal uptake of SO_2 is deduced from bulk stomatal canopy resistance measurements; leaf surface uptake represents the residual.

CONCLUSION

The development of our understanding of the exchange of pollutants beween vegetation and the atmosphere has highlighted the number of trace atmospheric gases, particles, and cloud droplets whose fluxes play an important role in ecosystem function.

The input of fixed nitrogen as gaseous NH_3, NO_2, and HNO_3, and as NO_3^- and NH_4^+ in precipitation has been shown to influence species composition in heathland and moorland plant communities. Recent research has shown the mechanism of exchange of the most important nitrogen sources (NH_3 and NO_2) between vegetation and the atmosphere. Using this information and simple models of the deposition pathway, it is now possible to model the inputs over heathland and possibly also for natural grasslands. For fertilized agricultural land, the processes regulating NH_3 emission by plants are not understood sufficiently well to permit estimates of the net nitrogen exchange with the atmosphere over a growing season.

Recent research has also shown that leaf surfaces are sites on which important chemical reactions between atmospheric trace gases occur. The co-deposition of SO_2 and NH_3 on moist surfaces of grass and heather canopies appears to modify the deposition rates of both gases, though the kinetics of the reaction have yet to be established. It seems probable that other important trace gases also undergo chemical transformations on leaf surfaces.

Stomatal uptake of the major phytotoxic gases SO_2, O_3, and NO_2 can be calculated using stomatal opening characteristics for a range of species and is a valuable technique for estimating dose. However, this assumes that the concentration of each of these gases is zero within substomatal cavities, while the actual concentrations of these pollutants within substomatal cavities are poorly understood.

Forest canopies have been shown to capture wind-driven fog and cloud droplets efficiently; as a consequence, pollutant inputs through this mechanism may represent a hazard for some forest species (e.g., red spruce and Norway spruce). In polluted regions, sites that are frequently exposed to orographic cloud may therefore experience species composition change, with short and more tolerant grass species replacing sensitive conifers. A knowledge of both the species' response to pollutants and the mechanisms of deposition, therefore, can be used to quantify the potential for changes in land use or species changes.

These examples of circumstances in which pollutants present on regional scales may force changes in biodiversity contrast with the very limited evidence for regional-scale biological effects of SO_2 or NO_2 directly. They show in particular that more attention should be directed toward natural plant communities, unlike pollution effects research of the last three decades. In this way, more subtle changes may be detected.

The range of trace gases and particles that are exchanged between the atmosphere and terrestrial surfaces appears to be growing. Not only are the primary pollutants and their oxidation products included but biogenic emissions that result from intensive agriculture (NH_3, N_2O) and HONO resulting from NO_2 deposition are also exchanged. It is also probable that volatile organic compounds from vegetation play a much more important role in photochemical oxidant production than was believed until recently. The research required to understand the role of surface-atmosphere exchange of pollutants, therefore, extends across a range of environmental problems and a long list of plant species. The goals set by major funding agencies currently are directed towards the understanding of regional budgets for the major sulfur- and nitrogen-containing pollutants. However, the wider scientific need to explain the roles of these pollutants in other chemical cycles within the atmosphere must not be overlooked.

ACKNOWLEDGMENTS

The author gratefully acknowledges support from the UK Department of the Environment under contract PECD 7/12/82 and by the Commission of the European Communities. The helpful comments and criticism from referees are also acknowledged.

REFERENCES

1. Adema, E.H. 1986. On the dry deposition of NH_3, SO_2, and NH_3 on wet surfaces in a small scale wind tunnel. Proceedings of the 7th World Clean Air Congress, Sydney, Vol. 2, ed. H.F. Hartmann, pp. 1-18. *Clean Air Society of Australia and New Zealand.*
2. Buijsman, E. 1987. Ammonia emission calculation: Fiction and reality. In *Ammonia and acidification*, ed. W.A.H. Asman and H.S.M.A. Diederen, pp. 13-28. EURASAP Symposium, Bilthoven, April 1987. Netherlands: National Institute of Public Health and Environmental Protection/National Institute of Environmental Sciences (RIVM/TVO).
3. Chamberlain, A.C. 1975. The movement of particles in plant communities. In *Vegetation and the atmosphere*, ed. J.L. Monteith, Vol. 1, pp. 115-201. London: Academic Press.
4. Choularton, T.W., M.J. Gay, A. Jones, D. Fowler, J.N. Cape, and I.D. Leith. 1988. The influence of altitude on wet deposition. Comparison between field measurements at Great Dun Fell and the predictions of a seeder-feeder model. *Atmos. Environ.* 22:1363-71.
5. Denmead, O.T., J.R. Simpson, and J.R. Freney. 1974. Ammonia flux into the atmosphere from a grazed pasture. *Science* 185:609-10.
6. Dollard, G.J., T.J. Davies, and J.P.C. Lundstrom. 1987. Measurements of the dry deposition rates of some trace gas species. In *Physico-chemical behaviour of atmospheric pollutants*, ed. G. Angeletti and G. Resteklli, pp. 470-79. Dordrecht: Reidel.
7. Duyzer, J.H., A.M.H. Bowman, H.M.S.A. Diederen, and R.M. von Aalst. 1987. *Measurements of dry deposition velocities of NH_3 and NH_4 over natural terrains*. Research Report No. R87/273. Netherlands: Netherlands Organization for Applied Scientific Research/Netherlands Institute of Environmental Sciences (TNO).
8. Duyzer, J.H., H.L.M. Verhagen, and J.H. Westrate. 1992. Measurement of the dry deposition flux of NH_3 on to coniferous forest. *Environ. Pollut.* 75 (1): 3-14.
9. Erisman, J.W., F.A.A.M. de Leeuw, and R.M. van Aalst. 1987. Depositie van de voor verzuring in Nederland belangrijkste componenten in de jaren 1980-1986 (Deposition of the most important acidifying components in The Netherlands in 1980-1986). Report 228473001, 57 pp. Bilthoven: RIVM.
10. Farquhar, G.D., P.M. Firth, R. Wetsetaar, and B. Wier. 1980. On the gaseous exchange of ammonia between leaves and the environment: Determination of the ammonia compensation point. *Plant Physiol.* 66:710-14.
11. Fowler, D., and M.H. Unsworth. 1979. Turbulent transfer of sulphur dioxide to a wheat crop. *Q. J. R. Meteorol. Soc.* 105:784.
12. Fowler, D., and J.N. Cape. 1983. Dry deposition of SO_2 onto a Scots pine forest. In *Precipitation scavenging, dry deposition and resuspension*, ed. H.R. Pruppacher, R.G. Semohin, and W.G.N. Slinn, pp. 763-74. New York: Elsevier.
13. Fowler, D., J.N. Cape, I.D. Leith, T.W. Choularton, M.J. Gay, and A. Jones. 1988. The influence of altitude on rainfall composition at Great Dun Fell. *Atmos. Environ.* 22:1355-62.
14. Fowler, D., J.N. Cape, and M.H. Unsworth. 1989. Deposition of atmospheric pollutants on forests. *Philos. Trans. R. Soc. Lond.* B 324:247-65.

15. Freedman, B., and T.C. Hutchinson. 1980. Smelter pollution near Sudbury Ontario, Canada, and effects on forest litter decomposition. In *Effects of acid precipitation on terrestrial ecosystems*, ed. T.C. Hutchinson and M. Haras, pp. 395-434. NATO ARI. New York: Plenum.
16. Galloway, J.N., and A. Gandry. 1984. The composition of precipitation on Amsterdam Island, Indian Ocean. *Atmos. Environ.* 18:2649-56.
17. Garland, J.A. 1978. Dry and wet removal of sulphur from the atmosphere. *Atmos. Environ.* 12:349-62.
18. Hallgren, J.E., S. Linder, A. Richter, E. Troeng, and L. Granat. 1982. Uptake of SO_2 in shoots of Scots pine: Field measurements of net fluxes of sulphur in relation to stomatal conductance. *Plant Cell Environ.* 5:75-83.
19. Hargreaves, K.J., D. Fowler, R.L. Storeton-West, and J.H. Duyzer. 1992. The exchange of nitric oxide, nitrogen dioxide and ozone between pasture and the atmosphere. *Environ. Pollut.* 75 (2): 53-60.
20. Heil, G.W., and W.M. Diemont. 1988. Raised nutrient levels change heathland into grassland. *Vegetatio* 53:113-20.
21. Hicks, B.B., and D.R. Matt. 1988. Combining biology, chemistry and meteorology in modelling and measuring dry deposition. *J. Atmos. Chem.* 6:117-131.
22. Hicks, B.B., M.L. Wesely, R.L. Coulter, R.L. Hart, J.L. Durham, R.E. Speer, and D.H. Stedman. 1983. *An experimental study of sulfur deposition to grassland in precipitation scavenging, dry deposition and resuspension*, ed. H.R. Pruppacher, R.E. Semonia, and W.E.N. Slinn, pp. 933-42. New York: Elsevier.
23. Hill, A.C. 1971. Vegetation: A sink for atmospheric pollutants. *J. Air Pollut. Control Assoc.* 21:341-46.
24. Hough, A.M. 1988. Atmospheric chemistry at elevated sites. In *Acid deposition at high elevation sites*, ed. M.H. Unsworth and D. Fowler, pp. 1-47. Dordrecht: Kluwer.
25. Huebert, B.J. 1983. Measurements of the dry deposition flux of nitric acid vapour to grasslands and forests. In *Precipitation scavenging, dry deposition and resuspension*, ed. H.R. Pruppacher, R.G. Semonin, and W.G.N. Slinn, pp. 785-94. New York: Elsevier.
26. Husar, R.B. 1980. *Sulphur in the atmosphere*. London: Academic Press.
27. Isaksen, I.S.A. 1987. *Tropospheric ozone, regional and global scale interactions*. Proceedings of the NATO Advanced Workshop on Regional and Global Ozone Interaction and Its Environmental Consequences. Lillehammer, Norway 1987. 425 pp. Dordrecht: Reidel.
28. Johansson, C. 1987. Pine forest: A negligible sink for atmospheric NO_x in rural Sweden. *Tellus* 39B:426-38.
29. Law, R., and T.A. Mansfield. 1982. Oxides of nitrogen and the greenhouse atmosphere. In *Effects of gaseous air pollution in agriculture and horticulture.*, ed. M.H. Unsworth and D.P. Ormerod, pp. 93-112. London: Butterworth.
30. Lee, J.A., M.C. Press, S. Woodin, and P. Ferguson. 1987. Responses to acidic deposition in ombrotrophic mines in the U.K. In *Effects of atmospheric pollutants on forests, wetlands and agricultural ecosystems*, ed. T.C. Hutchinson and K.M. Meema, pp. 549-60. Berlin: Springer-Verlag.
31. Lendzian, K.J., and G. Kerstiens. 1988. Interactions between plant cuticles and gaseous air pollutants. *Aspects Appl. Biol.* 17:97-104.
32. Leuning, R., M.H. Unsworth, H.H. Neumann, and K.M. King. 1979. Ózone fluxes to tobacco and soil under field conditions. *Atmos. Environ.* 13:1155-63.
33. Liss, P.S. 1971. Exchange of SO_2 between the atmosphere and natural waters. *Nature* 233:327-29.

34. Mansfield, T.A., and P.H. Freer-Smith. 1984. The role of stomata in resistance mechanisms. In *Gaseous pollutants and plant metabolism*, ed. M.J. Koziol and F.R. Whatley, pp. 131-46. London: Butterworth.

35. Marsh, A.R.W. 1978. Sulphur and nitrogen contribution to the acidity of rain. *Atmos. Environ.* 12:401-6.

36. Monteith, J.L. 1973. *Principles of environmental physics*. London: Arnold.

37. Nihlgard, B. 1985. The ammonia hypothesis: An additional explanation to the forest dieback in Europe. *Ambio* 14:2-8.

38. Payrissat, M., and S. Beilke. 1978. Laboratory measurements of the uptake of sulphur dioxide by different European soils. *Atmos. Environ.* 9:211-17.

39. Pruppacher, H.R., and J.D. Klett. 1980. *Microphysics of clouds and precipitation*. Dordrecht: Reidel.

40. Rennenberg, H. 1984. The fate of excess sulfur in higher plants. *Annu. Rev. Plant Physiol.* 35:121-53.

41. Sandroni, S. 1987. *Regional and long-range transport of air pollution*. 510 pp. Amsterdam: Elsevier.

42. Stull, R.B. 1988. *An introduction to boundary layer meteorology*. Dordrecht: Kluwer.

43. Sutton, M.A. 1990. *The surface-atmosphere exchange of ammonia*. Ph.D. thesis, Univ. of Edinburgh.

44. Sutton, M.A., J.B. Moncrieff, and D. Fowler. 1992. Deposition of atmospheric ammonia to moorlands. *Environ. Pollut.* 75 (1): 15-24.

45. Taylor, G.E., Jr., and D.T. Tingey. 1982. Sulfur dioxide flux into leaves of *Geranium carolinianum* L.: Evidence for a non-stomatal or residual resistance. *Plant Physiol.* 72:237-44.

46. Unsworth, M.H. 1981. The exchange of carbon dioxide and air pollutants between vegetation and the atmosphere. In *Plants and their atmospheric environment*, ed. J. Grace, E.D. Ford, and P.G. Jarvis, pp. 111-38. Oxford: Blackwell Press.

47. Unsworth, M.H., and D. Fowler. 1988. *Acid deposition at high elevation sites*. NATO ARI series. Dordrecht: Kluwer.

48. Van Breeman, N., and H.F.G. Van Dijk. 1988. Ecosystems effects of atmospheric deposition of nitrogen in the Netherlands. *Environ. Pollut.* 54:249-74.

49. Wesely, M.L., J.A. Eastman, D.H. Stedman, and E.D. Yelvac. 1982. An eddy-correlation measurement of NO_2 flux to vegetation and comparison to ozone flux. *Atmos. Environ.* 16:815-20.

50. Whelpdale, D.M. 1978. Large-scale atmospheric sulphur studies in Canada. *Atmos. Environ.* 12:661-70.

51. Wiman, B.L.B. 1988. Aerosol capture by complex forest architecture. In *Vegetation structure in relation to carbon and nutrient economy*, ed. J.T.A. Verhoeven, G.W. Heil, and M.J.A. Werger, pp. 157-183. The Hague: SPB Academic Publishing.

4

Monitoring Atmospheric Effects on Biological Diversity

Robert C. Musselman, Douglas G. Fox, Charles G. Shaw III, and William H. Moir

INTRODUCTION

Distinguishing natural successional and evolutionary changes in an ecosystem from those that occur because of subtle, anthropogenically driven processes is extremely difficult. Detailed information about ecosystem components is needed and a fundamental understanding of processes that dynamically link these components. The ecosystem components important in maintaining biodiversity are genetic variability, species composition, and the structure and function of these species in the ecosystem.

A sound approach to monitoring ecosystem changes requires a careful and comprehensive inventory of the presence, location, abundance and "health" of species; of the various ways these species interact to form communities; and of the inputs and exchanges of energy, water, and nutrients within and among these communities. Procedures for conducting such monitoring are described here. A case study of the Glacier Lakes Ecosystem Experiments Site (GLEES) monitoring program to evaluate atmospheric effects on an alpine and subalpine ecosystem is used to exemplify the nature of such comprehensive monitoring. The GLEES monitoring program measures both the atmospheric and the biotic components of the ecosystem. Current plans of the U.S. Department of Agriculture (USDA) Forest Service to monitor forest health in conjunction with the U.S. Environmental Protection Agency's (EPA) Environmental Monitoring and Assessment Program (EMAP) also are reviewed.

ISSUES IN MONITORING ATMOSPHERIC EFFECTS AND BIOLOGICAL DIVERSITY

Monitoring to determine effects of air pollutants on biodiversity must include both atmospheric and biological measurement. Before determining how to monitor atmospheric and biological components of ecosystems, we must discuss the concepts involved in monitoring of the atmosphere and the biosphere, atmospheric effects on biota, and indications of biodiversity.

Environmental Monitoring

Environmental monitoring is undergoing renewed emphasis as public awareness of the condition of our global environment increases. The primary objective of monitoring is to establish some baseline that indicates ecosystem condition and thereafter to document spatial and temporal changes. Baseline conditions need to be measured before changes can be determined. A monitoring program implies repeated measurements over time, perhaps decades. The following minimal set of considerations must be satisfied to establish a monitoring program.

Statistical Inference
Measurements must be representative in both time and space of the subject ecosystems. Thus monitoring plans need to address temporal and spatial variability. The design and sampling intensity needs to assure that evaluation of the status and conditions of the ecosystem relative to past conditions will be sensitive to real changes in the ecosystem. Frequency and intensity of sampling depend on monitoring objectives, system variability, experimental design, and available resources.

Representativeness
Sites selected for monitoring should appropriately represent the variability of some larger environment so that inferences on condition can be extrapolated to the larger area, be it forest, wetland, or desert. Unless sites are carefully selected without bias, the data collected may not relate to other similar systems, and inferences will not be valid. Plots should be randomly located within carefully selected representative ecosystems.

Uniqueness
The ethic of conservation biology is one of the most important concepts evolving in science today. The uniqueness of each ecosystem needs to be clarified by the sampling procedure. Particularly important is a species inventory and careful tracking of those identified as rare or endangered. Special sampling procedures to locate and document these species may be needed. Vascular plants and vertebrate animals are common in such inventories. Nonvascular plants and nonvertebrates are also important, though seldom inventoried. Changes in these taxa are often expected to be reflected in changes in higher trophic levels or in changes in ecosystem function, but this assumption needs further documentation.

Quality Control
To ensure the utility of collected data, we must be able to interpret data using documented techniques. In addition, data from various sites must be compatible and comparable. Just as taxonomists must arrive at a common understanding of the suite of unique characters that define a species, and chemists must agree on appropriate analytical methods and their associated precision and accuracy, so must those engaged in environmental monitoring agree on appropriate techniques and procedures. Even though caution must be exercised in changing monitoring methods, it is recognized that new and better monitoring techniques will develop. Research on improved monitoring techniques is encouraged. When it is

necessary to modify sampling protocols or analytical methods, overlap studies must be conducted to document comparability between and among various methods.

Analytical protocols and experimental techniques and procedures must be stringently documented to maintain continuity in monitoring. Technical staff may change, and even though written protocols should be easily understood and followed by new staff, overlap during staffing changes should be encouraged. Standardized protocols are necessary in the analysis and interpretation of monitoring data.

Atmospheric Effects

Atmospheric effects is the term used to describe how the biosphere is influenced by nonbiotic, above-ground surroundings. It is a comprehensive term that includes energy from the atmosphere as well as directly from the sun. It also encompasses water and other chemical exchanges between the atmosphere and the Earth's surface. Atmospheric effects can be delineated as follows (Musselman and Fox 1991):

First-Order Effects

These result from direct contact of the atmosphere with the biosphere, or with biotic tissue. These atmospheric effects influence both flora and fauna components of the biosphere. Examples include direct phytotoxic reactions to pollutants such as ozone, sulfur dioxide, and fluoride, as well as positive photosynthetic responses to increased CO_2 concentrations and increased growth from foliar fertilization by increased nitrogen oxides. Increased UV-B radiation on plant tissue and increased temperature effects on rate of physiological reactions would be considered first-order effects.

Second-Order Effects

Examples include indirect effects of an altered atmosphere on biota, or on the climate which subsequently affects the biota. Increased degree-day accumulations and redistributed precipitation patterns and resulting effects on soil moisture are examples. Premature senescence from ozone injury or delay of hardening because of late-season fertilization from atmospherically deposited nitrogen are also second-order effects.

Third-Order Effects

These include disturbances exacerbated by atmospheric changes. Fire, increased insect and disease occurrence, and biogeochemical alterations such as eutrophication have enormous potential to alter successional patterns of communities. Changes in community structure or fitness are third-order effects.

A high-quality environmental monitoring program should be able to identify first-, second-, and third-order effects of atmospheric deposition on ecosystems and, ideally, distinguish among them. First-order effects, since they are the first direct interactions of the atmosphere with the biosphere, are easier to detect and monitor.

Within the context of biodiversity, the utility of conventional dose/response testing is limited. Dose-response testing, while useful for examining plant processes and magnitudes of response of individual plants to atmospheric stresses, is difficult to control experimentally and examine at a community or ecosystem level. Although dose-response experiments are essential to enhance our understanding of species response to pollutants, the magnitude of the task using conventional dose-response experimental procedures appears overwhelming.

Biological Diversity

Biological diversity captures diversity at several interrelated levels. Indicators need to be selected and measured within each of these levels. Noss (1990) identifies the structural, functional, and compositional components to monitor at the different levels of organization. Biodiversity is typically examined at four levels of organization; genetic, species, community-ecosystem, and landscape.

1. *Genetic* diversity represents variation within a biological species, for example subspecies, ecotypic variation, additive and nonadditive variance, and heterozygosity.
2. *Species* diversity is a function of the number of species present and can be represented by the number of species contained within a specific area, the species richness. Measures of species diversity depend on the size of the area and, of course, on how carefully it is sampled.
3. *Community-ecosystem* diversity (alpha diversity) involves the number (richness) and distribution (evenness) of species within a community (Whittaker 1972). Age and spatial distribution within communities are components of diversity at the ecosystem level.
4. *Landscape* diversity (beta diversity) involves the numbers (richness) and patterns (evenness) of communities within a given area. Measures of landscape diversity using landscape indices are influenced by spatial and temporal scales (Turner and Gardner 1991). Beta diversity reflects the turnover of species along an environmental gradient. In addition to areas, size introduces concepts that deal with the steepness of the environmental gradient as well as the frequency and intensity of disturbances. Patchiness and connectivity are important parameters at the landscape level.

Biological diversity is of particular concern to Federal land management agencies. Land management agencies are charged with protecting plant and animal species, and often place special emphasis on those considered threatened and endangered. This management objective focuses on preserving diversity at the genetic and species levels. At the community and landscape levels, such agencies are concerned with land use and site conversion through management. At the landscape level, managers must address the loss of wetlands and deterioration of riparian areas, conversion or preservation of old-growth forests, urban expansion, and agricultural and community-oriented land use decisions.

Perhaps our greatest concern is with the inherent conflict between commodity- and diversity-oriented land management objectives. Several national land management agencies around the world, such as the USDA Forest Service, are charged with maintaining sustainable use of natural resources. Utilization often results in fragmentation and isolation, and creates barriers to dispersal of germ plasm. Sustainability suggests the need to set aside diverse lands such as wilderness tracts and natural areas. Questions exist regarding the intermingling of such refugia with commodity production lands. We need to know what size and spatial patterns of natural landscapes are needed to sustain viable populations, allow migrations, and ensure functional landscapes under potential global environmental changes.

Scale

The subject of interactions between various spatial and temporal scales is among the major contemporary issues in Earth science. Powerful tools exist to study spatial scale structure and function. At considerably larger spatial scales, remote sensing tools are developing rapidly. However, there remains a significant gap in both nature and intensity between information gained from remote sensing tools and the detailed information and science available to study the structure and function of individual organisms at specific plots. Recently, there have been attempts both to scale up from the organ and individual levels to population and community, and to scale down from the global to the regional to the landscape level. How individuals organize into populations, and how species interact to create unique communities in the landscape, is an area of increased study in the biological sciences.

Biogeochemical studies concentrate on inputs and outputs of various nutrients (Chen et al. 1983; Cosby et al. 1985). They provide insight into the health of systems on a watershed level. Similar overarching research frameworks unfortunately do not yet exist for examining interaction among communities and landscapes. At the other extreme, progress has been made in modeling global and regional response to atmospheric processes. However, large gaps in knowledge exist about the ways in which landscapes interact with atmospheric processes to form regional patterns. There is also a relationship between space and time scales. The level of organization is also important in the intensity of monitoring needed to measure change. For example, broad scale patterns may change relatively slowly, and thus require less frequent monitoring than would processes occurring at finer, more localized scales.

Research Sites

Sites selected for studying anthropogenic impacts on biological diversity should be carefully chosen to represent the average condition of the selected ecosystem. A comprehensive, holistic approach should be used for monitoring and research at these sites. Sampling should include a full suite of monitoring indicators selected to provide information at the various scales discussed above. Selected measurements can be made at less intensively monitored sites to determine the representativeness of the intensively monitored site.

Long-Term Comprehensive Monitoring

Monitoring of atmospheric and ecosystem processes must be conducted on a constant basis for decades because of the large variability in natural ecosystems and their random and unknown nature. Biological systems and atmospheric conditions constantly change naturally, both temporally and spatially. Both biological and atmospheric parameters vary stochastically. Long-term monitoring will be necessary to detect any pattern in change to either of these systems and linking atmospheric and biological changes will be difficult.

Implicit in this discussion is the need to collect long-term physical as well as biological data. Separating natural successional change from anthropogenic change or change in measurement technique requires a solid database of biological and physical conditions as well as a fundamental understanding of ecosystem processes and function. Detailed data on atmospheric and ecosystem processes is important for testing inferences regarding interaction of these two systems. Physical data should include measurements of atmospheric deposition (wet and dry), meteorological conditions, and biogeochemical information within soils, water, and vegetation. Biological data should include measurements or parameters that will allow us to evaluate changes in ecosystems' biodiversity, resiliency, and productivity. Monitoring should provide data on compositional, structural, and functional components of the biosphere (Noss 1990). Land management options should be considered, but not limiting, when one is determining the specifics of parameters to be measured.

Physical Environment and Biological Response

Responses of the biological system to components of the physical environment determine the nature and extent of ecosystem change. Certain physical components are more critical in particular ecosystems, such as cold temperatures and snowpack in alpine areas. Other physical factors may be more important at specific successional stages of development, such as moisture and temperature in reproduction and establishment of pioneer communities (Woodward 1989). Phenological stage of development and stage of maturity influence the sensitivity of some plants to air pollutants (Smith 1981). Although biota depend on the physical environment, ecosystem structures such as forest canopies can modify these conditions (FAO 1962).

Physical and biological systems can interact at the biogeochemical level. This is particularly important for aquatic ecosystems, whose biological components can respond quickly to small changes in atmospheric inputs. The response depends on the buffering capacity of the watershed and its surface waters. Changes in aquatic components of the ecosystem may be one of the first indications of an impact from atmospheric deposition.

The leaf is another site of intense interaction between physical and biological systems. Atmospheric deposition (wet, dry, and gaseous) interacts through direct surface contact and stomatal uptake. Leaves have some buffering capacity against both acidic and gaseous deposition. Knowledge of the physical and biological processes of interaction is weak, but an understanding of these processes is essential to measure and predict atmospheric effects on biological systems.

Theoretical constraints based on physical and chemical principles enable a detailed understanding of how energy, water, and nutrients are utilized within an ecosystem. Such models can predict system response in physical and chemical terms, but are not able to predict detailed biological responses. From this perspective, the expression of biological diversity and its response to physical and chemical alterations is largely unpredictable. There simply have not been sufficient experimental data collected to model many significant general relationships with environmental parameters.

Interdisciplinary Research

Ecosystem monitoring may not necessarily be directed to specific hypotheses. Baseline or benchmark data are necessary before change can be detected, or before hypotheses concerning such change can be formulated and examined. Research to determine biological responses to atmospheric change requires interaction among scientists from different disciplines; for example, ecology, plant physiology, taxonomy, chemistry, plant pathology, meteorology, soils, geology, hydrology, and limnology. These scientists must conduct their research and monitoring in a larger context, considering how such data will fit within the larger research goals. Data needs may transcend those required by each discipline. Thus individual scientists may need to collect data not directly relevant to their research. Conversely, scientists may need data from others outside their respective disciplines.

Absence of Models/Theory

Detailed and complex process-simulation models exist for tree physiology and forest growth; for the successional dynamics of tree species responses to disturbance; for the biogeochemical cycling of water, carbon, nitrogen, and other nutrients in a watershed context; and for atmospheric energy and mass transfer. A sampling of the models includes TREGRO (Weinstein, Beloin, and Yanai 1991) and MAESTRO (Wang and Jarvis 1990) for individual trees, FORET (Shugart 1984) and ZELIG (Smith and Urban 1988) for forest stands, MAGIC (Cosby et al. 1985) and ILWAS (Chen et al 1983) for watersheds and catchments, and models by Meyers and Paw U (1986), and Raupach (1987) for energy and mass transfer. However, the linkage is weak between the biological and physical/chemical aspects of the models, and considerable effort is now being expended on strengthening this (Landsberg et al. 1991). Modeling efforts scaling up from the tree level (Huston 1991), and scaling down from the global level (IGBP 1990) are being examined.

We can see an example of the lack of linkage between physical and biotic processes in the attempt to determine the effects of air pollutants on wilderness ecosystems. Dispersion models can give credible predictions of concentrations of a specific pollutant from a point source at a wilderness site (Ross and Fox 1991). Linking levels of concentration with biological responses such as photosynthesis or tree growth is difficult. However, there is an increasing interest and research in this linkage.

METHODS TO ASSESS AIR POLLUTION EFFECTS ON BIODIVERSITY

Biological diversity at a landscape level can be measured as the number of communities (richness component) and their spatial arrangement (evenness component). It is unlikely that exposure to short-term air pollution at current ambient concentrations would noticeably change either of these measures. However, pollution exposure over a long period of time, or pollution in combination with other aspects of climate change over the long term, may alter spatial arrangement as some species expand or contract at the expense of others. This might be detected only after decades of monitoring, but could occur rapidly for aquatic systems. Such species shifts resulting in community change are perhaps best monitored by aerial photography. More large-scale changes involving communities displacing other communities with strongly contrasting structures can be measured by remote sensing from orbiting satellites. A number of indices are available that describe spatial arrangements of communities (Turner 1990).

Biological diversity at the community level can be measured by a wide variety of techniques (Magurran 1988). As with diversity at the landscape level, these local measurements are usually summarized by indices that include both richness and evenness components. Most indices are composed of several taxa, any of which might be affected by pollution. Therefore, interpretations based on a changed or unchanged index are incomplete. To illustrate, decreasing biodiversity would take place as species, subspecies, or ecotypes become extinct (whether locally, regionally, or globally) without replacement by adapted species, subspecies, or ecotypes. Therefore, a diversity index will remain unchanged as species, subspecies, or ecotypes become extinct (loss of richness) but accompanied by a redistribution of survivors (gain in evenness), or by invasion of tolerant new species, subspecies, or ecotypes. Finally, an increase in biodiversity might be indicated in a monitoring system as new species, subspecies, or ecotypes invade just before sensitive taxa become locally extinct. This can happen by an increase in the "rare" component of the biodiversity index and negligible change in the evenness component. When evenness and richness are combined in a diversity measure, shifts between them may be of critical biological interest, but the index itself can remain unchanged. These examples illustrate that change or lack of change in an index as a measure of biological diversity cannot be interpreted without knowledge of what is happening to the various components of the index. The index itself reveals little. Peet (1974) discusses other difficulties in interpreting indices of biological diversity.

The effects of air pollution on sensitive species, and therefore on biological diversity, may be very subtle. Variation in populations or in the spatial displacement of affected communities may be relatively small and statistically undetectable. The changes induced by pollution may, in some instances, be overwhelmed by other causative factors. These include climatic variations and natural or man-caused disturbances such as fire, insects, diseases, and grazing.

Change as well as constancy in an index of biodiversity are clearly not easily interpreted. However, we are often not interested in the index of diversity so much as in the affected taxa. Biological diversity, whether at large or small spatial scales, is the mix of different communities, species, populations, or genotypes. It is the loss or constriction of a species, a population, or a habitat that concerns us. We propose that some monitoring effort be directed to known or potential sensitive communities, species, subspecies, or ecotypes in the landscape.

Simple floristic, faunistic, and plant community lists or surveys tallied over all the variations of a bounded landscape may be an effective monitoring scheme at present (Metzler 1980; Magurran 1988). These lists should be prepared by specialists in the taxa being inventoried. Additional data on abundance would be particularly helpful. Changes in the lists can be interpreted with corroborative information on changes in pollution loads and known physiological effects. We assume the lists will include surveys over a wide variety of habitats and successional stages. This will permit biological interpretations for those taxa that are deleted or added to the list.

Local extinctions, or even shifts in species composition, may have an important impact on ecosystems. Disappearance or decreases in abundance of widely distributed taxa, or disappearance or reduced abundance of taxa of known sensitivity to pollution, points toward pollution effects. However, it is often easy to misinterpret causal agents. Interpretation of the changed list requires knowledge of life history, ecological amplitudes, and limits of tolerance of the organisms. Conclusions drawn from changes are suggestive rather than conclusive. Collateral information from tissue analyses, measured changes in air chemistry, or background contamination levels is also needed to establish cause and effect. Lists that contain numerous species may already have known sensitivities established (including tolerances to pollutant buildup by accumulator or less sensitive species). Extinctions are difficult to document, in that it is difficult to sample to the degree that rare species are observed.

We recognize that the practicality of the list approach may be limited for some biological groups. For example, bird lists may be highly variable from year to year for reasons unrelated to the local area, especially in the case of migratory species. The systematics and taxonomy for certain arthropods may be incomplete or unknown for the area in question. While these limitations pose severe constraints on the interpretation of diversity from lists, the biologist may nevertheless find local lists useful as part of a broader monitoring system (Magurran 1988). For other taxonomic groups, such as lichens, floristic lists may present the most practical means for generalized pollution monitoring at present (Wetmore 1985, 1988).

Where practical, however, the list or survey approach (which is actually the biodiversity measure of richness) to monitoring is relatively easy, repeatable, and inexpensive. It requires expert knowledge, since it is desirable to sample all of the communities in an area. These experts provide quality control through standardized procedures of collecting, handling, labeling, and archiving voucher specimens (Wetmore 1988).

MONITORING ECOSYSTEMS

Conceptual Framework

Data collected from monitoring biodiversity must satisfy several conditions to be useful in determining or gauging system response to atmospheric changes. Measurements must be routine and repeatable, and the variability known and documented. Standard protocols must be developed, documented, and closely followed. Strict quality control procedures must be followed and documented. The monitoring design and sampling scheme must be carefully planned before the monitoring begins.

A statistician or biometrician with experience in biological sampling must be involved in the planning of the monitoring program. Many statistical concerns need to be addressed before one can design a sampling program to detect changes in biodiversity. The amount of change that needs to be detected, and how much change is necessary to be considered "real" must be determined. Estimates of population variance are necessary to determine sampling intensity. The type of data to collect and appropriate analysis techniques for determining changes in biodiversity are described in detail by Green (1979), Keith (1988), and Turner and Gardner (1991).

Before the monitoring begins, a database management program is needed that will be used to log, summarize, interface various data sets, and archive the data. This can be done on a personal computer using database management software. Field data collection with data-loggers or electronic data pads will aid in this process and is highly recommended. Monitoring activities do not end with data collection. Data must be analyzed, summarized, and reported. Database management is an important component of the monitoring program. It must include the ability to retrieve specific archived data years and perhaps decades later. The database must also contain annotations regarding limitations of the data, be able to accommodate updates and inclusion of additional variables, and be compatible with new software and hardware systems.

Guidelines to Document Current Condition

Monitoring is best accomplished under some conceptualization of the ecosystem, its elements, and the exchange processes that link the elements. For example, the measurement program could emphasize the role of the atmosphere as a source of energy, water, and nutrients for the ecosystem.

Using this conceptualization, Fox, Bernabo, and Hood (1987) developed a guide of various protocols needed for measuring the physical, chemical, and biological conditions of wilderness ecosystems. The guide documents sampling protocols for measurements of the atmospheric environment, visibility, soils and geology, aquatic chemistry, aquatic biology, and terrestrial vegetation. It was developed specifically for high-elevation, wilderness ecosystems; however, many of the experimental protocols can be used in other ecosystems.

The guidelines suggest a periodic revision to include new or updated monitoring techniques. Testing and evaluation of these protocols are ongoing, involving a comprehensive long-term ecological research program being conducted at a USDA Forest Service, Rocky Mountain Forest and Range Experiment Station field site in southeast Wyoming.

Chemical changes in the environment are perhaps the first signal that biological systems might be impacted by atmospheric deposition. This can be monitored by wet or dry deposition sampling or by surface water analysis. Aquatic systems respond rapidly to minor changes in water chemistry. Sampling protocols for both water chemistry and aquatic biota, specifically plankton, are routine and easy to follow; however, taxonomic experts such as phycologists should be consulted for species identification or verification. The relationship between atmospheric deposition and aquatic taxa has been documented (Schindler, Kasian, and Hesslein 1989). Aquatic sampling could be a first step in detecting atmospheric impact on ecosystems because these systems will likely respond earlier than would terrestrial populations, particularly in areas where surface waters have limited buffering capacity. Other early indications that ecosystems might be impacted by pollutants could be evidence from ambient concentrations of toxic pollutants. For example, monitoring of ambient ozone concentrations would give an indication of potential threat of ozone to sensitive terrestrial plants.

Glacier Lakes Ecosystem Experiments Site

An intensive monitoring program to determine effects of atmospheric deposition on alpine ecosystems has been established by the USDA Forest Service, Rocky Mountain Forest and Range Experiment Station, 60 km west of Laramie, Wyoming, in the Medicine Bow National Forest. The 600 ha alpine/subalpine ecosystem at 3,200-3,500 m elevation in southeastern Wyoming's Snowy Range, called the Glacier Lakes Ecosystem Experiments Site (GLEES) was established in 1987 for intensive monitoring of biological and physical processes. The monitoring program at the GLEES is designed to provide information on the effects of atmospheric change on alpine and subalpine ecosystems. This program requires an integrated approach to the study of these ecosystems, involving scientists from multiple disciplines including meteorology, soils, geology, hydrology, limnology, plant physiology, taxonomy, ecology, biogeochemistry, and snow chemistry and physics. The goal is to integrate these disciplines to determine interactions and movement of pollutants, chemicals, or gases between various components of the ecosystem. The expertise of individual scientists from different disciplines is necessary to integrate the diverse data available and necessary to understand the relationship between atmospheric and biotic processes. Ecosystem and atmospheric processes, and interactions between these two systems, are being studied at GLEES. Inputs and outputs, and energy and nutrient budgets, are also being examined. Process-based models incorporating both atmospheric and biotic systems will be developed to integrate individual studies into system-level descriptions of response to atmospheric change. These models will be necessary for predicting the effects of atmospheric change on such ecosystems. Long-term baseline atmospheric and biotic data from the GLEES will be used to parameterize and validate these models. Detailed topographic, soils, geologic, and vegetation maps of the GLEES area have been prepared. In addition, atmospheric, meteorological, biotic, hydrologic, and chemical data are collected routinely. Data monitored are fed directly to electronic data loggers, often solar-powered, and transferred to a personal computer using database management software.

The GLEES has snow machine access from 6 km in winter, and is accessible directly by road in the summer. It is characteristic of many high-elevation wilderness ecosystems in the western United States, having shallow, immature soils with low base saturation, small lakes with low acid-neutralizing capacity, and a short, cool growing season. The area has persistent high winds and low temperatures with frost conditions possible all year. Dominant habitats include alpine meadows, subalpine coniferous forest, krummholz spruce and fir, and rock and scree. The area is covered by snow for much of the year (8 or more months), and the extent of snow cover depends on prevailing winds. Vegetation patterns show a distinct relationship to snow cover. Once snow has melted, the area is subject to drought conditions.

Table 4-1 summarizes data currently collected at GLEES. This includes those minimum data suggested by Fox, Bernabo, and Hood (1987), plus an additional set of measurements for this intensively monitored ecosystem. The data are summarized in a GLEES interactive database. Meteorological data collected at the GLEES alpine habitat include air and soil temperature, relative humidity, precipitation, wind speed, wind direction, solar radiation, and wetness. Figure 4-1 illustrates a monthly summary of the meteorological data produced through the GLEES database management system. Pan evaporation is monitored during the growing season. Data are recorded at less than 1-minute intervals using solar-powered data loggers. At a GLEES subalpine forest site with electric power, meteorological data are collected, in addition to dry deposition (by filter pack), ozone concentrations, and carbon dioxide concentrations. Other measurements of precipitation, wet deposition chemistry,

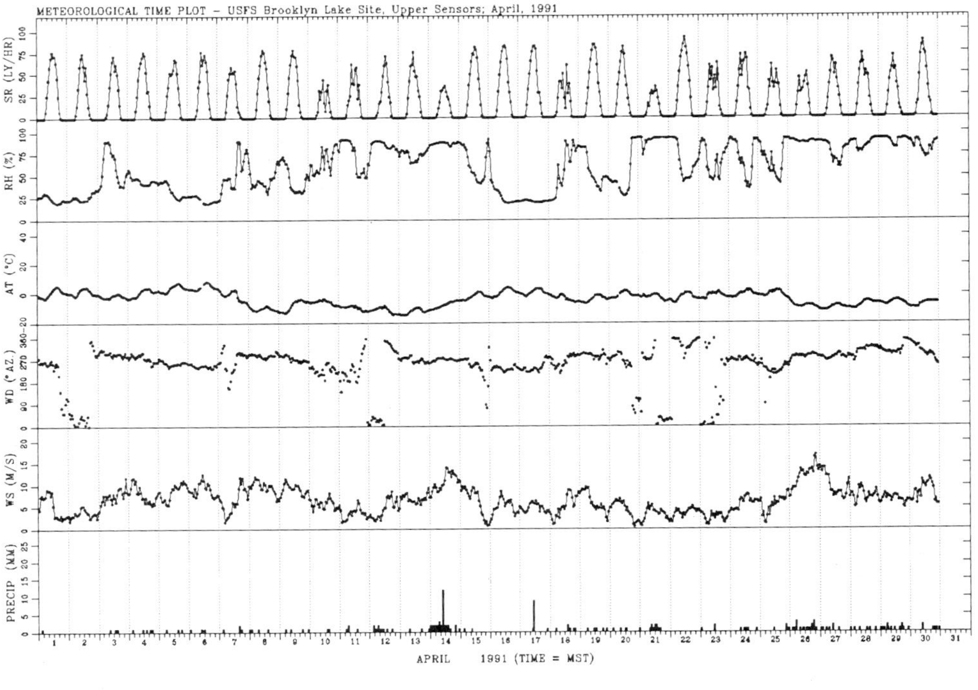

FIGURE 4-1 Summary of meteorological data from the GLEES Brooklyn monitoring site, April 1991. Solar radiation (SR), relative humidity (RH), air temperature (AT), wind direction (WD), and wind speed (WS) sensors at 30 m height. Precipitation (precip.) at 2 m height.

TABLE 4-1. Data Collected and Sampling Location at GLEES

	Meteorology	Air Quality	Water Chemistry	Snow	Vegetation	Soils & Geology
Data	Wind speed (WS)	Ozone, CO_2	Precipitation	Snow pits:	Mean % cover/	Mineralogy
	Vector WS	Coarse mass	Stream flow	Physical:	species	Soil type
	Wind direction	Fine mass	volumes	Layer top	DBH/species	Soil solution
	(WD)	H, Al, Si	pH (lab & field)	Layer bottom	Presence/absence	chemistry
	Vector WD	S, K, Ca, Mn	Conductance (lab	Layer density	Seed bank	pH, NH_4–N
	Temperature (T)	Ti, Fe, Ni	& field)	Snow crystal	Phenological	NO_3–N
	Delta T.	Cu, Zn, As, V	Ca, Mg, K, Na	type	patterns	Organic matter
	Rel. humidity	Se, Br, Pb	NH_4, NO_3, Cl	Grid location	Vascular plant	% Sand, % Silt
	Radiation	OMH[b]	SO_4, PO_4	Wetness	species list	% Clay
	Surface wetness	SO_4, SO_2	ANC, Si, Al	Chemistry: Same	Voucher	Textural class
	Precipitation	NO_3, HNO_3	Br, DIC & DOC	as water	specimens	CEC
	Soil temperature	NH_4, Na, K	Map location	chemistry	Aspect	Gravimetric
	Snow wetness	Ca, Mg, Cl		plus tracers	Elevation	moisture
	Snow depth	Optical adsorp.		Map location	Slope position	Moisture retention
	Map location	Map location			Map location	Map location
Sample Sites	Brooklyn[a]	Brooklyn[a]	Brooklyn[a]	Pits: 2 depths	150 permanent	Vegetation plots
	Glacier Lakes		6 lakes	4 aspects	plots	and various other
	Summit		21 streams	Snow surveys	1200 subplots	locations
				over 300 ha		

[a] Brooklyn monitoring site has meteorological monitoring at 15 m and 30 m height. NADP (National Atmospheric Deposition Program), NDDN (National Dry Deposition Network), and SFU (Stacked Filter Unit) for wet and dry deposition are located at the Brooklyn site.

[b] OMH = Organic matter from hydrogen

wind speed, wind direction, and incoming and outgoing radiation are monitored at various sites in the GLEES.

Hydrological data collected at the GLEES include streamflow volumes from two lake inlet streams and two lake outlets equipped with Parshall flumes. Lake, stream, soil water, and snow chemistry are also periodically sampled. A snow-core survey of the ecosystem is conducted at maximum seasonal accumulation where depth and density of snow are recorded. Two of the small watersheds within GLEES are sampled more intensely for snow depth and snow density. Snowpack data are gathered at monthly intervals from snowpits located on four different aspects.

Biotic data include information on both terrestrial and aquatic ecosystems. The GLEES contains three small alpine lakes, where phytoplankton and zooplankton are routinely monitored in conjunction with sampling of water chemistry. Approximately 90 permanent vegetation plots were established in two small watersheds, with plots randomly located within 100 m grid sectors overlaying the watersheds. Vascular plant species abundance data have been collected from these plots, along with environmental data including physical and chemical information on the soils, and on the plot's aspect, slope, and snow pack duration. A less intensive vegetation sampling grid was designed for the rest of the GLEES. Checklists of vascular plants, phytoplankton, and zooplankton have been assembled for the GLEES. Herbarium voucher specimens of the vascular plants are available at the University of Wyoming Herbarium and at the field laboratory. A checklist of nonvascular plant species is being assembled.

Linking biotic data with atmospheric data is difficult, because atmospheric data are available from only a few fixed sites at the GLEES. It is evident that micrometeorological conditions vary widely in the GLEES, and that vegetation responds to these different conditions. The extensive biological and environmental data collected from the permanent vegetation plots are being examined to relate biotic and physical components. A multivariate gradient analysis is being used to examine the relationships between vegetation and environment. This gradient analysis is a first step to linkage of biotic data with physical data expressed as environmental condition. Table 4-2 lists the community associations identified by the GLEES vegetation gradient study.

The environmental data available from the GLEES vegetation plots indicate that there is a wide range in micrometeorological conditions. Micrometeorological and microenvironmental conditions at the GLEES exhibit abrupt spatial changes, with associated changes in vegetation species and abundance. Meteorological conditions are measured at only a few fixed weather tower sites. Relating point-source meteorological tower data to micrometeorology at numerous permanent plot sites throughout the watershed is difficult because environmental and meteorological conditions are spatially discontinuous at GLEES due to terrain complexity. Additional weather stations have been situated at various locations throughout the GLEES to examine this variability, but weather stations are costly and the data obtained are only a small proportion of those needed. Sufficient equipment will never be available to monitor micrometeorological conditions at each permanent plot. Thus extrapolation of meteorological data between a few fixed atmospheric monitoring stations to estimate micrometeorological conditions at vegetation plots within the watershed, although difficult, is necessary in order to relate biotic response to the atmosphere. The gradient analysis suggests that vegetation habitats in the watersheds of GLEES depend on specific environmental parameters, particularly snowpack. Snow distribution, in turn, is highly dependent on terrain complexity and wind patterns.

TABLE 4-2. Characteristics of GLEES Community Types

Dominant Species	Community Type	Dominant Life	Topographic Form	Aspect Position	Species Richness
Eleocharis pauciflora/Carex aquatilis	Fen	Graminoid	Stream bottom	NE to SE	24
Salix planifolia/Carex aquatilis	Willow carr	Shrub/graminoid	Stream bottom	NW to E	37
Salix planifolia/Caltha leptosepala	Willow carr	Shrub/forb	Stream bottom	E to SW	39
Deschampsia cespitosa	Wet meadow	Graminoid/forb	Stream bottom	NE to S	46
Caltha leptosepala/Trollius laxus	Wet forb meadow	Forb	Stream bottom	SW	46
Abies lasiocarpa/Picea engelmanii/Salix planifolia	Riparian forest	Tree/shrub	Stream bottom	S to SW	59
Carex brevipes/Juncus drummondii	Upland meadow	Graminoid/forb	Lower slope to upper slope	E to SW	76
Abies lasiocarpa/Picea engelmanii/Vaccinium scoparium	Subalpine forest and krummholz	Tree/shrub	Lower slope to ridge top	NW to W	48
Abies lasiocarpa/Juniperus communis/Salix brachycarpa	Subalpine krummholz	Tree/shrub	Midslope	E to SE	24
Abies lasiocarpa/Salix brachycarpa	Alpine shrub thicket	Shrub	Upper slope to ridge top	NW to S	56
Carex rupestris	Turf	Graminoid/ cushion forb	Upper slope to ridge top	NW to SW	30

Methods are available to relate conditions at multiple points in a watershed with those recorded at a meteorological station. For example, we have constructed detailed wind speed, wind direction, and snow depth maps for the GLEES based on vegetative indicators of these meteorological and environmental factors. Wind-deformed trees indicate prevailing wind direction by the direction of their asymmetric deformation; they indicate relative wind speed by the extent of the deformation (Robertson 1987; Yoshino 1975). Empirical relationships have been developed between extent of deformation and actual wind speeds causing the deformation. Snow depth is indicated by lack of vegetative deformation from snow cover. The data measured with vegetative indicators are ground-checked with meteorological tower data and snow depth surveys. Results show a close match of wind speed and wind direction data from vegetative indicators with available meteorological data; and a close relationship of snow depth with snow survey data (Musselman et al. 1990) as indicated by the height at which deformation begins.

A major step in accomplishing GLEES goals has been the design and implementation of the GLEES interactive database management system. The database management system is implemented on PC hardware using commercially available database management and spread-sheet software. It allows all GLEES researchers ready access to all physical and biotic data, so they can speculate about relationships and formulate and test multiple hypotheses efficiently. One example of the associations that are facilitated by the GLEES interactive database is an examination of the relationship among vegetation occurrence, meteorological data, soil information, and snow distribution.

Forest Health Monitoring

The GLEES is an intensive multidisciplinary examination of atmospheric deposition effects on ecosystem processes. These changes in biodiversity can be detected and causes examined. On a wider scale, health of national forests also need to be monitored for changes. This monitoring must be conducted more broadly and for budgetary reasons must be less intensive. An example of such a program is the USDA Forest Service program to annually monitor the condition of forested ecosystems in the United States. The Forest Health Monitoring Program was developed in close coordination with and assistance of the EPA's Environmental Monitoring Assessment Program (EMAP), designed to monitor national resources of the United States. The Forest Health Monitoring Program described here is only one of eight resource categories monitored in a coordinated national program. The program is relatively uniform in measurement protocols so that data can be examined and compared on a national or a regional basis. The monitoring will provide data on current forest conditions, so that changes can be documented and meaningful inferences drawn about causes of change.

From a perspective of forest resource management, the goals of forest health monitoring are threefold:

1. To detect changes; to establish a baseline to determine if, when, and where changes are occurring; and to quantify those changes.
2. To evaluate possible causes of change. If a change is undesirable, unexpected, or unexplained, it is important to evaluate the cause of the change to decide if remedial action is indicated. If a change results from a planned manipulation, it is important to document the cause/effect relationship to evaluate the management practice. Natural and

desirable changes should be evaluated to enhance understanding of the resource for future management.

3. To increase our ability to anticipate or predict changes in forest resources. This is achieved by understanding the processes involved in regulating the function and controlling the structure of forest ecosystems. Forest ecosystems change rather gradually because they are dominated by long-lived individuals (trees). Therefore, many forest ecosystem processes can be understood only by repeated, long-term observations. Even the development of researchable hypotheses is often possible only after many years of observing the state of the ecosystem. Finally, a better understanding of the processes that control the functioning of forests will make it possible to better define baseline conditions.

The Forest Health Monitoring Program has three tiers of interrelated monitoring activities: detection monitoring, evaluation monitoring, and intensive-site ecosystem monitoring. (1) Detection monitoring records the condition of forest ecosystems, estimates baseline conditions, and detects changes from those baselines over time. (2) Evaluation monitoring determines the causes of detected changes, if possible, or hypothesizes causes that can be tested experimentally or with information from intensive-site ecosystem monitoring. (3) Intensive-site ecosystem monitoring provides high-quality, detailed information that will enable a rigorous assessment of cause/effect relationships and support experimental research on a small set of representative ecosystems. All three tiers are needed to fully explain the state of health of forest ecosystems.

A fourth component of the program applies to all three tiers of monitoring activity—research on monitoring techniques. Research is needed on topics such as sampling methods, sampling design, statistical analysis, assessment, remote sensing and other measurement procedures, and indicators of forest health. Research on monitoring techniques ensures that forest health monitoring will be effective.

Detection Monitoring

This type of monitoring consists of a network of permanent plots distributed throughout the forests of the United States, coupled with remote sensing observations and pest surveys. The sampling frame for the permanent plots will be existing Forest Service networks of inventory sample locations, augmented with additional sample locations as needed to represent all forest lands in the United States. From this augmented network, a subset of "sentinel plots" will be selected and visited annually. Sentinel plots will be selected to closely conform with the EMAP network of environmental monitoring locations. The amount of information collected on sentinel plots will be greater than that collected during regular forest inventories. Information from the sentinel plots will be pooled with information collected during routine forest pest surveys and other specifically focused monitoring activities. All information will be spatially linked to provide a more complete, annual estimate of forest condition.

Evaluation Monitoring

Evaluation monitoring usually will be activated by the results of detection monitoring. When detection monitoring reveals changes that represent areas of concern, a specific evaluation will be made to determine necessary followup activities. The details of evaluation monitoring cannot be specified in advance. By definition, evaluation monitoring will be implemented where unexplained changes have been detected, and it will be tailored to the specific nature of the problem. Activities could include additional targeted surveys, site-specific evaluation visits, more detailed temporary monitoring, and specific research studies.

Intensive-Site Ecosystem Monitoring

The goal of this monitoring is a more complete understanding of the mechanisms of change in forest ecosystems. Monitoring at this level provides data to better explain causal relationships and predict direction and rates of changes in forest condition. Ecosystem monitoring sites will represent key forest ecosystems throughout the United States. An example of research monitoring is the intensive monitoring conducted at the GLEES. Ten to 50 primary sites will be used to represent major forest ecosystems. These sites will be centers where very detailed information on all components of the forest ecosystem will be collected to supplement detection and evaluation monitoring efforts. The sites will support mechanistic research to identify, describe, or model tree and forest processes to (1) increase basic understanding of causal relationships, (2) provide explanations or projections of observations in the other levels of the forest health monitoring system, (3) help anticipate changes in forest health, and (4) provide the knowledge for developing management responses to unexpected changes.

In some cases, the availability of this detailed information might resolve questions that were raised but not answered by detection or evaluation monitoring. Information from intensive-site ecosystem monitoring will contribute to better understanding of intrasite variability as well as better understanding of relationships between detection monitoring indicators and other ecosystem characteristics. In this sense, information from intensive-site ecosystem monitoring has intrinsic value.

More importantly, intensive-site ecosystem monitoring will provide long-term data and the sampling infrastructure to support research on mechanisms and processes that shape forest ecosystems. In this sense, intensive-site ecosystem monitoring sites are similar to the National Science Foundation's Long-Term Ecological Research (LTER) sites. Information from ecosystem monitoring sites can be augmented with additional short-term measurements or it can be expanded by measuring new variables during specific studies. Such studies, along with ecosystem process modeling efforts, will be needed to enhance the value of the intensive-site ecosystem monitoring information in context with the more extensive aspects of the detection and evaluation monitoring data and analyses.

SUMMARY AND CONCLUSIONS

Monitoring atmospheric effects on biological diversity must include measurement of both atmospheric and biotic components of the ecosystem. Several issues must be addressed before a monitoring program can begin, including statistical design, representativeness of the monitoring sites to the ecosystem, the ability to measure the uniqueness of the ecosystem, and quality control of the data from collection to output summaries and archiving.

Biodiversity can be measured at different levels of organization. Typical classifications monitor biodiversity at the genetic, species, community, and landscape scales. Composition, structure, and function are attributes of biodiversity that should be measured at each of these levels. Monitoring intensity generally decreases as scale becomes larger. Atmospheric effects on biodiversity can be delineated as first-order, second-order, and third-order effects. First-order effects are direct effects from contact of the atmosphere with the biosphere. Second-order effects are those resulting in changes in environment or climate that in turn affect biota. Third-order effects are ecosystem disturbances resulting from climate change. First-order effects of air pollution on biodiversity are easier to detect than second-order effects. Third-order effects are extremely difficult to relate to atmospheric deposition causes.

A case study is presented of an intensive monitoring program to determine the effects of atmospheric deposition on terrestrial and aquatic ecosystem. The GLEES research includes an intensive monitoring program of meteorological, atmospheric, hydrological, snow and water chemistry, aquatic, and terrestrial biota information. The information is carefully quality-controlled and archived in an interactive database. A goal of the research program is to examine the interaction of atmospheric processes and ecosystem processes. The research requires a long-term commitment to monitoring as well as the interaction of scientists from several disciplines.

On a wider, less intensive scale, the USDA Forest Service's Forest Health Monitoring Program is described as an example of an approach to monitor ecosystem health on a national basis. The program involves detection monitoring to determine baseline condition, evaluation monitoring to determine causes of detected changes, and intensive-site ecosystem monitoring to examine the processes regulating the function and controlling the structure of forest ecosystems. Research on monitoring techniques is also included in the program.

Air pollution effects on biological diversity may be small compared with threats of atmospheric change on ecosystems. Perturbations resulting from climate change are expected to have particularly large impacts on biodiversity. Nevertheless, changes in biodiversity from air pollution can be detected if the monitoring program is carefully designed and conducted.

REFERENCES

1. Chen, C.W., S.A. Gherini, R.J.M. Hudson, and J.D. Dean. 1983. *The Integrated Lake-Watershed Acidification Study: Volume 1. Model principles and application procedures*. RP1109-5. EA-3221. Palo Alto, CA: Electric Power Research Institute.
2. Cosby, B.J., G.M. Hornberger, J.N. Galloway, and R.F. Wright. 1985. Modeling the effects of acid deposition: Assessment of a lumped-parameter model of soil water and streamwater chemistry. *Water Resour. Res.* 21:51-56.
3. Food and Agricultural Organization of the United Nations. 1962. The influence of the forest on the weather and other environmental factors. In *Forest influences*, pp. 83-137. Rome: FAO.
4. Fox, D.G., J.C. Bernabo, and B. Hood. 1987. *Guidelines for measuring the physical, chemical, and biological condition of wilderness ecosystems*. General Technical Report RM-146. Fort Collins, CO: USDA Forest Service, Rocky Mountain Forest and Range Experiment Station.
5. Green, R.H. 1979. *Sampling design and statistical methods for environmental biologists*. New York: John Wiley & Sons.
6. Huston, M.A. 1991. Use of individual-based forest succession models to link physiological whole-tree models to landscape-scale ecosystem models. *Tree Physiol.* 9:293-306.
7. IGBP Committee. 1990. *The International Geosphere-Biosphere Programme: A study of global change*. The Initial Core Projects. IGBP Secretariat. Stockholm: The Royal Swedish Academy of Sciences.
8. Keith, L.H., ed. 1988. *Principles of environmental sampling*. American Chemical Society. York, PA: Maple Press Company.

9. Landsberg, J.J., M.R. Kaufmann, D. Binkley, J. Isebrands, and P.G. Jarvis. 1991. Evaluating progress toward closed forest models based on fluxes of carbon, water and nutrients. *Tree Physiol.* 9:1-15.
10. Magurran, A.E. 1988. *Ecological diversity and its measurement*. Princeton: Princeton Univ. Press.
11. Metzler, K.J. 1980. *Lichens and air pollution: A study in Connecticut investigation.* Report of Investigations No. 9. State Geological and Natural History Survey of Connecticut, Natural Resources Center, Department of Environmental Protection.
12. Meyers, T.P., and Paw U, K.T. 1986. Testing of a higher-order closure model for modeling airflow within and above plant canopies. *Boundary-Layer Meteorol.* 37:297-311.
13. Musselman, R.C., and D.G. Fox. 1991. A review of the role of temperate forests in the global CO_2 balance. *J. Air & Waste Manage. Assoc.* 41:798-807.
14. Musselman, R.C., G.L Wooldridge, D.G. Fox, and B.H. Connell. 1990. Using wind-deformed conifers to measure wind patterns in alpine transition at GLEES. In *Proceedings-Symposium on whitebark pine ecosystems: Ecology and management of a high-mountain resource,* pp. 80-84. U.S. Department of Agriculture, Forest Service, Intermountain Research Station, Gen. Tech. Rep. INT-270.
15. Noss, R.F. 1990. Indicators for monitoring biodiversity: A hierarchical approach. *Conserv. Biol.* 4 (4): 355-64.
16. Peet, R.K. 1974. The measurement of species diversity. *Annu. Rev. Ecol. Sys.* 5:285-307.
17. Raupach, M.R. 1987. A Lagrangian analysis of scalar transfer in vegetation canopies. *Q. J. R. Meteorol. Soc.* 113:107-20.
18. Robertson, A. 1987. The use of trees to study wind. *Agric. J.* 11:127-43.
19. Ross, D.G., and D.G. Fox. 1991. Evaluation of an air pollution analysis system for complex terrain. *J. Appl. Meteorol.* 30 (7): 909-23.
20. Schindler, D.W., S.E.M. Kasian, and R.H. Hesslein. 1989. Biological impoverishment in lakes in the midwestern and northeastern United States from acid rain. *Environ. Sci. & Technol.* 23:573-80.
21. Shugart, H.H. 1984. *A theory of forest dynamics. The ecological implications of forest succession models.* New York: Springer-Verlag.
22. Smith, T.M., and D.L. Urban. 1988. Scale and resolution of forest structural pattern. *Vegetatio* 74:143-50.
23. Smith, W.H. 1981. *Air pollution and forests: Interactions between air contaminants and forest ecosystems.* New York: Springer-Verlag.
24. Turner, M.G. 1990. Spatial and temporal analysis of landscape patterns. *Landscape Ecol.* 4:21-30.
25. Turner, M.G., and R.H. Gardner. 1991. Quantitative methods in landscape ecology: An introduction. In *Quantitative methods in landscape ecology*, ed. M.G. Turner and R.H. Gardner, pp. 3-14. New York: Springer-Verlag.
26. Wang, Y.P., and P.G. Jarvis. 1990. Influence of crown structural properties on PAR absorption, photosynthesis, and transpiration in Sitka spruce: Application of a model (MAESTRO). *Tree Physiol.* 7:297-316.
27. Weinstein, D.A., R.M. Beloin, and R.D. Yanai. 1991. Modeling changes in red spruce carbon balance and allocation in response to interacting ozone and nutrient stress. *Tree Physiol.* 9:127-46.
28. Wetmore, C.M. 1985. *Lichens and air quality in Sequoia National Park.* Final Report, Contract CX0001-2-0034, National Park Service, 25 pp. + maps.

29. Wetmore, C.M. 1988. Lichen floristics and air quality. In *Lichens, bryphytes, and air quality*, ed. T.H. Nash III and V. Wirth, pp. 55-65. Bibliotheca Lichenologica 30. Berlin: J. Cramer.
30. Whittaker, R.H. 1972. Evolution and measurement of species diversity. *Taxon* 21:213-51.
31. Woodward, F.I. 1989. *Climate and plant distribution*. Cambridge: Cambridge Univ. Press.
32. Yoshino, M.M. 1975. *Climate in a small area: An introduction to local meteorology*. Tokyo: Univ. of Tokyo Press.

5

Action of Pollutants Individually and in Combination

Jenny Wolfenden, Philip A. Wookey, Peter W. Lucas, and Terry A. Mansfield

INTRODUCTION

This chapter will provide some information about the effects of major air pollutants on some physiological and biochemical processes in plants. Because it is impossible to review here all past research achievements and the diverse range of current studies in a comprehensive way, we have decided to focus attention on some selected processes and mechanisms—leaf level processes, biochemical and cellular perturbations—having implications for whole plant physiology.

It is not practicable to keep strictly within the context of biodiversity here because most of the studies that have provided reliable information on physiological and biochemical effects have been performed on crop plants. Many of our arable crops have been developed from species that in their natural habitats would be described as "ruderals," i.e., plants that are adapted to regimes in which disturbances are high but other stress factors are low. The life cycle of ruderals is usually annual or biennial, and if they have a perennial capacity it is likely to be short lived. Ruderals are able to make use of situations that are temporarily favorable for abundant plant growth. Our agricultural methods exploit them by placing them in such situations and providing them with the resources needed for rapid growth. Some of the main features of ruderals (e.g., the capacity for fast completion of the life cycle, high rates of dry matter formation, and abundant seed production) have been accentuated during the development of our modern cultivars.

The types of responses to air pollutants found in crops may be of greater or lesser significance in species adapted to quite different kinds of natural habitats. We have found that it is useful to refer to the types of strategies for growth and survival in contrasting habitats as defined by Grime (1979). These strategies are employed by:

1. "Competitor" species exploiting conditions of low stress and low disturbance. "Competition" is defined as "the tendency of neighboring plants to utilize the same quantum of light, ion of a mineral nutrient, molecule of water, or volume of space";

2. "Stress-tolerator" species exploiting conditions of high stress and low disturbance. "Stress" is defined as "the external constraints that limit the rate of dry matter production of all or part of the vegetation"; and
3. "Ruderal" species exploiting conditions of low stress and high disturbance. "Disturbance" is defined as "the mechanisms that limit the plant biomass by causing its partial or total destruction."

Although some controversy has been created by these definitions (largely based on views about a continuum of different strategies rather than the sharp boundaries that the definitions imply), few would deny that in many plant communities species are found with strategies that fit comfortably into Grime's categories. Different strategies represent a vital component of the biodiversity within a given community or region; therefore, if pollutants exert differential effects according to strategy, the balance of genotypes present would be changed and perhaps impoverished.

There have been very few attempts to perform studies of comparative physiology or biochemistry of polluted plants with different habitat strategies in mind. It is, however, possible to speculate in general terms on the basis of some well-established effects of pollutants. For example, stress tolerators tend to be long-lived perennials and possess features and adaptations that allow them to survive under conditions where only low productivity is possible. They usually are evergreens with leaves possessing xerophytic characteristics, and their physiological activities display a resilience to adverse climatic factors, e.g., temperature extremes and drought. Stress tolerators have well-developed storage systems that may be found in any part of the plant, i.e., roots, stems, or leaves. Many shrubs and trees are stress tolerators. Ruderals, on the other hand, are almost exclusively short-lived herbs, often with mesomorphic leaves. Physiologically they have a low resilience to adverse environmental conditions, principally due to a limited capacity for acclimation. Stress tolerators and ruderals differ significantly in their development of storage systems. Ruderals store little in vegetative structures, but their rapid production of seeds is an important attribute.

On the basis of these general differences between stress tolerators and ruderals, we can predict some differences in the nature of their responses to air pollutants. For example:

1. Exposure to a short-term acute dose of a pollutant such as sulfur dioxide (SO_2) and ozone (O_3) may cause more rapid damage to a ruderal than to a stress tolerator because the stomatal conductance of ruderals tends to be high. If it were not, the carbon acquisition required for rapid growth would be restricted. The xerophytic characteristics of the leaves of stress tolerators, on the other hand, often include low stomatal conductance which, as well as conserving water, also reduces the uptake of gaseous pollutants; and
2. Exposure to a low concentration of a pollutant for a long period may have a more serious impact on a stress tolerator than a ruderal because ruderals tend to respond to any form of stress by switching from vegetative to reproductive growth. Seed production is likely to be reduced but the effect of the stress on the success of the subsequent generation is minimized. Stress tolerators, on the other hand, have leaves that are metabolically functional at all times of the year when conditions are favorable. Uptake of a pollutant can continue for a long period but cumulative problems may occur. These can be accentuated by the low morphogenetic plasticity of stress tolerators, i.e., damaged tissues cannot be quickly replaced.

The pollution responses of competitors are of particular interest. These plants have the capacity for both a high rate of acquisition of resources and rapid growth, combined with a capacity for rapid reapportioning of photosynthate to produce the morphogenetic changes that are required to maintain the plant's ability to compete for resources. Especially important are adjustments in leaf and root development to maintain supplies of carbon and mineral nutrients required for growth. The competitive strategy also makes use of mechanisms for water conservation that interfere as little as possible with carbon acquisition. Effective stomatal control is of high priority, and this involves high conductances under favorable conditions and the capability to switch quickly to lower conductances when circumstances change.

There is now substantial experimental evidence showing that exposure to some air pollutants may not only disturb the morphogenetic plasticity of plants but also may interfere at a fundamental level with the ability of stomata to respond correctly when a plant encounters the first effects of water shortage in the soil. These effects may be of particular importance in terms of biodiversity if they exert selective pressures on the most highly successful competitor species.

In this chapter we shall first give an outline of some of the known detoxification mechanisms in plants. In most cases these are likely to be found in all species but their capacity may be different in plants from different habitats. Second, we shall describe briefly the consequences of additional atmospheric sources of nitrogen. Third, we shall give detailed attention to two types of effects at the physiological level, namely interference with stomatal functioning and allocation of assimilates. These effects are of particular interest for competitor species for the reasons outlined above.

DETOXIFICATION MECHANISMS

The severity of damage that a pollutant can inflict upon a plant depends on the extent of penetration into the tissues and on the reactivity of the pollutant and its products with plant components. Pollutant gases entering the stomatal cavities, such as SO_2, NO_x, and O_3, dissolve in extracellular fluid and react with plant material to produce ionic species and free radicals (Mansfield and Freer-Smith 1981; Peleg 1976) that are generally more reactive than the pollutant gases themselves. As a result of the reactions of these species with lipids and proteins in the cell walls and membranes, chain reactions are initiated, giving rise to more free radicals. Unless this potentially self-perpetuating process is blocked, it can lead to widespread damage to membrane systems, resulting in the breakdown of cell compartmentation and severe metabolic disorder. Among the more reactive products are active oxygen species, such as superoxide (O_2^-) and hydroxyl ($OH^\cdot$) radicals, which are also produced in the chloroplast during photosynthesis (Salin 1987). Plants have developed a range of mechanisms to remove such toxic molecules (Salin 1987; Larsson 1988, reviews). These protective mechanisms include carotenoids (e.g., β-carotene), phenolic compounds such as vitamin E (α-tocopherol), peptides (e.g., glutathione), enzyme systems (e.g., catalase, superoxide dismutase), polyamines, and organic buffering systems. Together these form a constituent part of the internal resistance of the plant to atmospheric pollutants.

Understanding the biochemical mechanisms of pollutant toxicity may bring us closer to discovering some of the processes that form the genetic basis for different degrees of sensitivity to pollutants. Enzyme induction, for example, or the regulation of alternative biochemical pathways, may be under relatively simple genetic control, some characters of which can be selected to improve tolerance. A number of studies have aimed at comparing the efficiencies of known detoxification processes in plants showing varying degrees of sensitivity to pollutants. An important development in this field has been the use of electrophoresis to identify isozymes of enzymes such as peroxidase, which have been considered useful markers of pollution sensitivity (Horsman and Wellburn 1975). Using this technique, numerous studies of forest tree species using both clonal and seed material have revealed genetic heterogeneity within populations and even within clones (Dochinger 1972) with relation to pollution sensitivity. Despite this variation, there have been many reports of isozyme differences between families and populations growing in different environmental conditions, and some of these have been used successfully in breeding programs to increase tolerance to air pollutants (reviewed by Mejnartowicz 1984). The approach of using a particular enzyme system as a tool for selection may be useful for plants that normally experience only one type of pollution or a mixture of pollutants having similar mechanisms of toxicity. It is often the case, however, that pollutant mixtures contain components that cause injury in different ways and may produce synergistic or mitigating effects. In such cases it is difficult to envisage successful tolerance selection based on a single enzyme system.

Ozone toxicity in plants is generally considered to occur via free radical oxidation of membrane components such as sulfhydryl groups and the double bonds of unsaturated fatty acids. There have been many reports of changes in activity and concentration of antioxidants and their associated enzymes following exposure of plants to O_3. In general, exposure to low concentrations of O_3 (less than 200 ppb) increases the activity of antioxidants and scavenging processes (Mehlhorn et al. 1986, 1987; Tanaka, Saji, and Kondo 1988), while higher concentrations are sometimes inhibitory to antioxidants, whose scavenging capacity probably becomes overloaded (Sakaki et al. 1983). In such situations, the protective mechanisms may even act as pro-oxidants, enhancing the effect of O_3 (Price, Lucas, and Lea 1990). The efficiency of these systems is considered to be an important factor determining the resistance of plants to O_3 and other air pollutants (Davies and Wood 1972; Guri 1983; Tanaka et al. 1985).

Recent advances in our understanding of O_3 toxicity to plants have pointed to an important involvement of unsaturated hydrocarbons such as ethylene in the mechanism of damage. Exposure to O_3 stress induces the formation of ethylene by the plant; the extent of this response has been suggested as a sensitive indicator of the O_3 susceptibility of different plant species (Tingey, Standley, and Field 1976). From more recent evidence it appears that endogenous ethylene is not only a result of O_3 stress, but also plays an important part in the injury mechanism. Taylor, Ross-Todd, and Gunderson (1988) found that by blocking the pathway of ethylene biosynthesis they could reduce O_3-induced stress to gas exchange processes in soybean. The precise role of endogenous ethylene in the injury process was further investigated by Mehlhorn and Wellburn (1987) and Mehlhorn, Tabner, and Wellburn (1990), who provided evidence of a free radical-forming reaction between O_3 and stress-induced ethylene. These findings suggest that plants growing under conditions of stress are likely to be more susceptible to O_3 injury, an hypothesis that is consistent with many of the observed interactive effects of O_3 and other environmental stresses.

Ethylene production may be a factor by which plants within a population exhibit a high degree of variation in their response to pollutants. In a long-term experiment where Norway spruce seedlings were exposed to 70 ppb O_3 for a 3-month period during summer, ethylene production in the needles increased significantly compared with trees grown in filtered air, and remained high for many weeks after the end of the experiment (Figure 5-1).

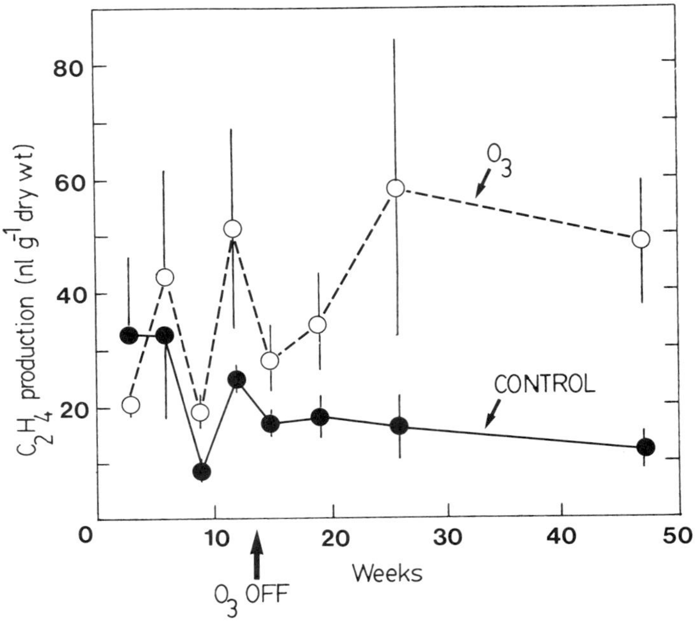

FIGURE 5-1. Ethylene production by Norway spruce needles during and after summer exposure to 70 ppb O_3. Means (± 1 SE) of 10 trees.

Source: Wolfenden, previously unpublished in this form.

The rate of ethylene production varied between individuals to a far greater extent in the O_3-treated trees than in the controls, perhaps indicating individual differences in sensitivity to ozone.

The role of ethylene also may be linked to the involvement of polyamines in plant response to O_3. Rowland-Bamford et al. (1989) reported that the polyamine spermidine increased in O_3-treated plants after a few days. Because the metabolic precursor of ethylene, S-adenosyl methionine (SAM) is also important in the biosynthesis of polyamines, it was suggested that a switch from ethylene production to polyamine production might be a way of reducing the danger of ethylene-enhanced O_3 damage. Further evidence in support of this idea comes from a study by Langebartels et al. (1991), who examined the response of polyamine and ethylene biosynthesis pathways in tobacco cultivars exposed to O_3. They found that the pathways were induced differently in sensitive and tolerant genotypes, and that the response was sufficiently rapid and intense to represent an important physiological control of lesion development. Little is known of the regulation of these pathways (Mijazaki and Yang 1987), but it is interesting to speculate on the possibility of a genetically controlled switch that allows the plant to reduce its ethylene production in response to stress and might thus confer an advantage during exposure to O_3. Polyamines also appear to offer protection from pollutants in a more direct way, perhaps as free radical scavengers. Bors et al. (1989) found that polyamines applied to the roots of tobacco plants accumulated in the leaves and reduced the sensitivity of the plants to O_3.

The toxic action of SO_2 arises partly from the direct effects of its solution products, SO_3^{2-} and HSO_3^-, on photosynthetic processes of photophosphorylation and ATP synthesis (Silvius 1975) and Rubisco activity (Ziegler 1972). These ions are, however, unlikely to accumulate within plant cells because they are rapidly oxidized to SO_4^{2-} in the chloroplast.

The efficiency of SO_3^{2-} detoxification has been related to the degree of SO_2 tolerance in cultivars of soybean (Miller and Xerikos 1978) and pea (Alscher, Bower, and Zipfel 1987; Jäger, Bender, and Grünhage 1985). Further toxic effects of SO_2 arise from free radical-generating chain reactions so that, as with O_3 exposure, scavenging mechanisms to remove active oxygen species are very important in SO_2-stressed plants. Increased activity of antioxidants often has been reported following SO_2 exposure (Grill, Esterbauer, and Klosch 1979; Chiment, Alscher, and Hughes 1986; Mehlhorn et al. 1986). In comparisons between two cultivars of pea and two species of deciduous trees, Jäger et al. (1985) found that SO_2 tolerance was related to the activities of superoxide dismutase and peroxidase, which were higher in tolerant plants. In other studies (Horsman and Wellburn 1975; Jäger and Klein 1977), the activity of glutamate dehydrogenase (GDH) has been found to respond to SO_2 more consistently than any other enzyme indicator.

Pollutant mixtures containing both SO_2 and NO_2 have been shown to be more detrimental to plants than either gas alone (Tingey et al. 1971; Ashenden and Mansfield 1978) and to have a more than additive effect on various enzymes (Horsman and Wellburn 1975). A biochemical explanation for the mechanisms behind this synergism was provided by Wellburn et al. (1981), who showed that SO_2 inhibits the normal pathway of NO_2 detoxification so that nitrite and nitrate ions are likely to accumulate to toxic levels in the chloroplasts, where they become involved in free radical reactions and enhance the damaging effects of SO_3^{2-} ions. In a long-term fumigation experiment using low concentrations of SO_2 and NO_2, Wellburn et al. (1981) compared the ratio of GDH to glutamine synthetase (GS) activity in clones of perennial ryegrass (*Lolium perenne*) and showed clear evidence of disturbance to nitrogen assimilatory pathways in plants exposed to these pollutant mixtures (Figure 5-2). While the enzyme ratio in all clones increased only slightly in NO_2, exposure to SO_2 resulted in a large increase in the SO_2-sensitive clone (S23), but not in the resistant clone. A mixture of both gases led to a significant increase in both the sensitive and resistant clones. In a third clone (Helmshore), which was SO_2 resistant and was isolated from a polluted industrial area, neither SO_2 nor the mixture had any effect on the ratio. It appears that the clone differed from the S_{23} resistant clone in its SO_2-resistant mechanism.

The effects of other combinations of pollutants on detoxification mechanisms have received relatively little attention; however, Mehlhorn et al. (1987) reported a larger increase in ascorbate peroxidase and glutathione reductase when pea plants were exposed to a mixture of SO_2, NO_2, and O_3 compared with a combination of any two of these gases. In spruce and fir trees, α-tocopherol and glutathione showed a more than additive response to a mixture of SO_2 and O_3 (Mehlhorn et al. 1986). The response of detoxification systems in trees to pollutant mixtures is complicated, however. While exposure to NO_x and nitric acid (HNO_3) has been shown to induce nitrate reductase activity in needles of Scots Pine (Wingsle et al. 1987) and red spruce (Norby, Weerisuriya, and Hanson 1989), this enzyme was inhibited in Norway spruce needles after exposure to mixtures of NO_2 with SO_2 or O_3 (Klumpp, Küppers, and Guderian 1989). This inhibition was most pronounced when all three gases were present, and the response was influenced by the age of the needles and the nutritional status of the plants. Nutrient effects may explain why the interaction of component ions in acidic mists is not always additive. In a long-term exposure of red spruce to different acidic mists, Chen, Lucas, and Wellburn (1990) found that sulfur concentration, rather than nitrogen content or pH of the mist, was the factor that most strongly influenced antioxidant levels and caused visible injury. NO_3^- or NH_4^+ ions mitigated this effect, perhaps by increasing the nutritional status of the plants.

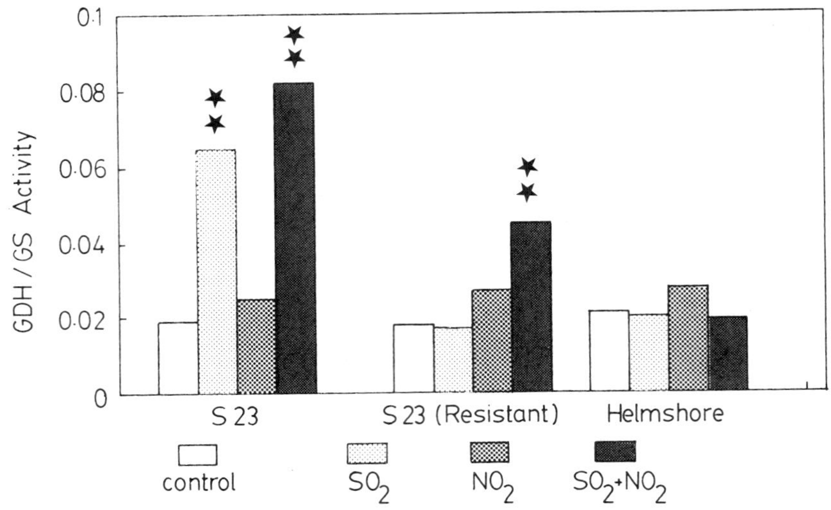

FIGURE 5-2. Glutamate dehydrogenase/glutamine synthetase activity in three clones of *Lolium perenne* showing different sensitivities to SO_2. Plants were exposed to 68 ppb SO_2, NO_2, or SO_2 + NO_2. Statistical significance denoted by: ** $P < 0.01$.

Source: tabulated data from Wellburn et al. 1981.

These supplementary factors make the identification of simple relationships between pollution tolerance and detoxification mechanisms very difficult when dealing with forest trees and other perennial plants. Antioxidant concentrations also are influenced by other environmental factors (Polle, Krings, and Rennenberg 1989) and undergo seasonal fluctuations that are considered to be important for winter survival, especially in conifers (Esterbauer and Grill 1978; Hausladen et al. 1989). The interference of pollutants with this normal seasonal variation has been observed (Hausladen et al. 1989) and could represent an additional factor governing the geographical limits of a species.

The entry of both gaseous pollutants and acidic mists into the plant is likely to result in accumulation of excess protons, derived from the oxidative reactions of pollutants and their products. Because most metabolic processes operate within a narrow pH range, changes in the pH of the cytoplasm and of the various subcellular compartments must be avoided. Plants have a large range of buffering systems and mechanisms for pH regulation (reviews by Nieboer et al. 1982; Nuticelli and Deamer 1982), including active membrane pumps, metabolic consumption of H^+ by organic acids such as malate, and inorganic buffers such as phosphate and bicarbonate. Individually these mechanisms have not been considered in relation to plant sensitivity to pollution, but a number of studies have investigated the overall buffering capacity of leaf tissue, especially of coniferous trees. Grill (1971) compared trees growing in polluted industrial areas with those in clean air and found that the buffering capacity of the whole needle homogenate was reduced in the former. The reduction in buffering at physiological pH was related to a decrease in the amounts of organic acids and phenolic groups, considered to be the main buffering components. Resistant trees were identified as those with a higher buffering capacity. In another study using Norway spruce, Scholz and Reck (1977) observed significant differences in the buffering capacity of different clones and suggested that this parameter may be useful in selecting trees resistant to pollution. It is important to realize, however, that measurement of the buffering capacity of whole leaf tissue provides only nonspecific information and tells us nothing of changes within different cellular compartments. More precise characterization of the mechanisms of pH control under pollution stress is required before reliable markers for tolerance selection can be identified.

Metabolic detoxification of pollutants can only proceed at a cost to the plant. Increased production of organic acids requires a change in carbon partitioning that is likely to be at the expense of other important processes. Increased rates of enzyme activity place an extra demand on the production of reductants such as ATP and NADPH. These higher demands may mean that although the plant is not directly harmed by the pollutant, it may grow more slowly and compete less well with others during periods when pollutant concentrations are low.

CONSEQUENCES OF ADDITIONAL ATMOSPHERIC SOURCES OF NITROGEN

NO_3^- and NH_4^+ ions are the most important nitrogen sources for higher plants; they are obtained mainly from the soil solution via the roots, although under natural conditions small amounts can enter the leaves from the atmosphere. In polluted environments, however, gaseous NO_X and NH_3 can provide additional sources of both NO_3^- and NH_4^+ for leaves. High rates of utilization of atmospheric NO_x and NH_3 have been reported: for example, Lockyer and Whitehead (1986) found that nearly half the nitrogen in the ryegrass *L. multiflorum* could be supplied by NH_3 taken into leaves. There are, however, potential problems of acid-base regulation in leaves when the rates of NH_3 and NO_X uptake are high (Wollenweber and Raven 1990). Although the absorption of NH_3 causes alkalinization, its subsequent assimilation can cause acidification because protons are released. Raven (1988) provides this equation for NH_3 assimilation:

$$3NH_3 + 45CO_2 + 32H_2O \longrightarrow C_{45}H_{72}N_3O_{32}^- + 45O_2 + H^+$$

Thus, the absorption and assimilation of NH_3 may lead to acid-base perturbations in leaf tissues. Problems also may occur when wet-deposited HNO_3 (derived from gaseous NO_x) is the nitrogen source. Raven (1988) provides this equation:

$$3HNO_3 + 45CO_2 + 35H_2O \longrightarrow C_{45}H_{72}N_3O_{32}^- + 51O_2 + H^+$$

Taking other factors into account, Raven concluded that when either NH_3 or HNO_3 are sources of nitrogen for shoots, they each lead to an excess of 0.22 mol H^+ per mol N assimilated. If ammonia enters the leaf in the form of NH_4^+, the excess is increased to 1.22 mol H^+ per mol N assimilated. However, the uptake of NO_3^- and its assimilation can lead to alkalinization:

$$3NO_3^- + 45CO_2 + 37H_2O \longrightarrow C_{45}H_{72}N_3O_{32}^- + 51O_2 + 2OH^-$$

If NO_3^- is the sole nitrogen source for leaves, an excess of 0.78 mol OH^- is produced per mol N assimilated. The precise balance of N-containing moieties entering the leaves thus plays a significant part in the complex processes of acid-base regulation.

Pearson and Stewart (1990) found that tree species differ greatly in their ability to assimilate nitrate in their leaves. Pioneer species such as birch and poplar have a much greater assimilatory capacity for nitrate than climax species such as oak and Sitka spruce. All species could, however, assimilate ammonium. In climax species, the capacity to assimilate NH_4^+ but not NO_3^- might lead to cytoplasmic acidosis.

The mechanisms for the assimilation of N-containing moieties by leaves are worthy of much more detailed attention in future work because differences in leaf nitrogen metabolism between species clearly could have important implications for biodiversity.

An example of the way in which excessive inputs of N can cause a differential competitive advantage among plants within an ecosystem is provided by the work of Heil and Bruggink (1987) and Heil et al. (1988). They have established that the changing nature of unmanaged heathland in the Netherlands, where the heath species *Calluna vulgaris* is being replaced by grass species (notably *Molinia caerulea*), is a result of the eutrophic effect of acidic rainfall and large inputs of nitrogen arising from intensive farming practices in the region. Both *Calluna* and *Molinia* are stress-tolerant species according to Grime (1979), but they have different growth patterns. *Calluna* is an evergreen and its long growing season can normally compensate for its slow growth rate, so that it competes successfully with the faster growing *Molinia* under normal nutrient-limiting conditions. A large increase in nitrogen supply, however, improves the competitive advantage of *Molinia*, increasing its growth rate so that it becomes the dominant species.

In ombrotrophic systems, plants such as *Sphagnum* moss are especially sensitive to atmospheric nitrogen inputs because they are physiologically adapted to maximize the uptake of normally scarce supplies of essential elements from the atmosphere. Studies by Press and Lee (1982) and Woodin, Press, and Lee (1985) have shown that nitrate reductase is induced rapidly when the leaves are supplied with nitrate, but the enzyme loses this capacity to be induced with successive additions of nitrate. This loss is attributed to an accumulation of ammonium in the tissues, which inhibits further enzyme activity. Thus, metabolism is closely coupled with nutrient supply, and the potential for perturbation of this system by atmospheric pollutants, therefore, is very high. In areas of intense NO_x pollution, the capacity of the leaves to metabolize the atmospheric inputs may be greatly exceeded; for example, it is likely that the observed decline in the number of species of *Sphagnum* in industrial areas like the southern Pennines in the United Kingdom is due to intolerance of this pollution. In transplant studies, when six *Sphagnum* species were grown and compared in a clean area and in the southern Pennines, only one species, *S. recurvum*, thrived in the polluted conditions (Ferguson and Lee 1983). *S. recurvum* is not a typical blanket bog species and is known to prefer nutrient-rich conditions.

SOME SPECIFIC PHYSIOLOGICAL EFFECTS: STOMATA, WATER RELATIONS, AND ASSIMILATE PARTITIONING

The processes of stomatal behavior, photosynthesis, and the translocation of assimilates are tightly coupled, and it is now known that air pollutants such as SO_2, NO_2, and O_3 can exert effects on each of them individually, with the potential for indirect impacts on linked processes. In spite of major advances in our understanding of these processes, largely through the use of controlled fumigation experiments, the implications of the results for field-grown crops and for natural and seminatural ecosystems have not received sufficient emphasis. Studies of the impacts of air pollutants on biodiversity, mediated via inter- or intraspecific variability in physiological responses, remain few in number, even though there is now ample justification for such work. Here we shall attempt to illustrate the ways in which pollution effects, initially on stomatal behavior and assimilate partitioning, could have far-reaching implications for biodiversity.

Stomatal pores in the leaves of higher plants provide a major pathway for the entry of atmospheric pollutants into the leaf (Garland and Branson 1977; Hosker and Lindberg 1982; Fowler 1985; Puckett 1990), and stomata themselves respond dynamically to the presence of pollutants in the atmosphere. Partial stomatal closure or wider openings often have been observed, the nature of the response being dependent upon the pollutants tested, their concentrations, the duration of exposure, environmental conditions (temperature, light intensity, humidity, and wind speed), and not least, upon plant species, cultivar, or clone (Unsworth and Black 1981; Darrall 1989).

Stomata can be thought of as the plant's first line of defense against air pollutants: closure in response to pollution may clearly represent a stress avoidance mechanism (Taylor 1978; Mansfield and Freer-Smith 1984), conferring resistance to the plant by preventing pollutants from entering the leaf. By contrast, enhanced stomatal opening may render the plant more susceptible to damage. Indeed, it is now well established that plants with high stomatal conductances may be less tolerant to SO_2 or O_3 singly, often as a consequence of greater uptake rates (Darrall 1989). Inter- and intraspecific variations in these responses to pollutants will—all other factors remaining equal—translate into variations in fitness that place some individuals at an advantage over others.

This principle has been illustrated by fumigation experiments with clones and cultivars of the same species. For example, Kimmerer and Kozlowski (1981) demonstrated that interclonal differences in the sensitivity of *Populus tremuloides* to SO_2 were related to stomatal conductance, more tolerant clones having lower conductances. Similar conclusions were reached by Pande (1985) in a study of the sensitivity of five spring barley *(Hordeum vulgare)* cultivars to 40, 80, and 120 ppb SO_2 applied over a 3-week period: stomatal conductance was lowest in cv. Midas, the most tolerant cultivar, and highest in cv. Koru, the most sensitive. Studies with O_3 further reinforce the view that stomata can fulfill a key role in determining sensitivity, as shown by Butler and Tibbitts (1979) in experiments with *Phaseolus* cultivars.

For pollutant mixtures, less information is available on the specific links between sensitivity and stomatal conductance, although many studies have demonstrated the major significance of cultivar and species variations in pollutant injury (reviewed by Ormrod 1982). Further studies are needed to establish the significance of stomatal behavior relative to other physiological or biochemical processes that could account for variations in sensitivity. Beckerson and Hofstra (1979) studied the effects of O_3 and SO_2 singly and in combination (150 ppb of each gas) on the development of visible injury on cucumber (*Cucumis sativus*), radish (*Raphanus sativus*), and soybean (*Glycine max*), and concluded that differences in stomatal conductances alone could not account for variations in sensitivity to the pollution regimes.

The work of Beckerson and Hofstra (1979) highlights the potential limitations of placing different clones, cultivars, or species into "sensitivity categories" on the basis of intrinsic stomatal conductances. It is particularly important to avoid visualizing conductance as a fixed value. The dynamic nature of stomata (Mansfield 1986) is central to their primary function, which is to balance two opposing priorities: maximization of CO_2

uptake for photosynthesis and prevention of excessive water loss by transpiration (Mansfield and Freer-Smith 1984). If the presence of gaseous pollutants disrupts the normal functioning of stomata, there clearly will be repercussions for photosynthesis, assimilate partitioning, and plant water status.

Research over the last 25 years has provided numerous examples of the sensitivity of stomata to the presence of pollutant gases (Majernik and Mansfield 1970; Biscoe, Unsworth, and Pinckney 1973; Ashenden 1979; Beckerson and Hofstra 1979; Neighbour, Cottam, and Mansfield 1988; Bennett, Lee, and Heggestad 1990), although fewer studies have attempted to examine the implications of altered stomatal functioning for plant survival that is mediated, for example, by changes in the ability of plants to withstand water deficits (see Darrall 1989). The danger of extrapolating the results of studies with well-watered plants to indicate how plants will function during water deficits has been clearly illustrated by the work of Unsworth, Biscoe, and Pinckney (1972). In studies of the stomatal responses of *Vicia faba* to short-term exposure to 100 ppb SO_2, Unsworth et al. (1972) observed an enhanced degree of stomatal opening (estimated to be equivalent to an 18% increase in transpiration) of well-watered plants. These effects alone suggest that exposure to short-term SO_2 episodes could render field crops more prone to water deficits; however, the same tests were also performed with plants from which water had been withheld for 2 days prior to fumigation. Again, exposure to SO_2 resulted in greater stomatal opening, although this was far more marked than in the well-watered plants, suggesting that the presence of SO_2 could exacerbate the effects of water deficits. Unfortunately, no information is available on the variation in these responses between cultivars.

In another study on the interactions between air pollutants and water deficits, Neighbour et al. (1988) exposed clonal silver and downy birch (*Betula pendula* and *B. pubescens*, respectively) to SO_2 and/or NO_2 at concentrations of each gas ranging from 20 to 65 ppb for a period of more than 3 weeks. Transpiration rates were increased markedly by exposure to the pollutant mixture, and there was clear evidence of a synergistic effect compared with the effects of each gas singly. Measurements of transpiration when water was withheld confirmed that plants exposed to this pollutant mixture succumbed to water-stress more rapidly than those exposed to clean air. There were interspecific differences in responses to the pollutants, although it is not clear whether clonal comparisons were significant. It is possible that differences in response between these two species could result in contrasting hybridization ratios in areas of contrasting pollution climate, although much longer term studies would be required to test this hypothesis.

A key example of within-species variability in the sensitivity of *B. pendula* to air pollutants has been provided by Whitmore and Freer-Smith (1982), although growth parameters were measured rather than stomatal functioning. Cuttings (one genotype) and plants grown from seed were exposed to a mixture of SO_2 and NO_2 (50 ppb of each gas) over a period of 80 days, and the increase in total plant dry weight was subsequently determined. The pollutant mixture caused significant reductions in growth of the cuttings and the seedlings, although the coefficients of variation were much higher for the seed stock (44.2% compared with 15.3% for the seedlings and cuttings, respectively), reflecting a very marked genetic heterogeneity within the seed population.

As was mentioned earlier, a fuller understanding of the implications of stomatal responses for biodiversity is dependent on a knowledge of other critical physiological processes. Alterations in carbohydrate assimilation and partitioning, occurring in response to air pollutants, could be especially significant in this respect. As with stomatal responses to pollutants, there is very diverse information in the literature on this subject, recently reviewed by Darrall (1989). Studies of assimilate partitioning have the potential to provide the researcher with a clearer idea of the implications for the whole plant of specific changes in stomatal processes, photosynthesis, and phloem loading. Inter- and intraspecific variability in assimilate partitioning (in response to air pollution and, indeed, many other abiotic or biotic stress factors) could translate into variations in the balance between root and shoot development which, in turn, may alter the nature of interspecific competition within a plant community (Hunt and Nicholls 1986; Grime et al 1986; Mansfield 1988). Such changes in root:shoot partitioning are known as "plastic" responses, and the significance of genetic factors (Ledig and Perry 1965), relative to environmental and/or biotic factors, remains a topic of much debate (Hunt and Nicholls 1986).

Both O_3 and SO_2 are known to be capable of causing reduced allocation of assimilates to root systems relative to shoots, with a consequent decrease in root:shoot ratio (Mansfield 1988, Darrall 1989). Bennett and Oshima (1977) found that root growth of crimson clover (*Trifolium incarnatum*) and annual ryegrass (*L. multiflorum*) was decreased more markedly than shoot growth during exposure to O_3. Studies using $^{13}CO_2$ to follow the translocation of assimilates in kidney beans (*Phaseolus vulgaris*) have shown that exposure to 200 ppb of O_3 over 4 days resulted in a 44% reduction in translocation to the root system (Okano et al. 1985). Similar observations have been made in fumigation experiments with SO_2 singly (e.g., Jones and Mansfield 1982; Freer-Smith 1984), with evidence of differing magnitudes of response for "sensitive" and "tolerant" lines of *Geranium carolinianum* (Taylor, Tingey, and Gunderson 1986).

There is now growing evidence that certain pollutant mixtures, notably SO_2 and NO_2, can act synergistically to produce reductions in the translocation of assimilates to root systems. This is particularly significant in view of the importance of such mixtures in the atmosphere (Unsworth and Black 1981; Ormrod 1982; Williams et al. 1989). Freer-Smith (1984) observed such synergism when black poplar (*Populus nigra*) was exposed to SO_2 and/or NO_2 at concentrations of 68 ppb of each gas for a period as short as 8 weeks. Further studies involving six tree species have demonstrated interspecific variability in responses to a mixture of SO_2 and NO_2 (62 ppb of each gas), with average reductions in root:shoot ratios ranging from 3 to 55% of control values for apple (*Malus domestica*) and grey alder (*Alnus incana*) respectively, following 22 weeks of exposure (Figure 5-3). Using $^{14}CO_2$, Gould and Mansfield (1988) were able to show that exposure of winter wheat (*Triticum aestivum*) to 80-100 ppb each of SO_2 and NO_2 caused a reduction in the translocation of assimilates from leaves to roots.

Reduced root:shoot ratios have been interpreted as a strategy whereby plants maintain photosynthetic area, at the expense of root systems, to compensate for pollutant-induced reductions in net assimilation rates (Mansfield 1988). Such responses may benefit plants when water or nutrients are in plentiful supply, but if these resources are limiting then plants with reduced root:shoot ratios may be at a competitive disadvantage. Changes in root:shoot partitioning could be particularly significant for plants adopting the "competitor" type of strategic specialization as proposed by Grime, Crick, and Rincon (1986). Such plants include those growing in stable habitats where there may be intense competition for light, water, and nutrients, and where plastic responses (e.g., changes in root:shoot ratio) form a basis for "active foraging" for resources.

Plasticity in developmental patterns may enable the plant to compensate in part for a reduction in biomass accumulation caused by pollution stress. This kind of compensatory mechanism has been observed particularly in relation to chronic O_3 exposure of sensitive

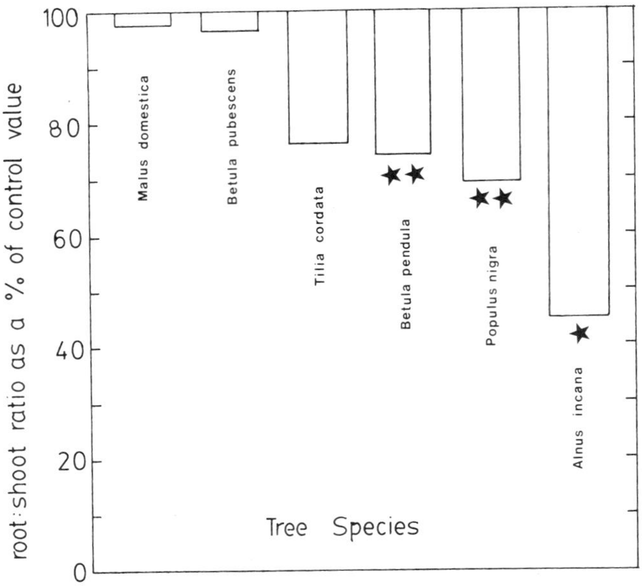

FIGURE 5-3. Reductions in root:shoot ratios (relative to charcoal-filtered control) of 6 tree species after 22 weeks of exposure to SO_2 + NO_2 at concentrations of 62 ppb of each gas. Statistical significance denoted by: * $P < 0.05$, ** $P < 0.001$.

Source: data recalculated from Freer-Smith 1984.

species like poplar, where pigment breakdown and a decline in net photosynthesis lead to premature leaf senescence (Reich 1983). In young leaves these effects are delayed to some extent by the operation of repair mechanisms, indicated by an increase in dark respiration, but the plant also compensates for the shortened lifespan of its leaves by increasing the rate of leaf production (Reich and Lassoie 1985). Although these adjustments are not sufficient to compensate completely for O_3 damage to photosynthesis, the ability to alter growth patterns probably represents a selective advantage to plants exposed to O_3 over long periods, and especially in the case of fast-growing plants with short life cycles or short growing seasons. Clearly, if there is some form of genetic control of plasticity, then the potential exists for air pollution to exert selective pressure on particular genotypes.

Further developments in our understanding of such selective processes will benefit from integrated experiments in which mixed plant communities are exposed to controlled pollution regimes. Processes such as stomatal behavior, photosynthesis, biomass accumulation, and root:shoot partitioning should be quantified for each plant species over extended time periods to allow any changes in community structure to be related to inter- or intraspecific variations in critical physiological processes. An example of such an approach is the study by Steubing et al. (1989) of the effects of SO_2, NO_2, and O_3 on herb layer species in a beech forest in Germany. They showed interspecific variations in responses to the pollution regimes, which translated into alterations in community structure. Such changes were evident after 4 years of SO_2 treatment (115 ppb applied for 4 hours per week during the growing season), while in the mixed pollutant treatment (SO_2,

NO_2, and O_3 at concentrations of 115, 51, and 98 ppb respectively), alterations in community structure became apparent after only 1 year. In their conclusions, Steubing et al. were able, albeit tentatively, to link stomatal responses of individual species to their sensitivity to pollution injury and, in some cases, to alterations in carbon assimilation rates. In a particularly sensitive species, *Allium ursinum*, it was found that carbohydrate metabolism was affected by exposure to the pollutants, with evidence for reduced translocation of assimilates away from leaves. Unfortunately, this study could not easily incorporate measurements of root development, so that the potential significance of the plasticity of root:shoot ratios could not be assessed.

Of special interest was the observation that sensitivity to air pollution could be related to the life form of each species. The most tolerant, *Hedera helix*, is an evergreen woody herb with scleromorphic leaves, while the most sensitive—*A. ursinum* and *Arum maculatum*—possess hygromorphic and mesomorphic leaves and are vernal geophytes. Although the authors provide stomatal density data only for *A. ursinum*, and there are no data for transpiration rates of *H. helix*, it seems reasonable to suggest that intrinsically low stomatal conductances of ivy may have conferred considerable resistance on this species by reducing the uptake of pollutants. As was stated by Steubing et al. (1989), "population development integrates all biochemical and physiological effects of pollution (and is, in addition, affected by competition)."

Continuing our theme of interrelationships between physiological responses, we conclude this section by considering the potential role of "chemical signalling" within plants (Davies et al. 1987). During the past decade it has become increasingly accepted that stomata can "sense" the water status of the soil by receiving and responding to a sesquiterpenoid compound, abscisic acid (ABA), produced in root tips in response to soil drying (reviewed by Davies and Mansfield 1988). Abscisic acid, arriving at the guard cells via the transpiration stream, can cause stomatal closure even in the absence of a fall in leaf turgor, which traditionally has been regarded as an essential precursor for stomatal closure (Zhang, Schurr, and Davies 1987). In this way, it is possible for a plant to reduce transpirational water loss during the early stages of soil water shortage, and thereby avoid the development of serious water deficits in the leaf tissues. The phrase "root-to-shoot communication" (Blackman and Davies 1985) has been coined to represent this process, with ABA being classed as a "stress hormone." Root-to-shoot communication provides another striking example of the integration and organization of physiological processes within plants.

Air pollutants have considerable potential to disrupt root-to-shoot communication, for example, by damaging the processes of stomatal functioning (and hence the ability to respond to ABA), by altering the balance of root and shoot development, or by inducing changes in the ability of the plant to synthesize, transport, and metabolize ABA. Indeed, in a recent study, Atkinson, Wookey, and Mansfield (1991) demonstrated that prior exposure of spring barley (*H. vulgare* cv. Klaxon) to a mixture of SO_2 and NO_2 (35 ppb of each gas) resulted in a reduced speed of stomatal closure in response to ABA which was supplied to the leaves via the transpiration stream (Figure 5-4). We still have much progress to make, however, if we are to understand the biochemical and mechanical bases for the observed effects. Unfortunately there is, as yet, very little further information available on the effects of air pollutants on root to shoot signalling, even though there are sound reasons to suspect that genetic heterogeneity in such responses may have profound implications for biodiversity.

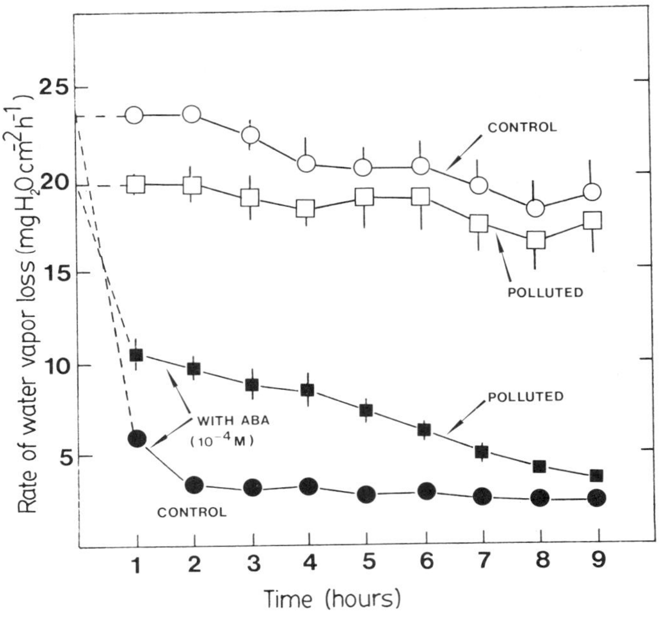

FIGURE 5-4. ABA response tests of spring barley (Hordeum vulgare L. cv. Klaxon) after 12 days of exposure to charcoal-filtered air (control) or SO2 + NO2 at concentrations of each gas of 24 and 33 ppb respectively. The results shown are the rates of water vapor loss per unit area (mean + 1 SE, n=10) from illuminated detached leaves supplied with either distilled water or 10-1 mol m-3 ABA solution. The youngest fully expanded leaf was always selected.

Source: Wookey, previously unpublished in this form.

POSTSCRIPT

Physiological and biochemical attributes contribute to biodiversity and can play a major role in the plastic responses that determine the survival of individuals of a species when one or more components of the environment change. Pollutant-induced disturbances to biochemistry may manifest themselves in different ways at the whole plant level, depending on environmental conditions, the nature of the stress, and the characteristics of the plant. On the one hand, the plant may suffer a reduction in vigor, putting it at a competitive disadvantage; this may occur, for example, if its defense mechanisms are disrupted so that it becomes less resistant to other stresses. Alternatively, the stress may be accommodated as a result of the stimulation of detoxification processes, or by switches in metabolism that constitute compensatory mechanisms and contribute to the overall stability of the plant system. There is now a substantial amount of knowledge of the ways in which individual air pollutants damage cells and tissues, and of the detoxification mechanisms that provide protection. Such mechanisms generally utilize metabolic pathways that have important roles in plants growing in unpolluted environments, such as the oxygen radical-scavenging systems that are essential in the protection of chloroplast membranes against harmful free radicals produced during photosynthesis. Rather less is understood about the detailed ways in which plants are stimulated to adjust their growth patterns to compensate for pollution damage, or of the way in which such changes may affect individuals or species in the long term. Whichever strategy is employed to cope with pollution stress, there will be some cost to the whole plant in terms of energy and resources. The balance between energy-demanding defense and repair processes and compensatory adjustments to growth patterns may determine the ultimate fate of the plant in a situation of long-term pollution exposure.

Air pollutants create selective pressures that are not normally encountered and can accentuate or subordinate heritable characters so that there are changes in the genetic makeup of populations. Air pollution that is widespread (e.g., O_3) may prove to be of more importance in the context of biodiversity than point sources that cause acute effects at particular locations. This is because genetic changes in populations over large geographical areas may have considerable ecological consequences. It is in this context that physiological and biochemical components of biodiversity will be of greatest significance.

In this chapter we have been able to define some aspects of metabolism and physiology that can be critically affected by air pollution. We have been able only to relate these speculatively to biodiversity because there has been insufficient research directed specifically to this question. However, some evidence (for example, that excessive loads of nitrogenous compounds from the atmosphere affect plants differentially in an ecosystem) does suggest that more attention should be given to this topic in future research.

ACKNOWLEDGMENTS

We are grateful to the Wolfson Foundation, the UK Department of Environment, and the Commission of the European Communities for their financial assistance during the preparation of this chapter.

REFERENCES

1. Alscher, R.G., J.L. Bower, and W. Zipfel. 1987. The basis for different sensitivities of photosynthesis to SO_2 in two cultivars of pea. *J. Exp. Bot.* 38 (186): 99-108.
2. Ashenden, T.W. 1979. Effects of SO_2 and NO_2 pollution on transpiration in *Phaseolus vulgaris* L. *Environ. Pollut.* 18:45-50.
3. Ashenden, T.W., and T.A. Mansfield. 1978. Extreme pollution sensitivity of grasses when SO_2 and NO_2 are present in the atmosphere together. *Nature* 273:142-43.
4. Atkinson, C.J., P.A. Wookey, and T.A. Mansfield. 1991. Atmospheric pollution and the sensitivity of stomata on barley leaves to abscisic acid and carbon dioxide. *New Phytol* 117:535-41.
5. Beckerson, D.W., and G. Hofstra. 1979. Response of leaf diffusive resistance of radish, cucumber and soybean to O_3 and SO_2 singly or in combination. *Atmos. Environ.* 13:1263-68.
6. Bennett, J.P., and R.J. Oshima. 1977. Effects of low levels of ozone on the growth of crimson clover and annual ryegrass. *Crop Sci.* 17:443-45.
7. Bennett, J.H., E.H. Lee, and H.E. Heggestad. 1990. Inhibition of photosynthesis and leaf conductance interactions induced by SO_2, NO_2, and $SO_2 + NO_2$. *Atmos. Environ.* 24A:557-62.
8. Biscoe, P.V., M.H. Unsworth, and H.R. Pinckney. 1973. The effects of low concentrations of sulphur dioxide on stomatal behavior in *Vicia faba*. *New Phytol.* 72:1299-1306.
9. Blackman, P.G., and W.J. Davies. 1985. Root to shoot communication in maize plants of the effects of soil drying. *J. Exp. Bot.* 36:39-48.
10. Bors, W., C. Langebartels, C. Michael, and H. Sanderman, Jr. 1989. Polyamines as radical scavengers and protectants against ozone damage. *Phytochemistry* 28 (6): 1589-95.

11. Butler, L.K., and T.W. Tibbitts. 1979. Stomatal mechanisms determining genetic resistance to ozone in *Phaseolus vulgaris* L. *J. Am. Soc. Hort. Sci.* 104:213-16.
12. Chen, Y-M., P.W. Lucas, and A.R. Wellburn. 1990. Relationships between foliar injury and changes in antioxidant levels in red and Norway spruce exposed to acidic mists. *Environ. Pollut.* 68:1-15.
13. Chiment, J.J., R.G. Alscher, and P.R. Hughes. 1986. Glutathione as an indicator of SO_2-induced stress in soybean. *Environ. Exp. Bot.* 26:147-52.
14. Darrall, N.M. 1989. The effect of air pollutants on physiological processes in plants. *Plant Cell Environ.* 12:1-30.
15. Davies, D.D., and F.A. Wood. 1968. The relative susceptibility of eighteen coniferous tree species to ozone. *Phytopathology* 62:14-19.
16. Davies, W.J., and T.A. Mansfield. 1988. Abscisic acid and drought resistance in plants. *ISI Atlas Sci. Anim. Plant Sci.* 1:263-69.
17. Davies, W.J., J.C. Metcalfe, U. Schurr, G. Taylor, and J. Zhang. 1987. Hormones as chemical signals involved in root to shoot communication of effects of changes in the soil environment. In *Hormone action in plant development—a critical appraisal*, ed. G.V. Hoad, J.R. Lenton, M.B. Jackson, and R. Atkin, pp. 201-16. London: Butterworth.
18. Dochinger, L.S., A.M. Townsend, D.W. Siegrist, and F. Bender. 1972. Responses of hybrid poplar trees to sulfur dioxide fumigation. *J. Air Pollut. Control Assoc.* 22:363-71.
19. Esterbauer, H., and D. Grill. 1978. Seasonal variation of glutathione and glutathione reductase in needles of *Picea abies*. *Plant Physiol.* 61:119-21.
20. Ferguson, P., and J.A. Lee. 1983. The growth of *Sphagnum* species in the southern Pennines. *J. Bryol.* 12:579-86.
21. Fowler, D. 1985. Deposition of SO_2 onto plant canopies. In *Sulfur dioxide and vegetation: Physiology, ecology, and policy issues*, ed. W.E. Winner, H.A. Mooney, and R.A. Goldstein, pp. 389-402. Stanford, CA: Stanford University Press.
22. Freer-Smith, P.H. 1984. The responses of six broad-leaved trees during long-term exposure to SO_2 and NO_2. *New Phytol.* 97:49-61.
23. Garland, J.A., and J.R. Branson. 1977. The deposition of sulphur dioxide to a pine forest assessed by a radioactive tracer method. *Tellus* 29:445-54.
24. Gould, R.P., and T.A. Mansfield. 1988. Effects of sulphur dioxide and nitrogen dioxide on growth and translocation in winter wheat. *J. Exp. Bot.* 39:389-99.
25. Grill, D. 1971. Die Pufferkapazität gesunder und rauchgeschädigter Fichtennadeln. *Z. Pflanzenkrankh. Pflanzenschutz* 78:616-22.
26. Grill, D., H. Esterbauer, and U. Klosch. 1979. Effect of sulfur dioxide on glutathione in leaves of plants. *Environ. Pollut.* 19:187-94.
27. Grime, J.P. 1979. *Plant strategies and vegetation processes.* Chichester, England: John Wiley & Sons.
28. Grime, J.P., J.C. Crick, and J.E. Rincon. 1986. The ecological significance of plasticity. In *Plasticity in plants*, ed. D.H. Jennings and A.J. Trewavas. London: Cambridge Univ. Press.
29. Guri, A. 1983. Variation in glutathione and ascorbic acid content among selected cultivars of *P. vulgaris* prior to and after exposure to ozone. *Can. J. Plant Sci.* 63:733-37.
30. Hausladen, A., N.R. Madamanchi, S. Fellows, R.G. Alscher, and R.G. Amundson. 1990. Seasonal changes in antioxidants in red spruce as affected by ozone. *New Phytol.* 115:447-58.
31. Heil, G.W., and M. Bruggink. 1987. Competition for nutrients between *Calluna vulgaris* L. Hull and *Molinia caerulea* (L.) Moench. *Oecologia* 73:105-7.
32. Heil, G.W., M.J.A. Werger, W. de Mol, D. van Dam, and B. Heijne. 1988. Capture of atmospheric ammonium by grassland canopies. *Science* 239:764-65.

33. Horsman, D.C., and A.R. Wellburn. 1975. Synergistic effect of SO_2 and NO_2 polluted air upon enzyme activity in pea seedlings. *Environ. Pollut.* 8:123-33.
34. Hosker, R.P., Jr., and S.E. Lindberg. 1982. Review: Atmospheric deposition of gases and particles. *Atmos. Environ.* 16:889-910.
35. Hunt, R., and A.O. Nicholls. 1986. Stress and the coarse control of growth and root-shoot partitioning in plants. *Oikos* 47:149-58.
36. Jäger, H-J., and H. Klein. 1977. Biochemical and physiological detection of sulphur dioxide injury to pea plants (*Pisum sativum*). *J. Air Pollut. Control Assoc.* 27:464-66.
37. Jäger, H-J., J. Bender, and L. Grünhage. 1985. Metabolic responses of plants differing in SO_2 sensitivity towards SO_2 fumigation. *Environ. Pollut.* 39:317-36.
38. Jones, T., and T.A. Mansfield. 1982. Studies on dry matter partitioning and distribution of ^{14}C-labelled assimilates in plants of *Phleum pratense* exposed to SO_2 pollution. *Environ. Pollut.* 28:199-207.
39. Kimmerer, T.W., and T.T. Kozlowski. 1981. Stomatal conductance and sulfur uptake of five clones of *Populus tremuloides* exposed to sulfur dioxide. *Plant Physiol.* 67:990-95.
40. Klumpp, A., K. Küppers, and R. Guderian. 1989. Nitrate reductase activity in needles of Norway spruce fumigated with different mixtures of ozone, sulfur dioxide, and nitrogen dioxide. *Environ. Pollut.* 58:261-71.
41. Langabartels, C., K. Kerner, S. Leonardi, M. Schraudner, M. Trost, W. Heller, and H. Sandermann, Jr. 1991. Biochemical plant responses to ozone. 1. Differential induction of polyamine and ethylene biosynthesis in tobacco. *Plant Physiol.* 95:882-89.
42. Larsson, R.A. 1988. The antioxidants of higher plants. *Phytochemistry* 27:969-78.
43. Ledig, F.T., and T.O. Perry. 1965. Physiological genetics of the shoot-root ratio. In *Proceedings of the Society of American Foresters,* pp. 39-43.
44. Lockyer, D.R., and D.C. Whitehead. 1986. The uptake of gaseous ammonia by leaves of Italian ryegrass. *J. Exp. Bot.* 37:919-27.
45. Majernik, O., and T.A. Mansfield. 1970. Direct effect of SO_2 pollution on the degree of opening of stomata. *Nature* 227:377-78.
46. Mansfield, T.A. 1986. The physiology of stomata: New insights into old problems. In *Plant physiology, a treatise, Vol IX: Water and solutes in plants,* pp. 155-224. Orlando, FL: Academic Press, Inc.
47. Mansfield, T.A. 1988. Factors determining root:shoot partitioning. In *Scientific basis of forest decline symptomatology,* ed. J.N. Cape and P. Mathy, pp. 171-81. Brussels: Commission of the European Communities.
48. Mansfield, T.A., and P.H. Freer-Smith. 1981. Effects of urban air pollution on plant growth. *Biol. Rev.* 56:343-68.
49. Mansfield, T.A., and P.H. Freer-Smith. 1984. The role of stomata in resistance mechanisms. In *Gaseous pollutants and plant metabolism,* ed. M.L. Koziol and F.R. Whatley, pp. 131-46. London: Butterworth.
50. Mehlhorn, H., G. Seufert, A. Schmidt, and K.J. Kunert. 1986. Effect of SO_2 and O_3 on production of antioxidants in conifers. *Plant Physiol.* 82:336-38.
51. Mehlhorn, H., and A.R. Wellburn. 1987. Stress ethylene formation determines plant sensitivity to ozone. *Nature* 327:417-18.
52. Mehlhorn, H., D.A. Cottam, P.W. Lucas, and A.R. Wellburn. 1987. Induction of ascorbate peroxidase and glutathione reductase activities by interactions of mixtures of air pollutants. *Free Rad. Comms.* 3:193-97.
53. Mehlhorn, H., B. Tabner, and A.R. Wellburn. 1990. ESR evidence for the formation of free radicals in plants exposed to ozone. *Physiol. Plant.* 79:377-83.

54. Mejnartowicz, L.E. 1984. Enzymic investigations on tolerance in forest trees. In *Gaseous pollutants and plant metabolism*, ed. M.L. Koziol and F.R. Whatley, pp. 381-98. London: Butterworth.
55. Mijazaki, J.H., and S.F. Yang. 1987. The methionine salvage pathway in relation to ethylene and polyamine biosynthesis. *Physiol. Plant.* 69:366-70.
56. Miller, J.E., and P.B. Xerikos. 1978. Residence time of sulphite in SO_2 "sensitive" and "tolerant" soybean cultivars. *Environ. Pollut.* 18:259-64.
57. Neighbour, E.A., D.A. Cottam, and T.A. Mansfield. 1988. Effects of sulphur dioxide and nitrogen dioxide on the control of water loss by birch (*Betula* spp.). *New Phytol.* 108:149-57.
58. Nieboer, E., J.D. MacFarlane, and D.H.S. Richardson. 1983. Modification of plant cell buffering capacities by gaseous pollutants. In *Gaseous pollutants and plant metabolism*, ed. M.L. Koziol and F.R. Whatley, pp. 313-30. London: Butterworth.
59. Norby, R.J., Y. Weerasuriya, and P.J. Hanson. 1989. Induction of nitrate reductase activity in red spruce needles by NO_2 and HNO_3 vapor. *Can. J. For. Res.* 19:889-96.
60. Nuticelli, R., and D.W. Deamer. 1982. *Intracellular pH: Its measurements, regulation, and utilization in cellular functions.* Kroc Foundation Series 15. New York: Alan R. Liss, Inc.
61. Okano, K., O. Ito, G. Takeba, A. Shimizu, and T. Totsuka. 1985. Effects of O_3 and NO_2 alone or in combination on the distribution of 13C-assimilate in kidney bean plants. *Jpn. J. Crop Sci.* 54:152-59.
62. Ormrod, D.P. 1982. Air pollutant interactions in mixtures. In *Effects of gaseous air pollution in agriculture and horticulture*, ed. M.H. Unsworth and D.P. Ormrod, pp. 307-31. London: Butterworth.
63. Pande, P.C. 1985. An examination of the sensitivity of five barley cultivars to SO_2 pollution. *Environ. Pollut.* 37:27-41.
64. Pearson, J., and G.R. Stewart. 1990. Susceptibility of trees to foliar inputs of nitrogen: A metabolic "vitality" theory. In *Abstracts of conference on acidic deposition: Its nature and impacts*, Glasgow, September 1990. p. 80. Edinburgh: Royal Society of Edinburgh.
65. Peleg, M. 1976. The chemistry of ozone in the treatment of water. *Water Res.* 10:361-65.
66. Polle, A., B. Krings, and H. Rennenberg. 1989. Factors modulating superoxide dismutase activity in needles of spruce trees (*Pices abies* L.). *Ann. Sci. For.* 46 (suppl.): 807s-810s.
67. Press, M.C., and J.A. Lee. 1982. Nitrate reductase activity of *Sphagnum* species in the southern Pennines. *New Phytol.* 92:487-94.
68. Price, A., P.W. Lucas, and P.J. Lea. 1990. Age dependent damage and glutathione metabolism in ozone fumigated barley: A leaf section approach. *J. Exp. Bot.* 41:1309-17.
69. Puckett, L.J. 1990. Estimates of ion sources in deciduous and coniferous throughfall. *Atmos. Environ.* 24 (A): 545-55.
70. Raven, J.A. 1988. Acquisition of nitrogen by the roots of land plants: Its occurrence and implications for acid-base regulation. *New Phytol.* 109:1-20.
71. Reich, P.B. 1983. Effects of low concentrations of O_3 on net photosynthesis, dark respiration, and chlorophyll contents in aging hybrid poplar leaves. *Plant Physiol.* 73:291-96.
72. Reich, P.B., and J.R. Lassoie. 1985. Influence of low concentration of ozone on growth, biomass partitioning and leaf senescence in young hybrid poplar plants. *Environ. Pollut.* 39:39-51.
73. Rowland-Bamford, A.J., A.M. Borland, P.J. Lea, and T.A. Mansfield. 1989. The role of arginine decarboxylase in modulating the sensitivity of barley to ozone. *Environ. Pollut.* 61:95-106.
74. Sakaki, T., N. Kondo, and K. Sugahara. 1983. Breakdown of photosynthetic pigments and lipids in spinach leaves with ozone fumigation: Role of activated oxygens. *Physiol. Plant.* 59:28-34.
75. Salin, M.L. 1987. Toxic oxygen species and protective systems of the chloroplast. *Physiol. Plant.*72:681-89.

76. Scholz, F., and S. Reck. 1977. Effects of acids on forest trees as measured by titration *in vivo*, inheritance of buffering capacity in *Picea abies*. *Water Air and Soil Pollut.* 8:41-45.

77. Silvius, J.E., M. Ingle, and C.H. Baer. 1975. Sulphur dioxide inhibition of photosynthesis in isolated spinach chloroplasts. *Plant Physiol.* 56:434-37.

78. Steubing, L., A. Fangmeier, R. Both, and M. Frankenfeld 1989. Effects of SO_2, NO_2, and O_3 on population development and morphological and physiological parameters of native herb layer species in a beech forest. *Environ. Pollut.* 58:281-302.

79. Tanaka, K., Y. Suda, N. Kondo, and K. Sugahara. 1985. O_3 tolerance and the ascorbate dependent H_2O_2 decomposing system in chloroplasts. *Plant Cell Physiol.* 26:1425-31.

80. Tanaka, K., H. Saji, and N. Kondo. 1988. Immunological properties of spinach glutathione reductase and inductive biosynthesis of the enzyme with ozone. *Plant Cell Physiol.* 29:637-42.

81. Taylor, G.E., Jr. 1978. Plant and leaf resistance to gaseous air pollution stress. *New Phytol.* 80:523-34.

82. Taylor, G.E., Jr., D.T. Tingey, and C.A. Gunderson. 1986. Photosynthesis, carbon allocation, and growth of sulphur dioxide in ecotypes of *Geranium carolinianum* L. *Oecologia* (Berlin) 68: 350-57.

83. Taylor, G.E., Jr., B.M. Ross-Todd, and C.A. Gunderson. 1988. Action of ozone on foliar gas exchange in *Glycine max* L. Merr.: A potential role for endogenous stress ethylene. *New Phytol.* 110:301-8.

84. Tingey, D.T., R.A. Reinart, V.A. Dunning, and W.W. Heck. 1971. Vegetation injury from the interaction of nitrogen dioxide and sulphur dioxide. *Phytopathology* 61:1506-11.

85. Tingey, D.T., C. Standley, and R.W. Field. 1976. Stress ethylene evolution: A measure of ozone effects on plants. *Atmos. Environ.* 10:969-74.

86. Tingey, D.T., and G.E. Taylor, Jr. 1982. Variation in plant response to ozone: A conceptual model of physiological events. In *Effects of gaseous air pollution in agriculture and horticulture*, ed. M.H. Unsworth and D.P. Ormrod. London: Butterworth.

87. Unsworth, M.H., P.V. Biscoe, and H.R. Pinckney. 1972. Stomatal responses to sulphur dioxide. *Nature* 239:458-59.

88. Unsworth, M.H., and V.J. Black. 1981. Stomatal responses to pollutants. In *Stomatal physiology*, ed. P.G. Jarvis and T.A. Mansfield, pp. 187-203. London: Cambridge Univ. Press.

89. Wellburn, A.R., C. Higginson, D. Robinson, and C. Walmsley. 1981. Biochemical explanations of more than additive inhibitory low atmospheric levels of SO_2 and NO_2 upon plants. *New Phytol.* 88:223-37.

90. Whitmore, M.E., and P.H. Freer-Smith. 1982. Growth effects of SO_2 and/or NO_2 on woody plants and grasses during spring and summer. *Nature* 300:55-56.

91. Williams, M.L., D.H.F. Atkins, J.S. Bower, G.W. Campbell, J.G. Irwin, and D. Simpson. 1989. *A preliminary assessment of the air pollution climate of the UK*. Warren Spring Laboratory Report LR 723 (AP). Warren Spring Laboratory, U.K. Department of Trade and Industry, Stevenage, U.K.

92. Wingsle, G., T. Näsholm, T. Lundmark, and A. Ericsson. 1987. Induction of nitrate reductase in needles of Scots Pine seedlings by NO_x and NO_3^-. *Physiol. Plant.* 70:399-403.

93. Wollenweber, B., and J.A. Raven. 1990. N acquisition from atmospheric NH_3 and acid-base balance in *Lolium perenne*. In *Abstracts of conference on acidic deposition: Its nature and impacts*, Glasgow, September 1990. p. 82. Edinburgh: Royal Society of Edinburgh.

94. Woodin, S., M.C. Press, and J.A. Lee. 1985. Nitrate reductase activity in *Sphagnum fuscum*, in relation to wet deposition of nitrate from the atmosphere. *New Phytol.* 99:381-88.

95. Zhang, J., U. Schurr, and W.J. Davies. 1987. Control of stomatal behavior by abscisic acid which apparently originates in the roots. *J. Exp. Bot.* 38:1174-81.

96. Ziegler, I. 1972. The effect of SO_3^{2-} on the activity of ribulose 1,5-diphosphate carboxylase in isolated spinach chloroplasts. *Planta* 103:155-63.

6

Air Pollution Interactions with Natural Stressors

Jeremy J. Colls and Michael H. Unsworth

INTRODUCTION

A complete review of solid, liquid, and gaseous air pollutants and their impacts on plant and animal species, in combination with the full range of natural stresses, is clearly outside the scope of this chapter. We focus here on plants; on the four primary phytotoxic gases, ozone (O_3), nitric oxide (NO), nitrogen dioxide (NO_2), and sulfur dioxide (SO_2); and on the principal natural stresses due to mechanical loading, shortage of water, pests and diseases, nutrient deficiency, and low temperatures. The great majority of published material on air pollution responses is of course on single species or agricultural bicultures such as grass/clover mixtures. Even for those studies in which multiple natural species have been involved, such as forest ecosystem damage caused by smelter emissions in North America (e.g., Gordon and Gorham 1963), the pollutant load has often been so high, and the damage so acute, that the results cannot be extended to chronic pollutant effects, let alone to their interactions with other stresses. Consequently, our deductions on biodiversity impacts are based primarily on the interpretation of environmental and physiological data from rather simple biological systems. Superimposed on this already complex situation of pollutant exposure we have a set of relatively new concerns brought about by increasing atmospheric carbon dioxide and UV-B radiation, and by possible changes in the distribution of global climate patterns that we have previously considered stable. If we remember, as background to this discussion, that around one-half of all known species inhabit tropical rain forests about which we have virtually no information on air pollution or its effects (Wilson 1988), then the sheer scale of the problem becomes apparent.

NATURAL STRESSES

Of course, even stress physiologists do not necessarily agree on a definition of natural stress. We tend to follow the definition of Austin (1989), that plants are considered to be under stress when they experience a relatively severe shortage of an essential constituent or an excess of a potentially toxic or damaging substance. The term "substance" here must include physical parameters such as heat and water, as well as chemical ones such as nutrients and pollutants. The major proviso to this definition in the current context is that the maintenance of species diversity must take as a priority the survival of species, not merely their maximum productivity.

Air pollutants are known to affect many aspects of plant development and reproduction. Some of these effects, such as changes in stomatal conductance and photosynthesis, are very short-term, and the integrated consequences for the plant cannot easily be predicted. Thus Sanders, Clark, and Colls (1990) found that stomatal conductance and photosynthetic rate of green bean (*Phaseolus vulgaris* cv. Lit) fell rapidly at the start of ozone exposure in open-top chambers, but recovered subsequently to their initial value. Long-term responses may be subtle and require detailed measurements to elucidate; work at Nottingham (Sanders et al. 1990) showed that although a mixture of ambient air pollutants had no significant effect on the number of leaves on field beans (*Vicia faba* L.) or on their total area, the younger leaves tended to have larger areas in charcoal-filtered treatments than in nonfiltered treatments (Figure 6-1). Such imperceptible differences probably have implications for the effects of pollutants in combination with natural stresses such as pest invasion and particle deposition, but the significance of these cannot yet be quantified.

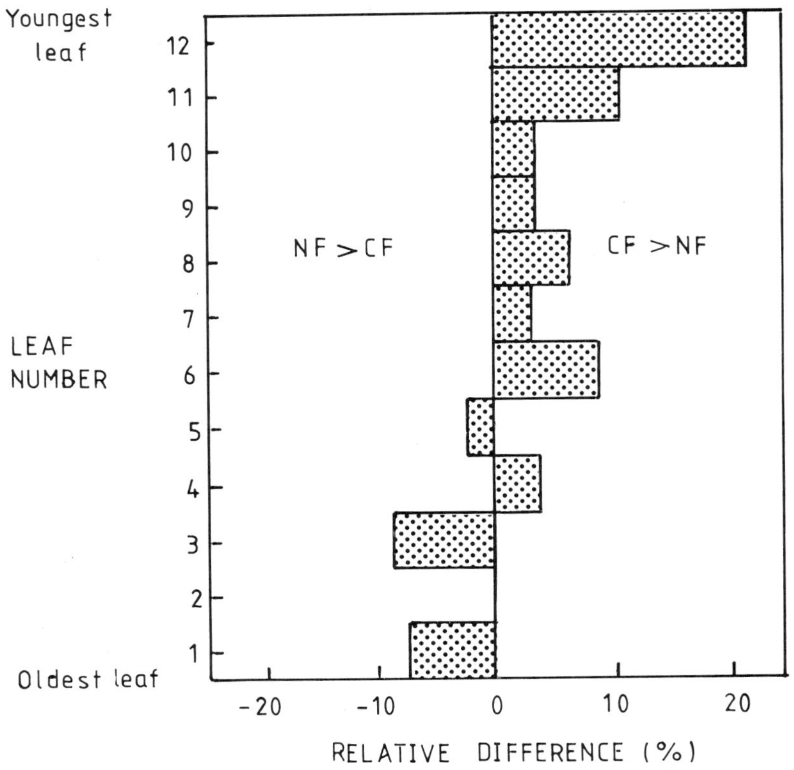

FIGURE 6-1. The difference in leaf area of individual leaves on *Vicia Faba* plants from nonfiltered (NF) and charcoal-filtered (CF) treatments at 49 days after emergence.

Source: Sanders, Clark, and Colls 1990.

MECHANICAL LOADING

Mechanical stress caused by air movement can take a variety of forms and have a range of consequences, depending on the plant architecture and wind profile. Thus responses can range from the stripping of leaves without stem damage (Moore and Osgood 1985), to the bending of stems in cereal lodging (Pinthus 1973) or the entire uprooting of trees by wind throw (Woodward 1987). Little research has been done on pollutant interactions with wind damage with the exception of lodging, although there is new work in progress in Scotland. B.R. Werkman (pers. communication 1991) has exposed Norway spruce seedlings to acid mist (pH 2.5) and ozone (140 ppb) with and without a wind of 16 ms^{-1}. Although the acid mist alone reduced the frost hardiness of the seedlings below that of the control (pH 5.0), the combined effect of the acid mist and wind was to increase the frost hardiness.

Effects of pollutants on lodging were observed when Baker, Fullwood, and Colls (1990) exposed winter barley (*Hordeum vulgare* cv. Igri) to sulfur dioxide in an open-field gas release system. Plants were grown under SO_2 concentrations that were raised by nominal amounts of 200, 100, and 50 ppb throughout the season. At the end of the season, lodging was experienced in all treatments, but was worse in the 100 and 200 ppb treatments than in the 50 ppb or control (ambient) treatments. The structural strength and anatomy of plant stems were investigated. The internode strength of stems from the two more severe SO_2 treatments was significantly reduced, and this was related to changes in the structure of sclerenchyma cells.

Any predictions of ecosystem responses to both wind force and air pollutants must, for reasons given above, be speculative at this stage. Pollutants can alter leaf areas, stem strengths, and root biomass—thus the severity of all three major forms of wind damage may be changed.

SHORTAGE OF WATER

Until recently, the prevailing view of transpiration was of an hydraulic system in which water flowed from soil to atmosphere through a resistance network, at a rate controlled primarily by the soil water potential, the saturation vapor pressure deficit, and the stomatal or canopy resistance. The plant responded to short-term water stress by reducing the stomatal aperture in order to maintain turgor, and to long-term water stress by changing the root depth and density so as to maximize extraction. Since the deposition of gaseous pollutants was seen as an essentially passive process in which the pollutant flux was controlled by stomatal conductance, there was a clear-cut mechanism by which water stress could interact with pollutant deposition.

A substantial change in perspective has been brought about by improved understanding of the role of chemical messengers, in particular abscissic acid (ABA), in influencing stomatal conductance (Neales et al. 1989). It appears, for example, that if a small proportion of root tips are water stressed the stomata will close, even when a plentiful supply of water is available to the majority of roots. Furthermore, the stomatal closure is triggered by root-sourced ABA that has moved to the leaf epidermis. This new perspective has profound implications for aspects of pollutant/water stress interaction. Any experimental work on such interactions should be re-examined for the possibility that spurious differences in root water environment influenced the pollutant uptake. This could happen with pot plants in which the root tips grew beyond the pots or with plants growing in soil in which physical inhomogeneities such as stones or cavities caused local drainage variations. In a specific experiment on this topic, we grew green beans (*P. vulgaris* cv. Lit) in open-top chambers. Some plants were grown in pots, others in the soil; all treatments were irrigated to keep them close to field capacity. Five ozone concentrations were applied in the chambers, corresponding to charcoal-filtered air (CF), nonfiltered air (NF), and NF air to which three ozone elevations were added. The 7-hour seasonal mean concentrations were 8, 21, 27, 33, and 38 ppb, respectively. Three different responses were observed. First, we found significant differences between the pot- and soil-grown plants even within the NF treatment. Pot-grown *P. vulgaris* grew better initially, although the soil-grown plants had caught up by 30 days after emergence and achieved a higher maximum number of leaves per plant. Leaves of pot-grown plants senesced more rapidly toward the end of the season, leaving fewer leaves per plant by harvest. Nevertheless, the yield from pot-grown plants was 14 percent higher than from soil-grown. Second, vegetative aspects such as leaf dry weight showed less sensitivity to ozone in pot-grown plants than in soil-grown plants although the growth stages in the two cultivation treatments were unaffected by ozone. Third, there was an important difference in the reproductive performance. Soil-grown plants at the lowest ozone addition (27 ppb as compared with CF 8 ppb; 7-hour means) showed significant increases in the number of pods per plant and seed yield per plant, whereas pot-grown plants showed decreases. Thus the difference in cultivation practice gave a different sign to the response. Pots have been used in many of the experiments in which the combined responses of water and pollutant stresses on plants have been investigated. Results from such experiments should be interpreted with some caution when applied to soil-grown plants.

Field-based experimental work on water and pollutant interactions has mainly involved crop experiments with monocultures or with, for example, grass/clover swards. Results on the effects of water stress in modifying pollutant responses have often been inconsistent. For example, Heggestad and Lesser (1990) found that water stress did not alter the yield reduction expected from soybean under ozone exposure. Miller et al. (1989) reinforced this result. The National Crop Loss Assessment Network (NCLAN) program conducted a specific series of experiments on the soil moisture/ozone interaction. It was concluded (Heagle et al. 1988) that responses could range between (1) no measurable change in O_3 sensitivity due to water stress; (2) reduced O_3 sensitivity; and (3) increased sensitivity in both additive and synergistic modes. The issue is at the forefront of pollution effects research, because most research is done under carefully controlled conditions (including irrigation) whereas most agricultural crops are intermittently water stressed. Within NCLAN, experiments in which water stress was varied were conducted on cotton, soybean, clover/fescue, alfalfa, and barley. The simplistic view that water stress would act to close stomata and reduce ozone deposition was shown to be appropriate under some conditions but not under others. There is currently no full understanding of the reasons for these variations in observed responses. In light of this lack of insight into either the mechanisms or the consequences of water stress affecting pollutant response of simple systems, we can only speculate on the possible interactions in more complex plant communities. Mansfield et al. (1986) discussed the changes in the epidermal layer caused by pollutants—changes that might in turn affect drought responses. In what has now become a common measure of pollutant impact on stomatal function, Neighbour et al. (1988) determined the weight loss of excised leaves to show that the closure of stomata from polluted plants was impaired. Thus we might anticipate that, in times of water stress, species having an intrinsically higher water-use efficiency and showing least impairment of stomatal function under pollutant exposure would be favored in a competitive environment.

In a specific study of the effects of both water and ozone stresses on competition between species, Heagle et al. (1989) exposed a community of ladino clover (*Trifolium repens* L. cv. Regal) and tall fescue (*Festuca arundinacea* Schreb. cv. Kentucky 31) to ozone at two levels of soil moisture stress. They found a statistically significant interaction between ozone and water stress. This was thought to be because the fescue grew better than the clover in all except the well-watered, charcoal-filtered treatments, and the clover was more sensitive than the fescue to both water and ozone stresses.

Although ozone is the dominant gaseous air pollutant in the United States, many other regions of the world have concurrent concentrations at similar magnitude of four gases—O_3, SO_2, NO, and NO_2. Lucas (1990) explored the interaction between severe water stress and simultaneous exposure to two of these gases. Timothy grass (*Phleum pratense* L.), which is widespread in Europe both as lawn cover and as a forage crop, was grown in vertical soil tubes. The plants were exposed for 40 days to either 30 ppb NO_2 + 30 ppb SO_2, or 60 ppb NO_2 + 60 ppb SO_2, or 90 ppb NO_2 + 90 ppb SO_2. Following this pretreatment, the plants were allowed to grow at ambient conditions: one-half in well-watered soil and the remainder without any further watering. After 23 days, leaf area and shoot biomass were reduced at the two highest concentrations of SO_2 + NO_2, and the greatest reduction occurred in the water-stressed treatment. This was attributed to a change in their water-use efficiency.

PESTS AND PATHOGENS

Lechowicz (1987) summarized plant responses to air pollutants that could affect their resistance to pests and pathogens. The key points were (1) for SO_2 exposure, more mass was allocated to leaves and less to stems; (2) no clear patterns existed for O_3 or NO_2; (3) for SO_2 or O_3 exposure, root and leaf carbohydrate concentration sometimes increased and sometimes decreased; and (4) for SO_2, O_3, and NO_2 exposure, plant reproduction was suppressed.

As with other natural stresses, a very wide range of resource capture and allocation responses may be found; this naturally makes it difficult to predict how the modified plant system will respond to the additional stress of pest or pathogen attack. The picture is further confounded by the difficulty of deciding whether changes to aspects of the plant such as root biomass and leaf area are actually perceived by the plant as a stress.

Many observations have been made of population increases of herbivorous insects on plants damaged by air pollutants (see, e.g., Alstad, Edmunds, and Weinstein 1982). More recently, systematic experiments have revealed more about the fascinating interdependencies that are involved. In this chapter we concentrate our attention on the aphid, many species of which are widespread agricultural pests, and on cereal pathogens. Dohmen (1988) measured the mean relative growth rate (MRGR) of *Aphis fabae* Scop. growing in clip cages on field bean (*V. faba* L. cv. Sutton). Exposures of the plants to SO_2 (90 to 130 ppb), NO_2 (200 ppb), or O_3 (85 ppb) were made for periods of between 2 and 7 days inside ventilated Perspex chambers. The aphids were then placed on the plants for 3 days, after which MRGR was determined. There were significant increases in MRGR for aphids from the plants exposed to SO_2 and NO_2, but decreases for those from the O_3-exposed plants. The former influence on MRGR was attributed to the effect of nitrogen and sulfur in enhancing plant metabolic processes, possibly increasing the nutrient supply to the aphids.

A complementary series of experiments was carried out at Lancaster, England, by Warrington (1987). This time the MRGR of the pea aphid (*Acyrthosiphon pisum*) was measured over a range of SO_2 concentrations to establish a dose-response characteristic. The aphids were allowed to feed for 4 days on plants that had been exposed to between 20 and 300 ppb of SO_2. The percentage change in MRGR increased linearly with SO_2 concentration up to about 11 percent at 100 ppb, before declining again at higher concentrations (Figure 6-2). This stimulant effect on aphid growth was again attributed to changes in nitrogen metabolism in plants, in particular to higher concentrations of free and total amino acids. Houlden et al. (1990) investigated the responses of three cereal aphids, three legume aphids, and two species feeding on brassica plants, when the plants had been previously exposed to 100 ppb of SO_2 or NO_2 for 7 hours. They found that the pollutant exposure increased the MRGR by between 7 and 75 percent in all cases but one; the MRGR for *A. pisum* on *V. faba* was negative under both SO_2 and NO_2 exposure. Perhaps surprisingly, there was no consistent difference between NO_2 and SO_2 in affecting MRGR. The powerful conclusion from these experiments is that significant increases in aphid growth—and hence in fecundity and population—seem to be a general result of both SO_2 and NO_2 exposure.

This comparatively straightforward result for SO_2 and NO_2 does not hold for O_3. For example, Whittaker et al. (1989) studied the MRGR response of two aphid species feeding on pea (*Pisum sativum* L.) or dock (*Rumex obtusifolius* L.) and exposed to O_3 concentrations in the range 21–206 ppb. They found changes of between +24 and –6 percent in MRGR, and no evidence of a dose-response relationship. In fact they concluded that the same O_3 treatment could produce opposite responses in similar insects feeding on different food plants, or vice versa. This variability was confirmed by Braun and Fluckiger

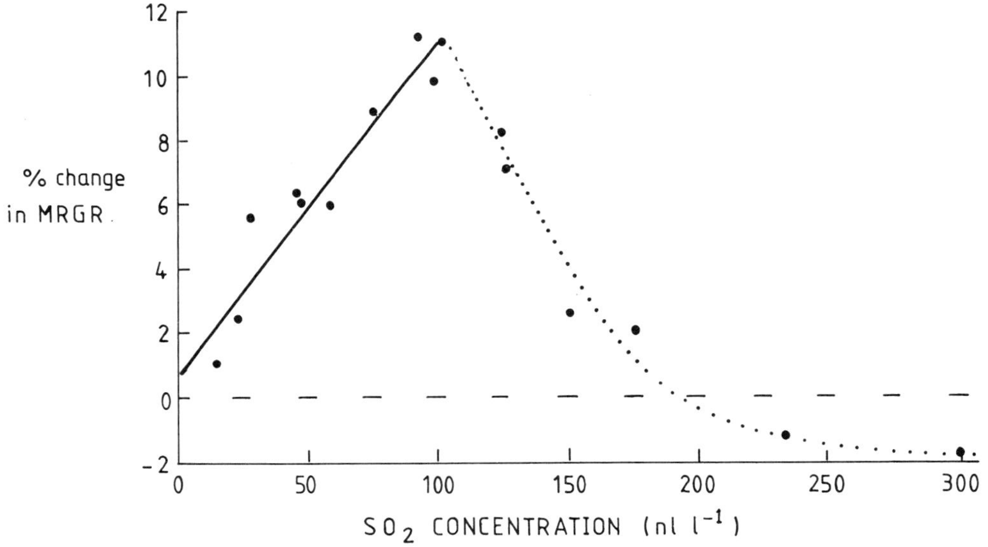

FIGURE 6-2. Relationship between SO_2 concentration and percentage change, relative to controls, in MRGR of pea aphids feeding on pea plants.

Source: Warrington 1987.

(1989), who found that *A. fabae* Scop. was depressed by ambient ozone on *P. vulgaris* L. but that *Phyllaphis fagi* L. was stimulated by the action of the same air pollutant on *Fagus sylvatica* L. Hence invertebrate diversity may be enhanced or reduced by pollutant action.

Leaf pathogenic fungi such as leaf blotch have a large impact on cereal yields; the secondary effects of air pollution on plant susceptibility to these diseases have therefore been topics of increasing interest. Tiedemann et al. (1990) examined the responses to ozone of five wheat pathogens and three barley pathogens. The ozone treatments were at maximum concentrations of 60, 90, and 120 ppb for 7 days before inoculation. In general, they found that the 7-day ozone pretreatment gave a significant enhancement of pathogen-related leaf damage, and that intrinsic pathogen viability factors such as spore production were also stimulated in these cases. Conversely, in no case did they find that ozone pretreatment conferred resistance on the host plant to parasite invasion. The authors linked these effects to the well-known acceleration of senescence caused by ozone, and to evidence that *Septaria* species, for example, spread more vigorously on senescent than on young leaves.

In contrast to the stimulation described above, Dohmen (1987) found that O_3 depressed the vigor of a brown rust (*Puccinia recondita* f. sp. tritici) inoculated onto young wheat plants. The inverse problem was investigated by Damicone et al. (1987), who showed that soybeans infected with the soil-borne pathogen *Fusarium oxysporum* were more sensitive to ozone damage than were uninfected controls.

Turning now to sulfur dioxide, we have recent information from both chamber and field-release experiments. Lorenzini, Farina, and Guidi (1990) inoculated field bean with the rust fungus *Uromyces viciae-fabae*. Exposure to SO_2 did not decrease germination ability nor alter the morphological parameters of the fungus urediospores. However, the overall effect of the rust on the host plant varied with SO_2 concentration and time of fumigation. No reliable method for predicting effects could be established.

The great majority of experimental work has been carried out in open-top or laboratory chambers, and open-field pest interaction studies are rare. McLeod (1988) exposed field crops of barley and wheat in successive years to sulfur dioxide at concentrations in the range 10–60 ppb. Pathogen infection was compared with that in companion control plots. Two fungal pathogens infected the barley, two more infected the wheat, and one infected both. It was concluded that all treatment SO_2 concentrations reduced the incidence of brown rust and leaf blotch in barley; plants exposed to SO_2 in this way developed and maintained a higher green leaf area index and achieved a higher crop dry weight prior to final harvest. For the wheat crop, powdery mildew was increased by SO_2 while eyespot and brown root rot were reduced.

NUTRIENT DEFICIENCY

All plants need a balanced supply of nutrients for normal growth. Variations in the nutrient requirements and in responses to nutrient stress are likely to influence biodiversity in natural ecosystems. There has been considerable discussion and experimental investigation of the effects of acidic deposition on soils and on trees growing in them; there is much less information on other plant species. Acidic deposition could alter the nutrient status of soils and impose a stress on plants in a number of ways.

Soil Acidification

It is well established that in some forest soils in Scandinavia, Germany, and the United Kingdom, acidity has increased greatly over the last 40 years (Berden et al. 1987; Billett et al. 1990). It is much less clear whether the measured increases in acidity have resulted from acidic deposition or from forest growth, for example the accumulation of base cations in the tree biomass or the replacement of hardwoods with conifers. At some sites careful analysis appears to show convincingly that acidic deposition has contributed to the acidification, but at many other sites it seems that changes in soil pH can be accounted for by changes in ground vegetation and forest canopy.

Decreased Nitrogen Mineralization

It has been proposed that acidic inputs and an increase in soil acidification would suppress the mineralization of organic nitrogen in soils because the necessary micro-organisms are intolerant of very acid conditions. This suggestion does not seem to be supported by experiments; it appears that significant reductions in rates of nitrogen mineralization are found only with simulated acidic rain of pH < 3.0, and such conditions are very rare (Stroo and Alexander 1986). In the longer term the unusually slow rates of litter decomposition reported in some forests in polluted areas must alter the carbon:nitrogen ratio in the soil, and this may have long-term implications for nitrogen availability. Certain specialized ecosystems such as the ombrotrophic *Sphagnum* mires studied by Woodin and Lee (1987) show interesting responses to atmospheric deposition of nitrogen. *Sphagnum* is effectively isolated from the soil nutrient supply and has adapted to use both dry and wet deposited airborne nitrogen. In a polluted environment, the supply of nitrogen is supraoptimal and may result in the loss of inorganic nitrogen retention by the *Sphagnum* carpet.

Base Cation Deficiencies

Considerable evidence from field and laboratory investigations shows that there can be increased leaching of base cations when soils are exposed to simulated acidic rain or sulfur dioxide (e.g., Morrison 1983). The amount of leaching and the consequences for soil nutrient status depend on the chemistry of the deposition and on soil properties. Rainfall acidity needs to be below pH 3.0 before there are major increases in leaching and significant impacts on soil base saturation. More important are likely to be the effects of large inputs of ammonia or inputs of neutral solutions containing large concentrations of mobile acid anions, especially when the soils concerned are already acidic (Hultberg et al. 1990). Ammonia inputs may lead to rapid soil acidification if nitrification produces nitrate, which cannot be used by the soil plant system. This response has been established in The Netherlands (Draaijers et al. 1989) (Figure 6-3), and is likely to be an important factor influencing biodiversity in many areas where intensive agricultural production (involving releases of ammonia) occurs close to natural ecosystems. A recent survey in the United Kingdom (Pitcairn, Fowler, and Grace 1991) has revealed several areas where substantial changes in natural species diversity appear to be linked to this mechanism. Although the evidence is largely circumstantial, it does appear that in areas of the United Kingdom subject to annual nitrogen deposition loads in excess of 30 kg ha^{-1}, some species such as *Calluna* have declined whereas *Brachypodium* has flourished.

Although it is well known that "Type One" forest decline in central Europe is associated with low foliar concentrations of magnesium (Rehfuess 1987), there has been considerable

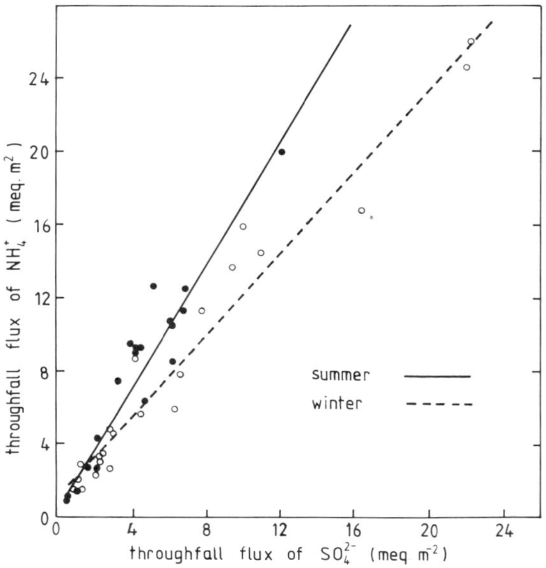

FIGURE 6-3. Relationship between the throughfall flux of SO_4^{2-} and the throughfall flux of NH_4^+ at one location for both summer (solid symbols, *n*=21) and winter (open symbols, *n*=22). In winter, five rain periods had missing values.

Source: Draaijers et al. 1989.

argument over whether this deficiency is a cause of the decline or is a secondary response. Although the fundamental reasons for the reduction in magnesium status of foliage are still uncertain it seems clear that pollution is implicated (Roberts, Skeffington, and Blank 1989). There seem to be two linked effects in the damaged regions: (1) deposition of acidic anions (SO_4 and NO_3) leads to increased leaching of Mg and (2) increased atmospheric inputs of nitrogen may have acted as a fertilizer, increasing growth so that it exceeds the ability of acidic soils to supply magnesium by mineralization.

Imbalance of Base Cation Uptake

Plants take up the majority of their base cation requirements from soil solutions. In areas with high deposition of ammonia, the dominant cation in the soil solution is ammonium, and this effectively suppresses concentrations of magnesium, calcium, and potassium in soil solution. In these conditions experimental studies have clearly demonstrated that plants may show cation deficiencies linked to an imbalance of uptake (van Dijk et al. 1990). There is some suggestion that similar deficiencies occur in forests in the uplands of the United Kingdom where there are relatively large atmospheric inputs of nitrogen. It has been suggested that the excessive supply of nitrogen has induced a deficiency of phosphorus and that this deficiency limits further growth, so that the continuing excess nitrogen deposition passes through the ecosystem into water courses (Stevens, Hornung, and Hughes 1989).

Aluminum Toxicity

One of the earliest hypotheses to explain observed reductions in forest health such as needle loss and yellowing was that pollutant input to acid soils resulted in large soil solution concentrations of aluminum, leading to root damage and reduced nutrient uptake (Ulrich, Mayer, and Khann 1980). Although several field and experimental studies have generally supported this hypothesis, it is clear that not all forms of aluminum in soil solution are toxic to roots, and so it seems likely that several aspects of the inorganic and organic chemistry of solutions need to be taken into account. Although the direct toxicity of aluminum as a factor in forest decline is still far from certain, the suggestion that high levels of aluminum in some soils may inhibit uptake of base cations can also not be excluded (Ryan, Gessel, and Zososki 1986).

LOW TEMPERATURES

Air pollution and cold stress may interact in two quite distinct ways. First, exposure to pollution may alter the sensitivity of plants to a number of different forms of cold stress. Second, the reduction in metabolism associated with plants growing in the cold can limit the ability of plants to detoxify air pollutants, and may allow concentrations to accumulate in foliage to an extent that eventually damages plants. The subject was reviewed thoroughly by Davison, Barnes, and Renner (1988) and so will be only briefly treated here, with some emphasis on recent studies in the United Kingdom concerning the sensitivity of red spruce.

Winter conditions produce three types of stress for plants: desiccation, cold or freezing temperatures, and photo-oxidation of pigments. The least information is available on the effects of pollution on photo-oxidation, which occurs when a combination of high irradiance and low temperatures permits a buildup of free radicals in leaf tissue, and these free radicals then attack chlorophyll (Oquist 1983). It has become clear from several pieces of air pollution research, noticeably by Wellburn and his group at Lancaster United Kingdom, that free radical scavenging is important for the protection of leaf tissue from a range of air pollutants (Mehlhorn and Wellburn 1987). Further research to investigate the pollution/photo-oxidation interaction would be well justified.

Desiccation of evergreen species can be very damaging in winter if there are weather conditions conducive to evaporation and the supply of water through the roots is restricted (Tranquillini 1982). It is generally believed that stomata are closed in these conditions and consequently that desiccation results from cuticular transpiration. There is good evidence that air pollution alters the rate of cuticular development, changes the structure of cuticular waxes, and impairs stomatal functioning (Crossley and Fowler 1986; Mansfield and Freer-Smith 1984). To support this circumstantial evidence that air pollution alters risks of desiccation, a substantial body of observations, especially by Huttunen, Havas, and Laine in Finland (1981) shows significant desiccation injury to conifers growing near sites of SO_2, NO_2, and HF pollution.

The sensitivity of plants to chilling and freezing injury is not necessarily well correlated with sensitivity to desiccation. It is generally believed that damage to cell membranes is the main cause of chilling and freezing injury. Ozone, SO_2, and NO_2 are all known to interact with membranes and consequently may alter the sensitivity to cold injury of this type. Equally, the processes of hardening have a metabolic cost, and it would be expected that air pollution stress alters the ability to harden and deharden.

The strongest evidence for air pollution influencing freezing injury exists for SO_2. In the field, observations in central Europe demonstrated that the frost sensitivity of spruce trees was increased after exposure to rather high concentrations of SO_2 (Materna 1984). Increased sensitivity to freezing injury has been reported for grasses by Davison and Bailey (1982), and increased frost injury was observed on winter wheat and winter barley exposed in the field to SO_2 released in a field fumigation system (Baker, Unsworth, and Greenwood 1982). A complex manifestation of cold stress was detected by Baker and Fullwood (1986). They exposed winter barley (*H. vulgare* cv. Igri) to SO_2 in a field release system throughout the growing season. In the spring, the crop was sprayed with a mixture of two contact herbicides, a fungicide, and a growth regulator. Shortly afterwards, leaf scorching appeared, which was related in both severity and extent to SO_2 concentration and was probably cold-induced.

In the early 1980's it became increasingly clear that red spruce growing at high elevations in the northern Appalachian Mountains of the United States were declining (Johnson and Siccama 1983). An early hypothesis to explain the decline was that the trees at these high elevations received large inputs of pollutants, especially from intercepted cloud water, which led to especially large inputs of nitrogen that might alter their frost hardiness (Friedland et al. 1984). The U.S. Department of Agriculture Forest Service established a Spruce-Fir Cooperative to undertake a wide range of coordinated research on the decline problem. In the United Kingdom, a consortium of groups based at the Institute of Terrestrial Ecology, Edinburgh, and the Universities of Lancaster and Nottingham combined to address two hypotheses as part of the U.S. Spruce-Fir Cooperative. The hypotheses were

1. Exposure of red spruce to acidic cloud water and air pollutants alters frost hardening.
2. Exposure of red spruce to SO_2 and NO_2 in winter causes foliar accumulation of toxic substances that alter spring growth.

The research results were reviewed recently by Unsworth (1991). Exposure to acidic mists disturbed autumn frost hardening of red spruce seedlings (Fowler et al. 1989; Cape et al. 1991). The results showed that acidic mist delayed the onset of autumn hardening, but once hardening started it proceeded at a constant rate irrespective of the chemistry of cloud water treatment (Figure 6-4). Although it was not possible to study the effects of each chemical ion in isolation, the evidence showed that the disturbances were caused by sulfate rather than nitrate ions and suggested that acidity per se had no effect on frost hardening.

Exposure of red spruce to ozone throughout the summer altered the autumn frost hardening only when the ozone treatment was in a system that also allowed the plants to receive nitric oxide (Neighbour, Pearson, and Mehlhorn 1990). In general the role of nitrogen oxides has been conspicuously ignored in North American research; this result suggests that important interactions may have been missed in several studies of the effects of ozone.

The second hypothesis was addressed by growing red spruce seedlings in freezing cabinets to which SO_2 and NO_2 could be added over winter. The results demonstrated that the seedlings accumulated sulfite and nitrite in their extra cellular fluids; there was little evidence that oxidation to less toxic sulfate and nitrate took place. There were indications that these accumulations were associated with delayed flushing of buds in the following spring (Wolfenden, Pearson, and Francis 1991). There are surprisingly few monitoring data

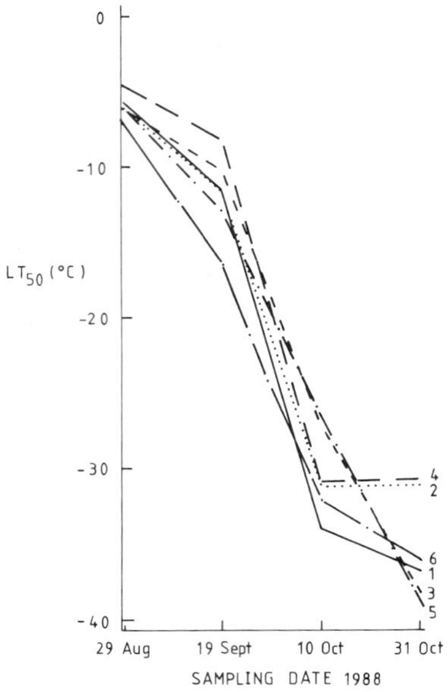

FIGURE 6-4. The development of frost hardiness in shoots of seedling red spruce (*Picea rubens*) exposed to polluted mist, expressed as the LT_{50} (temperature that killed 50% of shoots). Mist treatments were: 1. ammonium nitrate (1.6 mol m^{-3}); 2. ammonium sulfate (1.6 mol m^{-3}); 3. deionized water "control"; 4. sulfuric acid + ammonium nitrate (1.6 mol m^{-3} + 1.6 mol m^{-3}); 5. nitric acid (3.2 mol m^{-3}); sulfuric acid (0.5 mol m^{-3}).

Source: Cape et al. 1991.

from high-elevation sites in the United States of the concentrations of SO_2 and NO_2 in winter. It is clear from the small amount of data existing that in these conditions, when the oxidizing power of the atmosphere is limited, there can be substantial long-range transport of these pollutant gases. The role of winter SO_2 and NO_2 as a source of pollutant stress to remote ecosystems requires further investigation.

PLANTS AND RESOURCE CAPTURE

The analysis of the response of plants to the environment, developed from various aspects of research at Nottingham over the last two decades, gives a framework into which the effects of pollution and their consequences for biodiversity may be set. The growing plant captures resources from the atmosphere and the soil. The plant absorbs carbon dioxide from the atmosphere and intercepts light; from the soil it takes up nutrients and water. The processes of canopy expansion and root exploration are essential for this resource capture. The rate of canopy expansion is strongly dependent on temperature; there is little indication from our field studies and from open-top chambers that air pollution significantly alters the main phase of canopy expansion. However, the maximum amount of light intercepted and the duration for which the canopy remains green have been shown in studies with open-top chambers to depend on ozone concentration (Unsworth, Lesser, and Heagle 1984). Although there have been no detailed studies of root growth and root efficiency in response to air pollution in natural soils, there is good evidence that the amount of assimilate partitioned to roots is significantly reduced at quite low concentrations of many pollutants. It is important to discover the implications of this distortion of root shoot ratio for the balance of resource capture, and to discover whether, for example, the rate of capture of one of the four basic resources imposes a limit on plant growth.

When water is limiting, the analysis of plant growth in terms of resource capture takes a rather different form. The rate of expansion of the root system in relation to the rate of water extraction is of critical importance for the success of the plant and for the expansion of above-ground parts. The amount of dry matter produced by a plant is often proportional to the transpired water, and important conservative relations can be found when saturation deficit of the air is taken into account. Although these relations are conservative for particular genotypes there is considerable difference between species. There is scope for exploring these types of analysis in more complex plant communities. We do not know of any work that has exploited these approaches to investigate how plants respond to air pollution when water is limiting, but such approaches might help to clarify the apparent conflicts of results that have been reported in the literature for crops exposed to ozone.

CONCLUSIONS

Researchers on pollutant effects are only now getting into a position from which they can predict the responses of the very simplest plant communities (stress-free monocultures) to the simplest pollutant loads (single gaseous pollutants). Furthermore, this capability has been achieved empirically rather than from a theoretical understanding. Thus there is little real prospect of predicting interactions between the range of natural stressors and pollutant mixtures for complex ecosystems over long periods. What we can do, however, is apply any general principles from previous research to these more difficult topics. If air pollution reduces the root effectiveness of some species more than others, then those species will be more susceptible to water or nutrient deficiency. If air pollution reduces the stomatal aperture of some species more than others, then they may conserve water if drought stressed, overheat if heat stressed, or fall behind in a high CO_2 environment. In any one ecosystem, many conflicting and competing processes such as these may mold the long-term community structure.

REFERENCES

1. Alstad, D.N., G.F. Edmunds, and L.H. Weinstein. 1982. Effects of air pollutants on insect populations. *Annu. Rev. Entomol.* 27:369-84.
2. Austin, R.B. 1989. Prospects for improving crop production in stressful environments. In *Plants under stress*, ed. H.G. Jones, T.J. Flowers, and M.B. Jones, pp. 235-48. Cambridge: Cambridge Univ. Press.
3. Baker, C.K., and A.E. Fullwood. 1986. Leaf damage following crop spraying in winter barley exposed to sulphur dioxide. *Crop Prot.* 5 (5): 365-67.
4. Baker, C.K., A. Fullwood, and J.J. Colls. 1990. Lodging of winter barley (*Hordeum vulgare* L.) in relation to its degree of exposure to sulphur dioxide. *New Phytol.* 114:191-98.
5. Baker, C.K., M.H. Unsworth, and P. Greenwood. 1982. Leaf injury on wheat plants exposed in the field in winter to SO_2. *Nature* 299:149-51.
6. Berden, M., S.I. Nilsson, K. Rosen, and G. Tyler. 1987. Soil acidification—extent, causes, and consequences. Solina, Sweden: National Swedish Environmental Protection Board.

7. Billett, M.F., F. Parker-Jones, E.A. Fitzpatrick, and M.S. Cresser. 1990. Forest soil chemical changes between 1949/50 and 1987. *J. Soil Sci.* 41:133-45.
8. Braun, S., and W. Fluckiger. 1989. Effect of ambient ozone and acid mist on aphid development. *Environ. Pollut.* 56:177-87.
9. Cape, J.N., I.D. Leith, D. Fowler, M.B. Murray, L.J. Sheppard, D. Eamus, and R.H.F. Wilson. 1991. Sulphate and ammonium in mist impair the frost hardening of red spruce seedlings. *New Phytol.* 118:119-26.
10. Crossley, A., and D. Fowler. 1986. The weathering of scots pine epicuticular wax in polluted and clean air. *New Phytol.* 103:207-18.
11. Damicone, J.P., W.J. Manning, and S.J. Herbert. 1987. Growth and disease response of soybeans from early maturity groups to ozone and *Fusarium oxysporum*. *Environ. Pollut.* 48:117-30.
12. Davison, A.W., and I.F. Bailey. 1982. SO_2 pollution reduces the freezing resistance of ryegrass. *Nature* 297:400-402.
13. Davison, A.W., J.D. Barnes, and C.J. Renner. 1988. Interactions between air pollutants and cold stress. In *Air Pollution and Plant Metabolism*, ed. S. Schulte-Hostede, L.W. Blank, N.M. Darrall, and A.R. Wellburn. London: Elsevier Applied Science.
14. Dohmen, G.P. 1987. Secondary effects of air pollution: Ozone decreases brown rust disease potential in wheat. *Environ. Pollut.* 43:189-94.
15. Dohmen, G.P. 1988. Indirect effects of air pollutants: Changes in plant/parasite interactions. *Environ. Pollut.* 53:197-207.
16. Draaijers, G.P.J., W.P.M.F. Ivens, M.M. Bos, and W. Bleuten. 1989. The contribution of ammonia emissions from agriculture to the deposition of acidifying and eutrophying compounds onto forests. *Environ. Pollut.* 60:55-66.
17. Fowler, D., J.N. Cape, J.D. Deans, I.D. Leith, M.B. Murray, R.I. Smith, L.J. Sheppard, and M.H. Unsworth. 1989. Effects of acid mist on the frost hardiness of red spruce seedlings. *New Phytol.* 113:321-35.
18. Friedland, A.J., R.A. Gregory, L. Karenlampi, and A.H. Johnson. 1984. Winter damage to foliage as a factor in red spruce decline. *Can. J. For. Res.* 14:963-65.
19. Gordon, A.G., and E. Gorham. 1963. Ecological aspects of air pollution from an iron-sintering plant at WaWa, Ontario. *Can. J. Bot.* 41:1063-78.
20. Heagle, A.S., L.W. Kress, P.J. Temple, R.J. Kohut, J.E. Miller, and H.E. Heggestad. 1988. Factors influencing ozone dose-yield response relationships in open-top field chamber studies. In *Assessment of crop loss from air pollutants*, ed. W.W. Heck, O.C. Taylor, and D.T. Tingey. London: Elsevier Applied Science.
21. Heagle, A.S., J. Rebbeck, S.R. Shafer, U. Blum, and W.W. Heck. 1989. Effects of long-term ozone exposure and soil moisture deficit on growth of a ladino clover tall fescue pasture. *Phytopathology* 79:128-36.
22. Heggestad, H.E., and V.M. Lesser. 1990. Effects of ozone, sulphur dioxide, soil water deficit, and cultivar on yields of soybean. *J. Environ. Qual.* 19:488-95.
23. Houlden, G., S. McNeill, M. Amina-Kano, and J.N.B. Bell. 1990. Air pollution and agricultural aphid pests I: fumigation experiments with SO_2 and NO_2. *Environ. Pollut.* 67:305-14.
24. Hultberg, H., Y.-H. Lee, U. Nystrom, and S.I. Nilsson. 1990. Chemical effects on surface-, ground- and soil-water of adding acid and neutral sulphate to catchments in southwest Sweden. In *The surface waters acidification programme*, ed. B.J. Mason. Cambridge: Cambridge Univ. Press.
25. Huttunen, S., P. Havas, and K. Laine. 1981. Effects of air pollutants on the wintertime water economy of the Scots pine (*Pinus sylvestris*). *Holarct. Ecol.* 4:94-101.

26. Johnson, A.H., and T.G. Siccama. 1983. Acid deposition and forest decline. *Environ. Sci. & Technol.* 17:294a-305a.
27. Jones, H.G., and M.B. Jones. 1989. Introduction: Some terminology and common mechanisms. In *Plants under stress*, ed. H.G. Jones, T.J. Flowers, and M.B. Jones. Cambridge: Cambridge Univ. Press.
28. Lechowicz, M.J. 1987. Resource allocation by plants under air pollution stress. *Bot. Rev.* 53:281-300.
29. Lorenzini, G., R. Farina, and L. Guidi. 1990. The effects of sulphur dioxide on the parasitism of the rust fungus *Vromyces vicia-fabae* on *Vicia faba. Environ. Pollut.* 68:1-14.
30. Lucas, P.W. 1990. The effects of prior exposure to sulphur dioxide and nitrogen dioxide on the water relations of Timothy grass (*Phleum pratense*) under drought conditions. *Environ. Pollut.* 66:117-38.
31. Mansfield, T.A., and P.H. Freer-Smith. 1984. Effects of urban air pollutants on plant growth. *Biol. Rev.* 56:343-68.
32. Mansfield, T.A., W.J. Davies, and M.E. Whitmore. 1986. Interactions between the responses of plants to pollution and other environmental factors such as drought, light, and temperature. In *Working party III on effects of air pollution on terrestrial and aquatic ecosystems*. Brussels: Commission for the European Communities.
33. Materna, J. 1984. Impact of atmospheric pollution on natural ecosystems. In *Air pollution and plant life*, ed. M. Treshow. New York: John Wiley & Sons.
34. McLeod, A.R. 1988. Effects of open-air fumigation with sulphur dioxide on the occurrence of fungal pathogens in winter cereals. *Phytopathology* 78:88-94.
35. Mehlhorn, H., and A.R. Wellburn. 1987. Stress ethylene formation determines plant sensitivity to ozone. *Nature* 327:417-18.
36. Miller, J.E., A.S. Heagle, S.F. Bozzo, R.B. Philbeck, and W.W. Heck. 1989. Effects of ozone and water stress, separately and in combination, on soybean yield. *J. Environ. Qual.* 18:330-36.
37. Moore, P.H., and R.V. Osgood. 1985. Assessment of sugar cane crop damage and yield loss by high winds of hurricanes. *Agric. For. Meteorol.* 35:267-79.
38. Morrison, I.K. 1983. Composition of percolate from reconstructed profiles of two Jack Pine forest soils as influenced by acid input. In *Accumulation of air pollutants in forest ecosystems*, ed. B. Ulrich and J. Pankrath. Dordrecht: Reidel.
39. Neales, T.F., A. Masia, J. Zhang, and W.J. Davies. 1989. The effects of partially drying part of the root system of *Helianthus annus* on the abscisic acid content of the roots, xylem sap, and leaves. *J. Exp. Bot.* 40:1113-20.
40. Neighbour, E.A., D.A. Cottam, and T.A. Mansfield. 1988. Effects of sulphur dioxide and nitrogen dioxide on the control of water loss by birch (*Betula* spp.) *New Phytol.* 108:149-57.
41. Neighbour, E.A., M. Pearson, and H. Mehlhorn. 1990. Purafil-filtration prevents the development of ozone-induced frost injury: A potential role for nitric oxide. *Atmos. Environ.* 24A:711-15.
42. Oquist, G. 1983. Effects of low temperature on photosynthesis. *Plant Cell Environ.* 6:281-300.
43. Pinthus, M.J. 1973. Lodging in wheat, barley, and oats. *Adv. Agron.* 25:208-25.
44. Pitcairn, C.E.R., D. Fowler, and J. Grace. 1991. Changes in species composition of semi-natural vegetation associated with the increase in atmospheric inputs of nitrogen. Final Report to Nature Conservancy Council on Project TO705057:1. Institute of Terrestrial Ecology, Edinburgh, UK.
45. Rehfuess, K.E. 1987. Perceptions of forest diseases in central Europe. *Forestry* 60:1-11.
46. Roberts, T.M., R.A. Skeffington, and L.W. Blank. 1989. Cause of Type 1 spruce decline in Europe. *Forestry* 62:179-222.

47. Ryan, P.J., S.P. Gessel, and R.J. Zososki. 1986. Acid tolerance of Pacific Northwest conifers in solution culture II. Effect of varying aluminium concentration at constant pH. *Plant Soil* 96:259-72.
48. Sanders, G.E., A.G. Clark, and J.J. Colls. 1990. *Phaseolus vulgaris* and ozone: An open-top chamber study. In *Environmental research with plants in closed chambers*, ed. H.D. Payer, T. Pfirrmann, and P. Mathy. CEC Air Pollution Research Report 26. Brussels: Commission for the European Communities.
49. Sanders, G.E., N.D. Turnbull, A.G. Clark, and J.J. Colls. 1990. The growth and development of *Vicia faba* in filtered and unfiltered open-top chambers. *New Phytol.* 116:67-78.
50. Stevens, P.A., M. Hornung, and S. Hughes. 1989. Solute concentrations, fluxes, and major nutrient cycles in a mature Sitka spruce plantation in Beddgelert Forest, North Wales. *For. Ecol. Manage.* 27:1-20.
51. Stroo, H.F., and M. Alexander. 1986. Available nitrogen and cycling in forest soils exposed to simulated acid rain. *Soil Sci. Soc. Am. J.* 50:110-14.
52. Tiedemann A.V., P. Ostlander, K.H. Firsching, and H. Fehrmann. 1990. Ozone episodes in southern Lower Saxony (FRG) and their impact on the susceptibility of cereals to fungal pathogens. *Environ. Pollut.* 67:43-59.
53. Tranquillini, W. 1982. Frost-drought and its ecological significance. In *Physiological plant ecology 2, Encyclopedia of plant ecology*, Vol. 2b, ed. O.L. Lange, P.S. Nobel, C.B. Osmand, and H. Zeigler. Berlin: Springer-Verlag.
54. Ulrich, B., R. Mayer, and P.K. Khann. 1980. Chemical changes due to acid precipitation in a loess-derived soil in central Europe. *Soil Sci.* 130:193-99.
55. Unsworth, M.H. 1991. Air pollution and vegetation: Hypothesis, field exposure, and experiment. *Proc. R. Soc. Edinb. B* 97:139-53.
56. Unsworth, M.H., V.M. Lesser, and A.S. Heagle. 1984. Radiation interception and the growth of soybeans exposed to ozone in open-top field chambers. *J. Appl. Ecol.* 21:1059-80.
57. van Dijk, H.F.G., M.H.J. De Louw, J.G.M. Roelofs, and J.J. Verburgh. 1990. Impact of artificial, ammonium enriched rainwater on soils and young coniferous trees in a greenhouse. Part II: Effect on the trees. *Environ. Pollut.* 62:317-36.
58. Warrington, S. 1987. Relationship between SO_2 dose and growth of the pea aphid (*Acyrthosiphon pisum*) on peas. *Environ. Pollut.* 43:155-62.
59. Werkman, B.R. 1991. Informal presentation at the meeting of the Committee for Air Pollution Effects Research. London: Imperial College.
60. Whittaker, J.B., L.W. Kristiansen, T.N. Mikkelsen, and R. Moore. 1989. Responses to ozone of insects feeding on a crop and a weed species. *Environ. Pollut.* 62:89-101.
61. Wilson, E.O., ed. 1988. *Biodiversity*. Washington, DC: National Academy Press.
62. Wolfenden, J., M. Pearson, and B.J. Francis. 1991. Effects of over-winter fumigation with sulphur and nitrogen dioxides on biochemical parameters and spring growth in red spruce. *Plant Cell Environ.* 14:35-45.
63. Woodin, S.J., and J.A. Lee. 1987. The fate of some components of acidic deposition in ombrotrophic mires. *Environ. Pollut.* 45:61-72.
64. Woodward, F.I. 1987. *Climate and plant distribution*. Cambridge: Cambridge Univ. Press.

Part III

Consequences of Air Pollution Exposure on Biodiversity

7

Genetic Diversity of Plant Populations and the Role of Air Pollution

George E. Taylor, Jr., and Louis F. Pitelka

INTRODUCTION

Biodiversity is a general term used to characterize variability in biotic resources occurring at the level of populations, species, communities, and ecosystems (see White and Nekola, Chapter 2, this volume). Biodiversity can be simplified to the issue of genetic diversity, and this manuscript addresses that diversity at the level of plant populations and the role that air pollution may play.

When challenged with any environmental stress, plant populations can exhibit four different responses. The first and most common is one in which the population exhibits no response because the individuals are resistant to the environmental stress. The second and most severe case is the situation in which the population is extremely sensitive, resulting in mortality of all individuals and local extinction. In the remaining two categories of response, individuals within the population are affected by the stress, but the mechanism of response and the genetic consequences differ. In the third response category, the stress is physiologically accommodated, and growth and reproductive success of the individuals are unaffected (physiological accommodation). In the fourth category, the stress differentially affects members of the population, with some individuals exhibiting better growth and reproductive success due to genetically determined traits. Over several generations, this latter situation results in the progressive elimination of sensitive individuals and a shift in the genetic structure of the population towards greater resistance (microevolution). For stresses that are intermediate-to-low in intensity and prolonged in exposure duration (i.e., chronic stress), the most likely responses are physiological accommodation and/or microevolution, with only the latter affecting biodiversity.

Genetic diversity is responsive to biotic and abiotic stresses of natural origin (e.g., Turesson 1922; Clausen, Keck, and Hieser 1940) and to a lesser extent to stresses of anthropogenic origin (e.g., Bishop and Cook 1981). This conclusion underlies a basic tenet of ecological genetics—most species exhibit variation in physiology and morphology that consistently covaries with the environment. When investigated under carefully controlled conditions, this phenotypic variation is often genetically determined rather than environmentally induced, which is consistent with the hypothesis that the environment selects for alleles that enhance survival and reproductive success in a given location. Whereas the same alleles may exist throughout the species' distribution, their frequency and association with other gene complexes are specific to the local gene pool.

Although air pollution is regarded as an ecological stress in many areas (see Fowler, Chapter 3, this volume; Wolfenden et al., Chapter 5, this volume), its potential to function as a selective agent at the level of plant populations is only now being recognized more widely by the ecological genetics community (Bradshaw and McNeilly 1981; Pitelka 1988; Scholz, Gresorius, and Rudin 1988; Taylor, Pitelka, and Clegg 1991). There is no reason to believe that air pollution stress should not elicit shifts in the genetic structure of plant populations, and some notable examples support this conclusion (Dunn 1959; Bell, Ashmore, and Wilson 1991). However, it is important to recognize that air pollution exhibits some unusual attributes in terms of evolution that influence the extent and role that microevolution may play.

The objective of this manuscript is to address air pollution stress as a selective factor operating on genetic diversity at the level of plant populations. The specific objectives are fourfold:

1. Describe the processes controlling microevolution;
2. Discuss the evidence that air pollution can influence the genetic structure of plant populations;
3. Discuss the ecological and biological consequences of air pollution-induced changes in biodiversity; and
4. Discuss future changes in air quality and the potential for changes in genetic diversity.

Because the issue of air pollution as an evolutionary agent has been discussed in several recent reviews (Pitelka 1988; Taylor, Pitelka, and Clegg 1991), this aspect (objectives 1 and 2 above) is only briefly discussed. Instead, emphasis is placed here on the ecological and biological consequences of changes in biodiversity due to air pollution and the potential consequences of future changes in atmospheric chemistry (objectives 3 and 4).

PROCESSES CONTROLLING MICROEVOLUTION

Throughout a species' distribution, a mosaic of different environments exists such that the ecological factors prevalent in one location may not be significant in another. With respect to the species, genotypic variation in morphology, physiology, and biochemistry is pervasive. The result is predictable: in any given locality, natural selection favors organisms possessing heritable traits maximizing fitness to the unique ecology of the local environment. Conspecific populations thriving in other areas possess different assemblages of habitat-specific traits. The pattern that emerges over time is a nonrandom array of genotypes throughout the species' distribution reflecting, in part, variation in the environment. The species' gene pool in any given generation is structured into pockets of locally attuned genetic variation.

The evolution of population resistance to any environmental stress depends on two factors: (1) the availability of genetically determined, plant-to-plant variation in response to the stress and (2) natural selection for resistance. The interplay between these two processes dictates not only the speed with which populations can evolve resistance but the limits to what natural selection can achieve. The salient point is that evolution at the level of populations can be substantial, but there are significant limits to what evolution can achieve (Bradshaw and McNeilly 1991a). Each of these three factors is discussed below.

One of the hallmark features of the plant kingdom is the wealth of genetic variation at all levels of organization (i.e., molecular to whole-plant growth and reproduction). While the extent of variability is species- and population-dependent, most species possess an immense pool of intrinsic variability. This variation results in an array of individuals with differential fitness potential depending on the local environment, and it is this array upon which natural selection operates. However, the critical issue is not just the existence of genetically unique individuals but rather variants whose genotypes are differentially sensitive to the environmental stress. In reality, only a small percentage of the variation within most populations challenged by a novel stress influences the fitness of those individuals. Accordingly, it is essential to identify that component of the variability that is relevant to the particular stress.

The other prerequisite to microevolution is the presence of an environmental stress that effectively selects for alleles conferring resistance. The prevalence of strong natural selection pressure within many terrestrial landscapes is evident by the remarkable discontinuities in species' distributions due to stresses of natural and anthropogenic origin. At ecotones within landscapes, selection coefficients approach unity (i.e., high level of mortality) as species are unable to colonize terrain that is often times only several meters distant. In less obvious situations, natural selection is well below unity and differentially reduces the biotic potential of individuals, depending on their genetic predisposition to the stress.

The intensity of the selection coefficient is extremely important because it dictates how rapidly alleles that confer resistance increase in frequency within the population. Any stress that affects growth, survival, and reproduction exerts a selection coefficient, but it is experimentally difficult to demonstrate an effect unless the coefficient exceeds 0.2 (i.e., selection reduces reproductive performance by 20%) in outcrossing species (Bradshaw and McNeilly 1991b). The theoretical role that the selection coefficient can play is illustrated in Figure 7-1, which shows the number of generations needed to alter the frequency of an allele given selection coefficients of 0.1, 0.5, and 0.9. The inverse relationship between the generation time and selection coefficient indicates that changes in genetic frequency are most probable and dramatic in those environments in which the performance of sensitive individuals is reduced by at least 50%. As discussed later for air pollution stress and evolution, the magnitude of the selection coefficient is a key feature.

While both genetic variation and the selection coefficient individually influence the process of microevolution, the interplay between the two determines the speed with which populations evolve resistance and the limits to which natural selection can improve fitness of the population. Clearly, evolution is limited, and the limits are unique to species and stress scenarios. If the appropriate variation does not exist or its genic expression is constrained by pleiotropic effects, evolution may not result in ecotypically differentiated populations. Similarly, if the selection coefficient is moderate-to-low and highly episodic, the probability of significantly changing the frequency of alleles is low. Equally important constraints arise from the species' population biology including the mode of reproduction (e.g., sexual versus asexual), form of sexual reproduction (e.g., degree of inbreeding versus outcrossing), and generation time (e.g., annual, perennial).

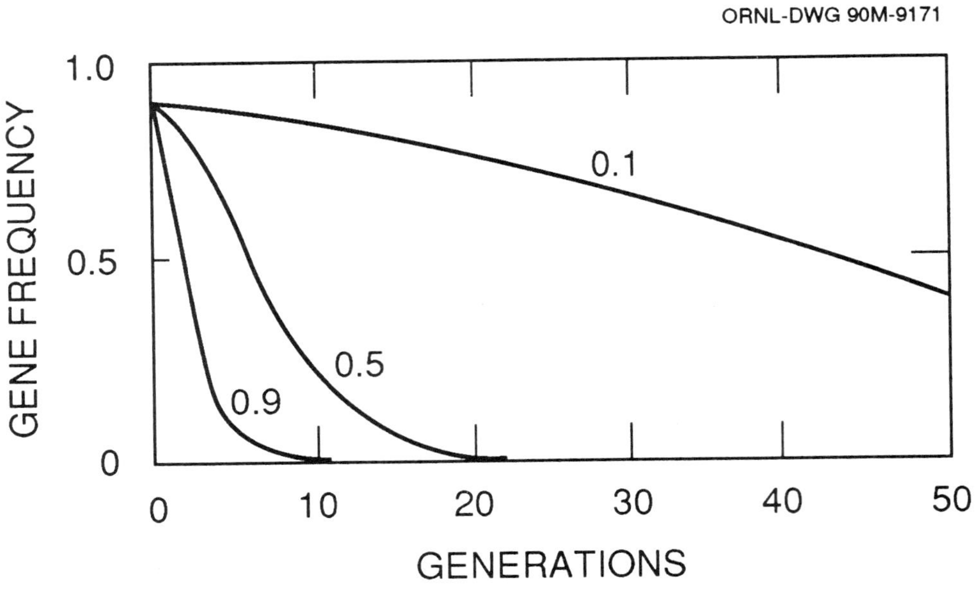

FIGURE 7-1. The theoretical relationship between the frequency of a stress sensitive gene in a population and number of generations given three different scenarios for selective pressure operating with selection coefficients of 0.9, 0.5, or 0.1. The graph demonstrates the disparity in time that it takes for the sensitive gene to change frequency from near unity given performance reductions of 10, 50, and 90%.

Source: Bradshaw and McNeilly 1991.

While microevolution is a characteristic of all plant and animal species, there are a number of attributes in the plant kingdom that facilitate microevolutionary responses (Bradshaw 1972) to air pollution:

1. The porosity of the foliar surfaces for gas exchange dictates that plants are more tightly coupled to the gaseous pollutant exposure regime;
2. The site of gaseous pollutant deposition in the leaf interior is proximal to the metabolic machinery that governs the plant's ability to assimilate carbon dioxide in net photosynthesis, which in turn is linked to productivity;
3. Most species exhibit a wealth of intrinsic, heritable variation in physiological and morphological traits;
4. Plants are stationary, more subject to the vagaries of the local environment, and unable to resort to an escape avoidance mechanism; and
5. Dispersal of gamete and seed is strongly leptokurtic so that progeny remain localized and thereby increase the probability of inbreeding.

These traits suggest that among biological species the probability of observing microevolution in response to air pollution stress is greater in the plant than the animal kingdom, although there are some noteworthy exceptions (Kettlewell 1955; Roush and McKenzie 1987).

AIR POLLUTION AS AN EVOLUTIONARY FACTOR

As an evolutionary stress, air pollution exhibits several noteworthy attributes, and the first is one of distribution. Many conventional studies of microevolution, particularly in situations involving anthropogenic stresses, have focused on spatial dimensions in which stress effects are evident in areas approaching meters in length. The most notable examples are responses to heavy metal (Antonovics, Bradshaw, and Turner 1971) or herbicide contamination (Whitehead and Switzer 1963). In these environments, selection coefficients can range from zero to unity in distances less than a meter, and as a consequence the effects of the stress are restricted to a very small part of a species' distribution. In contrast, most air pollution stress is distributed over much larger geographical areas and in many cases is regarded as being either regional (e.g., ozone, acidic precipitation, ultraviolet-B radiation) or global (e.g., carbon dioxide) in distribution. The significance is that the stress may not solely impact a few populations but rather affect large segments of the species' distribution.

Coupled to the distribution of air pollution stress is the intensity and exposure dynamics of stress that underlies the selection coefficient. Air pollution is commonly episodic, with periods of intermediate stress followed by respites of minimal to no stress. As a consequence, the selection coefficient is highly variable in space and time rather than at a constant intensity level. For most regionally and globally distributed pollution stresses, it is likely that selection coefficients are less than 0.3, meaning that the mean performance of the population is reduced by a maximum of 30%. This affects not only the probability of evolving resistance but also the physiological mechanism of resistance.

While some forms of air pollution are truly novel chemicals, most air pollutants are not toxicologically unique in their mode of action. First, many air pollutants are naturally occurring, and their effects on vegetation are simply a function of exceeding a threshold of toxicity. Notable examples are ozone, sulfur dioxide, and ultraviolet-B radiation (Caldwell 1979). Secondly, toxicity for many dissimilar chemicals is frequently expressed biochemically in a common mode of action, most notably the increase in oxidative state of the system via the generation of free radical species (e.g., superoxide anion), as induced by ozone, ultraviolet-B radiation, or hormones (Tingey and Andersen 1991).

From an evolutionary perspective, this last feature is particularly important because the requirement of mutation rate, being the only source of heritable variation, is removed, which otherwise would significantly impede evolution. Accordingly, the wealth of intrinsic variation within populations serves as the immediate source of genetic variation. Equally importantly, this feature allows for selection for resistance to multiple pollutants to occur more easily as long as the mechanism of resistance is common. The alternative and less expedient process is the coevolution of resistance by traits inherited at different loci.

An example of the intrinsic, plant-to-plant variation in response to air pollution stress is demonstrated by the diversity of foliar injury from sulfur dioxide in the herbaceous annual *Geranium carolinianum* L. (Taylor and Murdy 1975). In a variety of populations collected from seed and grown in controlled environments to maximize phenotypic response to the stress, exposure to the pollutant resulted in highly variable responses within each of the eight populations and a 24% mean coefficient of variability. Even after selection for resistance over a 30-year period, the populations experiencing the most intense selection coefficients remained highly variable, and while the mean responses were significantly offset from that of the control populations (less foliar injury and greater resistance), the populations continued to exhibit a high degree of variability (Figure 7-2). Similar patterns of variability in response to many forms of air pollution stress are well documented and are

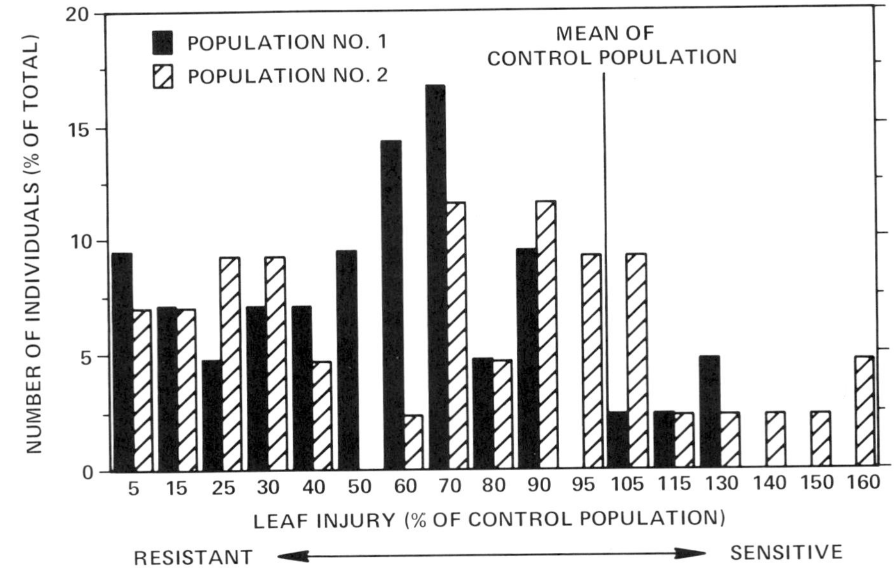

FIGURE 7-2. Intrinsic intraspecific variation in plant response to sulfur dioxide in two populations of *Geranium carolinianum* L. in which the evolution of resistance has been documented following 30 years of exposure to the pollutant. Injury is categorized as foliar necrosis and expressed as a percentage of the mean response of a control population. The net effect of selection pressure was to shift the distribution of individuals towards greater resistance.

common across a variety of taxa and for most physiological and growth parameters (Bell, Ashmore, and Wilson 1991).

Another unusual evolutionary aspect of air pollution stress is that air pollutants can be either toxic in their mode of action or provide a nutrient that is otherwise limiting to growth, with either providing selection pressure. The former includes heavy metals, ozone, and ultraviolet-B radiation. The latter includes carbon dioxide and wet and dry deposition of nitrogen and sulfur oxides. The rationale for the latter is that a species niche is determined by an array of biological, chemical, and physical factors, and the genotype is configured to maximize fitness within that matrix. Any change in the chemical component, irrespective of whether it ameliorates a chronic shortage of an element, may place the species in a new competitive environment in which it is at a selective disadvantage.

The final unusual aspect of air pollution is the distinction between direct and indirect effects. Air pollutants exert direct effects in which the pollutants chemically perturb the biochemistry of the plant, resulting in characteristic symptoms. Foliar mottling indicative of ozone exposure on broadleaf and coniferous species is one of the most classic examples. It is evident that chronic levels of air pollution may also exert their effects on a plant's physiology and growth in an indirect fashion by (1) subtly altering the organism's ability to compete for limited resources (e.g., light) or withstand other stresses of biotic (e.g., pest and pathogens) or abiotic (e.g., drought) origin and (2) producing other toxic chemicals that in turn provide the stress (e.g., enhanced levels of soil aluminum or depleted base cations due to acidic deposition) (Johnson and Taylor 1989; see Colls and Unsworth, Chapter 5, this volume). Examples are the enhanced winter hardiness in high-elevation spruce trees induced by exposure to acidic precipitation (Fowler et al. 1990), change in foliar water use efficiency in trees due to elevated levels of carbon dioxide or ozone (Neighbour, Cottam, and Mansfield 1988), and alteration of leaf angle and community composition from ultraviolet-B radiation (Gold and Caldwell 1983). From an evolutionary perspective, the significance is that selection for resistance will arise indirectly and only in situations in which the interacting stress is present.

ECOLOGICAL AND BIOLOGICAL CONSEQUENCES OF AIR POLLUTION-INDUCED CHANGES IN BIODIVERSITY

Even though evolution in response to air pollution is probable in areas of intermediate-to-high pollution stress, there are a number of ecological and biological issues that affect the (1) speed and frequency with which adaptation rises, (2) physiological mechanism of resistance, and (3) significance of microevolution in context of ecology and biodiversity. Collectively, these provide a framework in which to judge the probability that microevolution will occur and the potential significance of the process in applied and basic research.

Cost of Evolving Resistance

Microevolution in response to air pollution stress may carry with it negative physiological costs that make the plant less competitive or able to tolerate other stresses when the pollutant is absent, as illustrated in Figure 7-3 (Roose, Bradshaw, and Roberts 1982;

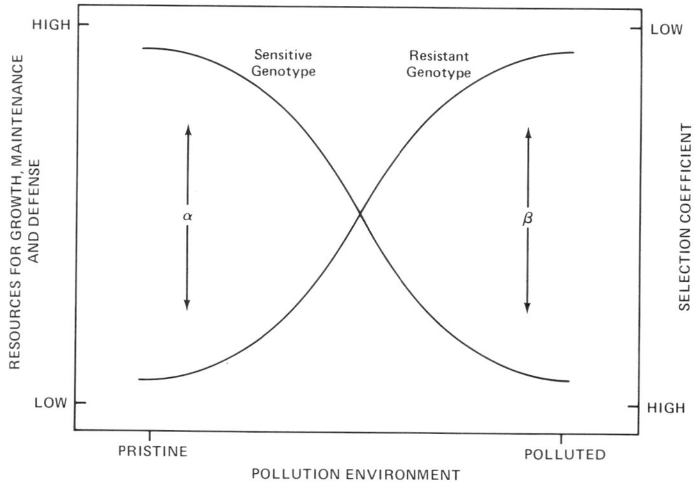

FIGURE 7-3. Theoretical relationship between the air pollution stress and the (i) allocation of carbon resources for growth, maintenance, and defense and (ii) selection coefficient in both sensitive and resistant genotypes. In pristine environments (low stress due to air pollution) sensitive genotypes are at an advantage over their resistant counterparts as shown by the low selection coefficient and the high allocation of resources to growth, maintenance, and defense. The converse arises in polluted environments in which resistant genotypes are in a selective advantage. The difference between the performance of the genotypes in pristine environments (α) is a measure of the cost of being resistant whereas the corresponding parameter in polluted environments (β) is the cost of being a sensitive genotype relative to a resistant counterpart.

Pitelka 1988; Winner et al. 1991). It is generally accepted that response to stress affects the plant's carbon economy in general, and most often is manifest in an increase in carbon allocation from growth to maintenance energy (Amthor 1988). The mechanisms of resistance may be accountable to stress avoidance, stress detoxification, or repair and compensation for injury. The energetic or growth costs of different mechanisms of resistance are likely to vary (Roose, Bradshaw, and Roberts 1982). For instance, if resistance to a gaseous air pollutant is due to a decline in stomatal conductance (Mansfield and Freer-Smith 1984), the potential cost would be a reduction in total carbon gain via carbon dioxide assimilation. However, if low stomatal conductance was a preexisting and necessary adaptation to other natural stresses in the environment, then the actual cost of being more resistant to air pollutants could be negligible. When the resistance mechanism is one of tolerance (e.g., Taylor, Tingey, and Gunderson 1986), it is more probable that there is an energetic cost associated with the adaptation, particularly if enzyme systems are necessary to detoxify or sequester a pollutant. But as in the case of avoidance, traits that confer tolerance to the pollutant may also be the basis for adaptation to other natural stresses. It is possible that relatively few fundamental mechanisms underlie the responses to many outwardly dissimilar stresses (Parsons 1988). An understanding of the biochemical and physiological basis of tolerance to a pollutant is necessary in order to determine the extent to which the adaptations carry with them a physiological cost.

The best evidence for costs of adaptation to pollutants comes from competition experiments or observations that resistance is selected against in the absence of the pollutant (Pitelka 1988). Without information about the mechanisms underlying resistance, such observations must be interpreted carefully because other differences between resistant and sensitive genotypes could explain the reduced success of the resistant genotype when the pollutant is not present.

The ramifications for biodiversity and ecosystem integrity of the costs of evolving resistance are uncertain. If plants of all species in an ecosystem are equally sensitive to pollution stress and natural selection results in the same mechanisms of resistance, the cost of resistance is likely to have no net effect at the community level. However, given the intrinsic variation within and among species, the probability is small that physiologically identical adaptive responses would evolve. Dissimilar mechanisms of resistance are more probable, and since each would have its own fitness costs, a shift in the competitive interactions among species would favor the species with the minimum cost of resistance. This would involve tradeoffs between different levels of cost rather than between the presence and absence of costs, and it is unlikely that the effect would be large compared with the many other variables that also determine competitive interactions.

It is possible that plants incur costs due to the evolution of resistance to air pollutants such that their ability to tolerate natural stresses (e.g., drought or winter temperature) is influenced (Fowler et al. 1990). To the extent that common responses and adaptations to different natural and pollutant stresses exist, a diminished chance of survival would depend on the stress and species.

Finally, it is proposed that species possessing resistance to a pollutant suffer in competition with other species if the pollutant stress is removed. The extent to which this happens and is a cost of adaptation is a function of (1) how quickly the transition to an unpolluted environment occurs, (2) how rapidly selection against the trait operates within resistant populations, and (3) the extent of the competitive advantage among species. Only if the cost of resistance is very high (i.e., the competitive advantage large) while the rate of selection against resistant individuals is low would species replacement occur. This phenomenon has been well documented in heavy metal tolerance among grasses (Antonovics, Bradshaw, and Turner 1971).

Role of Habitat Fragmentation

While the potential for air pollution to function as an evolutionary force at the level of populations is evident, it is important to recognize the pervasive, interactive consequences of habitat fragmentation. One of the most obvious effects of anthropogenic activities on terrestrial landscapes is the site-specific change in vegetation relative to that which existed prior to human intervention, with the more natural continuous distribution of landscapes being replaced by a mix of natural and managed ecosystems. These appear as a patchwork of landscapes with very discrete populations and communities varying significantly in size and species' associations (Turner 1987).

Array of communities within continental landscapes has profound implications for the genetic structure of species in a manner analogous to that well documented in island biogeography. Fragmentation of species into discrete, small patches or populations exerts several effects. The first is the reduction in population size, increasing the potential for population bottlenecks to affect genetic diversity in both a random and nonrandom manner. The second is gene dispersal: because pollen and seed dispersal tends to be leptokurtic, exchange of alleles among populations is restricted by fragmentation, thereby increasing the probability by chance alone that spatial isolation will result in unique genetic configurations for a variety of physiological and morphological traits. The net effect at the species level is that clinal variation in genetic structure is rivaled in importance by the variability due to fragmentation of the landscape. In one respect, this increases the probability that the genetic structure will change rapidly in an area in which air pollution-induced selection coefficients are maintained and the appropriate variation is present upon which selection can operate. On the other hand, the reduction in population size dictates that some populations are likely not to have the required intrinsic variation in response to air pollution and thus cannot evolve resistance to an equivalent degree.

Population Demographics in Response to Air Pollution Stress

Most areas experiencing air pollution stress are ones in which only a fraction of the total species in the community are negatively impacted, and for most of these sensitive species the stress is chronic rather than acute. This situation contrasts with many of the more widely publicized studies in which intense air pollution stress simplified ecosystems via removal of most of the terrestrial plant community (Gordon and Gorham 1960; Guderian 1977; Woodwell 1970). In these highly stressed regions, the demographics of the local populations progressed to complete mortality, and the change was visually apparent. The more current and common situation is different: given chronic levels of air pollution stress and a low selective pressure, most sensitive populations (particularly if they are annuals) are likely to experience no net change in population size. The performance of sensitive individuals within the population will decline in importance, and they will contribute less to the gene pool of subsequent generations. Conversely, the growth and reproductive success of resistant individuals will increase. There is no net change in population demographics and yet the population has experienced an increase in the frequency of resistant individuals.

This situation is not well documented for air pollution stress, but an analogy from the literature on heavy metal tolerance is appropriate (Farrow, McNeilly, and Putwain 1981). Over a 5-year period, the net population size of *Agrostis canina* showed little variation in number but only because the cumulative gains in resistant individuals equaled the cumulative losses of their sensitive counterparts (Figure 7-4).

Loss of Genetic Diversity Due to Air Pollution

There is legitimate concern among biologists about the depletion of genetic diversity through the loss of species, cultivars, and populations resulting from different kinds of human activities, and the principal concern is the role that such losses may play in limiting the future evolutionary potential of species. If significant and permanent losses of genetic variability occur as a result of air pollution, it is a matter of concern. The evolution of resistance to air pollutants is hypothesized to contribute to the loss of genetic variability within plant species (Bergmann and Scholz 1989; Karnosky 1991). However, several issues require that this conclusion be readdressed, including the mechanism by which air pollution causes a loss of genetic diversity, the extent to which evolution of resistance within a population results in losses of genetic variability, and the evolutionary implications of declines in variability.

There are four mechanisms by which any stress could result in the loss of genetic diversity (Figure 7-5). If the level of stress is so acute that entire species or populations become locally extinct, there is little question that significant losses of variability occur. This situation would tend to occur only locally, near point sources of intense pollution (Gordon and Gorham 1960). The extent to which this would represent a loss of diversity

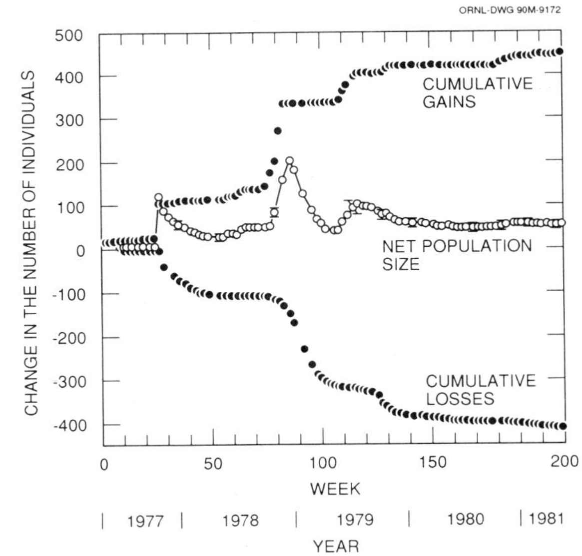

FIGURE 7-4. The change in population demographics over a 5-year period in the grass *Agrostis canina* on a copper mine. The graph illustrates the absence of any net change in population size, which is accountable to the paired cumulative increases and decreases over the same time period due to individuals being recruited and dying, respectively (Farrow, McNeilly, and Putwain 1981).

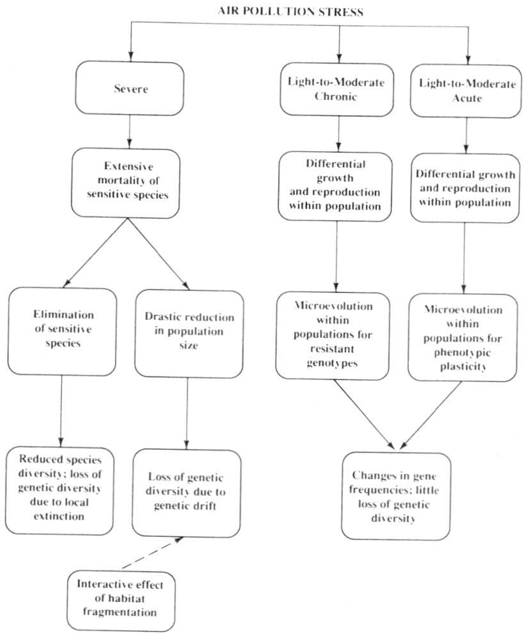

FIGURE 7-5. Flow diagram illustrating the potential effects of air pollution stress on biodiversity of plant species and populations with the effect dependent on the severity of the stress, availability of heritable variation in response to the pollutant, and habitat fragmentation.

within a widely distributed species depends on how genetically unique the populations were. Moreover, as discussed by Holt (1990), "worrying about microevolution is moot when (stress) is so extreme as to force rapid extinction."

The second mechanism by which losses of genetic variability can occur is through drift, whereby a population is abruptly reduced in size. If only a few individuals survive to establish a new population, they may carry only a fraction of the original genetic variability [for an exception, see Carson's example (1990) of genetic variance increasing following a population bottleneck]. In such situations, some of the loss of genetic variability is accountable to selection for resistance because those alleles conferring resistance and those that are linked genetically or physiologically will persist, while those conferring sensitivity will be lost. However, if only a few individuals comprise the founder population, it is likely that most of the losses of genetic variability will be random (i.e., founder effect). The important point is that the majority of the losses are not a direct consequence of the evolution of resistance but rather result from the massive level of mortality (Figure 7-4). As in the case of population extinction, losses of variability following a bottleneck will only occur in extreme and localized situations.

The third means by which genetic variability can be influenced by air pollution is through selection for resistance (Figure 7-5). While there is some evidence for this (Bergmann and Scholz 1985), it is important to recognize that other explanations exist for the correlation between low levels of genetic variability and high levels of pollution stress (Barrett and Bush 1991). The consequences of selection for resistance are not necessarily clear (Parsons and Pitelka 1991). Selection can result in losses of genetic variability in that alleles or gene complexes associated with sensitivity will become less common. However, it is unlikely that such alleles will be lost entirely from a population unless selective mortality is very high (i.e., approaching the second situation above). It is also not clear that changes in gene frequencies or modest losses in variability in any way limit the future evolutionary potential of the population. There is nothing unique about selection for resistance to air pollution or the associated losses of variability. Some losses of variability accompanying selection must occur in all species as a result of selection, and these have not resulted in reduced evolutionary potential.

The fourth way in which genetic diversity could be affected by air pollution operates at the community level and is based on the concept that preadaptation dictates that some species are more resistant to the pollutant than others (Figure 7-5). If pollutant exposure results in mortality or even simply reduced growth or reproduction, sensitive species will gradually replace their more resistant counterparts. This is especially likely when the generation time of the sensitive species is so long that evolution of resistance is slow compared with the time necessary for competitive replacement. If the resistant species were already components of the community rather than being opportunists, then the loss of sensitive species means that the species diversity is reduced and thus the overall genetic diversity of the community.

Significant losses of genetic diversity are only likely to occur as a result of population extinction or genetic drift associated with intense pollution in relatively restricted areas. It is unclear that selection for resistance per se will result in any significant or permanent losses of genetic diversity. In recent years, scientists, land managers, and policy makers have become more concerned about regional, low-level pollution from such agents as ozone, acidic deposition, carbon dioxide, and heavy metals. While selection for resistance to pollution may be occurring in such situations, there is little experimental evidence for concluding that genetic diversity is actually threatened (Parsons and Pitelka 1991).

Phenotypic Plasticity of Physiological Traits

If an environmental stress is variable in its intensity, the most optimal adaptive strategy is one in which the organism's physiology changes rapidly to match the speed and severity of the fluctuating stress. This form of adaptation is one of phenotypic plasticity and contrasts with the more conventional or "constitutive" form of adaptation in which the trait is always present (Bradshaw and Hardwick 1989). Given that many forms of air pollution stress are episodic, it is hypothesized that rapid changes in phenotype at the physiological level are very common modes of adaptation. Unfortunately, little research has been done to address this mechanism of response to air pollution.

Adaptations based on phenotypic plasticity are well documented for a variety of environmental factors including drought (e.g., stomatal closure, leaf abscission), predation (e.g., phytoalexins), nutrient deficiency (e.g., root exploration), submergence (e.g., leaf morphology), temperature (phenology), and solar radiation (e.g., rapid response of net carbon dioxide assimilation to sun flecks). The capacity to vary the phenotype is subject to genetic control (more so at the interspecific level) and natural selection confers a selective advantage on alleles that underlie these responses (Sultan 1987). From an evolutionary perspective, the rationale for selection favoring such adaptations lies in the costs of evolving resistance whereby constitutive adaptations confer selective advantage only during periods of stress and are at a disadvantage during nonstress periods. Accordingly, the capacity to respond rapidly with a variable phenotype is a solution both in the presence and absence of the stress.

The argument for phenotypic plasticity in response to air pollution is not well documented but is very likely to be far more common than constitutive responses. Whereas a number of candidate mechanisms are proposed (Table 7-1), one of the most likely is a decrease in stomatal conductance. The responsiveness of stomata to air pollution stress is well documented and closure is a common avoidance mechanism (Taylor 1978). Assuming this responsiveness is subject to genetic control, alleles allowing variable stomatal porosity in response to episodically high levels of pollution would have a selective advantage both in terms of conferring resistance during episodes and avoiding the negative costs of constitutive resistance during nonstressed periods (e.g., constant low stomatal porosity).

The capacity of the phenotype to physiologically respond to a change in intensity of air pollution stress requires a finely tuned mechanism whereby the stress is perceived and the plant's physiology and biochemistry rapidly adjusted. The theoretical work of Koshland, Goldbeter, and Stock (1982) provides a framework for such a stress perception and communication system: the most effective sensory and regulatory system is one in which a small change in an effector concentration elicits a very large and rapid change in physiology (Figure 7-6). The chemical form of the effector is not specified but could be an ion (e.g., calcium), electrical potential gradient across membranes, or hormone. The most effective systems are classified as "ultrasensitive," in which a major change in physiology is achieved with a very small change in effector concentration. Conversely, sluggish responses (hyper- or subsensitive) and ones with less evolutionary potential require far greater changes in effector concentration.

One of the most probable ultrasensitive response systems is stress ethylene, which is known to (1) be responsive to air pollution, (2) exhibit the necessary ultrasensitivity to

TABLE 7-1. Examples of Air Pollution-Induced Changes Indicative of Phenotypic Plasticity at the Physiological and/or Biochemical Level

Stress	Mechanism	Time Scale
Ozone	Phenological Shifts	Days/Weeks
	Stomatal Closure	Minutes/Hours
	Free Radical Scavenger	Minutes/Hours/Days
Ultraviolet-B Radiation	Leaf Orientation	Days/Weeks
	Pigmentation	Days/Weeks
Sulfur Dioxide	Stomatal Closure	Hours/Weeks
	Hydrogen Sulfide Emissions	Minutes/Hours
Elevated Carbon Dioxide	Phenological Shifts	Days/Weeks
	Stomatal Closure	Hours/Days
Nitrogen Deposition	Nitrate Reductase	Hours/Days
Ethylene	Phenological Shifts	Days/Weeks
	Stomatal Closure	Minutes/Hours
	Leaf Orientation	Hours/Days
Heavy Metals	Rhizosphere Precipitation of Metal	Days/Weeks
	Leaf Sequestration of Metal	Days/Weeks
	Stomatal Closure	Days/Weeks

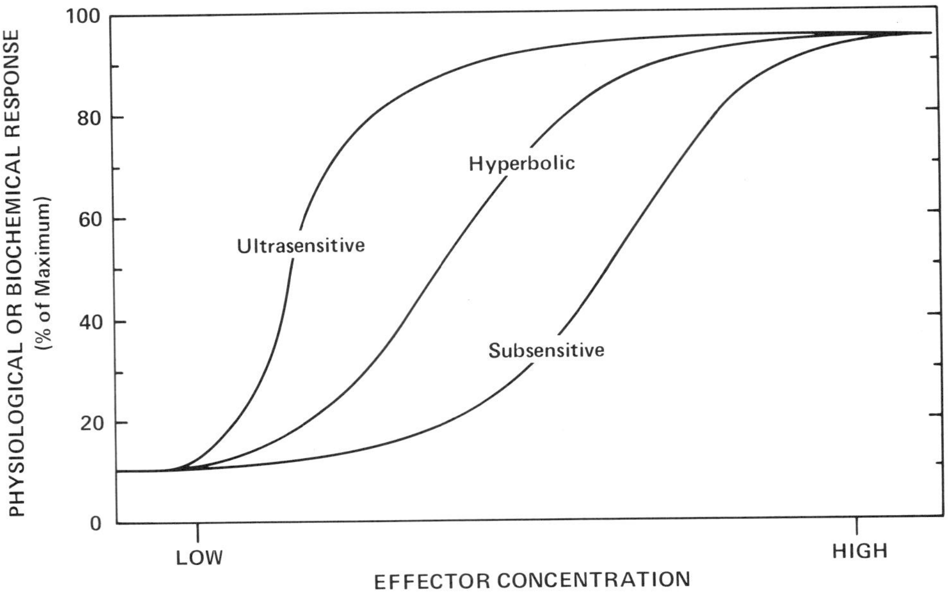

FIGURE 7-6. Relationship underlying the basis for an adaptation based on phenotypic plasticity in which an alteration in the organism's physiology or biochemistry is triggered by a very small change in effector concentration. From an evolutionary perspective, a trait that exhibits the characteristics of an ultrasensitive response would have a selective advantage over other adaptations (e.g., hypersensitive, subsensitive) if the stress is episodic in intensity.

Source: Koshland, Goldbeter, and Stock 1982.

elicit rapid responses at levels of organization ranging from biochemistry to physiology to whole plant growth and development, and (3) exhibit the traits Koshland, Goldbeter, and Stock (1982) identify as being most evolutionarily advantageous. Stress ethylene increases dramatically in the presence of air pollution (Tingey, Standley, and Field 1976), and the increase is evident within a matter of minutes (Hogsett, Raba, and Tingey 1981). Exogenously applied ethylene results in marked changes in stomatal physiology and net carbon dioxide assimilation, and the effects are indicative of a hormonally mediated effect rather than one of phytotoxicity (Gunderson and Taylor 1988). Given the role of stress ethylene in leaf gas exchange of carbon dioxide and water vapor, polyamine biochemistry, leaf senescence, and antioxidant formation, it is proposed that this system exhibits the necessary criteria for a phenotypically plastic adaptation in response to air pollution stress.

Biodiversity, Air Pollution, and Wilderness Areas

While air pollution appears to threaten genetic diversity principally in highly polluted situations, the effects of regionally distributed air pollutants on the genetic structure of plant populations in national parks and wilderness areas is of special concern. These are areas set aside in perpetuity by State and Federal agencies to preserve the natural resources, and any change resulting from human activities is unnatural and should be prevented (Graber 1983). Evolutionary response to an air pollutant would be considered undesirable regardless of whether it has any effect on the integrity of the species or ecosystem.

Although this is a policy issue concerning societal values rather than scientific issues (see McKelvey and Henderson, Chapter 13, this volume), it is relevant in the context of this discussion.

It is useful to examine the significance of the evolution of resistance to air pollution from the perspective of how it might help to preserve biodiversity, both genetic diversity within populations and species diversity in ecosystems. It is not necessary to accept the premise that some level of air pollution is acceptable in order to recognize that plants in many parts of the world have been and will continue to be exposed to various types of air pollutants. The complete elimination of air pollutants is unlikely in the foreseeable future (McLaughlin and Norby 1991). Artificial selection or genetic engineering of resistant genotypes may offer opportunities for actually supplementing the natural evolutionary process. In cases where air pollution cannot be reduced in the near future and certain species are threatened, artificial selection or genetic engineering may be a means of managing the ecosystem and preserve species diversity (Karnosky 1991).

Methodological Aspects of Detecting Changes in Biodiversity in Response to Air Pollution

The methodological problems of detecting air pollution-induced changes in the genetic structure of plant populations are not trivial. The previous discussion outlined many of the caveats that must be considered in order to accurately assess changes in biodiversity, and most of these caveats are based on drawbacks attributable to methodology.

Equally important is the recognition that adaptations based on phenotypic plasticity are far more likely than constitutive ones, and yet only the latter have been investigated. In part, this is due to the difficulty in experimentally investigating plasticity responses: by definition, these cannot be evaluated using the traditional "reciprocal transplant" techniques because they are stress dependent. Consequently, nontraditional approaches must be devised to tease out their potential regulatory role in conferring resistance while avoiding the costs.

Finally, the probability of finding a population whose gene pool is affected by a chronic-level stress is not necessarily dependent on the same criteria used in traditional studies in ecological genetics, particularly those addressing intense selection coefficients operating over a distance of meters. The reason for this disparity lies in the interaction between air pollution as a regional phenomenon (without clearly defined gradients) and the role of habitat fragmentation in establishing pockets of individuals that are somewhat genetically isolated from other populations. Because of genetic drift and founder effects, it is highly probable that at least some of these pockets have genetically diverged due to air pollution. However, it is equally probable that the small population size (with its associated finite availability of genetic variation) and the repetitive nature of human intervention have precluded evolution or have resulted in random changes unrelated to any pollution gradient. Being able to identify populations in which air pollution stress has altered the genetic structure unambiguously is confounded significantly by the degree to which other anthropogenic processes fragment the landscape.

One way of simplifying the process of identifying populations with a high probability of air pollution-induced changes in genetic structure is illustrated in Figure 7-7 (W. Hogsett, pers. communication. 1991). The regional distribution of ozone in the eastern United States extends from the mid-Atlantic states westward to the Ohio Valley and northward into the Great Lakes region. This regional stress is juxtaposed with the geographical distribution of

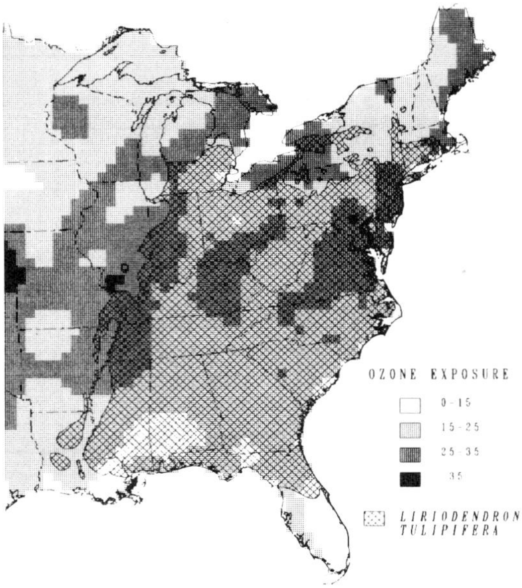

FIGURE 7-7. Joint distribution of ozone stress in the eastern United States for the summer of 1985 with the geographical distribution of *Liriodendron tulipifera* (W. Hogsett, pers. communication, 1991). The ozone exposure potential is a relative index that calculates an ozone value based on an inventory of ozone precursor emissions, ozone photochemical production as a function of temperature and meteorological stagnation, dispersion and decay, and transport as a function of wind speed and direction.

Liriodendron tulipifera, an ozone-sensitive tree species. What is not evident is the degree of habitat fragmentation on a fine scale, which results in the species being relegated to islands of varying size in urban, suburban, and rural landscapes. It is proposed that the probability of finding genetically altered populations is highest in areas in which the (1) stress is greatest and (2) species is at the geographical limits of its distribution, which for *L. tulipifera* is along either the northern or western segments of its distribution. Conversely, the least probable site with an equivalent ozone exposure would lie in the more interior regions of the species' distribution (e.g., Middle Atlantic States). This argument is based on the principle that individuals along the margins of the species' distribution are more likely to operate on a carbon production to maintenance energy ratio that approaches unity, and that any additional anthropogenic stress would place a premium on an adaptation that maintains that ratio. Conversely, in the more interior areas of the distribution, the production to maintenance energy ratio in many individuals greatly exceeds unity so that the same degree of air pollution stress is not likely to carry an equivalent selective pressure.

FUTURE CHANGES IN AIR QUALITY AND POTENTIAL CHANGES IN GENETIC DIVERSITY

Anthropogenic activities will continue to influence air quality in the future although the intensity, distribution, and forms of pollution will differ. In many of the more highly industrialized countries, progress is likely in reducing the levels of toxic air pollutants, notably sulfur and nitrogen oxides and chlorofluorocarbons. However, a different situation is likely in the lesser developed countries: air pollution will collectively exceed that of the industrialized nations, and the intensity will range from acute to chronic, with effects on local and regional scales. The most probable pollutants are oxides of sulfur and nitrogen from fossil fuel consumption, heavy metals, ozone, and air toxics (McLaughlin and Norby 1991).

The other aspect of air pollution stress is the recognition that some pollutants are truly global in distribution, and levels are likely to increase over the next several decades. For example, carbon dioxide concentration is predicted to double within the next 4 to 6 decades from its current concentration of 340 ppm. Knowing that intrinsic genetic variation exists and that it differentially influences plant physiology, growth, and reproductive success (Wulff and Alexander 1985), it is hypothesized that elevated levels of carbon dioxide may operate as a selective factor (Strain 1991). Equally important is the anticipated increase in ultraviolet-B radiation due to stratospheric ozone depletion. As with elevated levels of carbon dioxide, ultraviolet-B radiation is already documented to have played a role in plant evolution (Caldwell 1979) and its potential increase will place a premium on genetically determined traits that either avoid or tolerate the stress.

The threats to genetic diversity posed by air pollution must be evaluated in the broader context of all threats to biodiversity. For instance, subtle and slow changes in the genetic structure of populations resulting from air pollution may be irrelevant if the greenhouse effect results in climate change as great and rapid as that predicted by some general circulation models. With some predictions, climate zones in northern temperate latitudes could shift 600 to 1,000 km to the north. Because of the relation between climate and vegetation, vegetation zones would be predicted to shift accordingly (Emanuel, Shugart, and Stevenson 1985), as would the ranges of individual species (Davis 1989). Narrowly distributed species, whose predicted new ranges are entirely disjunct from their current ranges, might not be able to migrate fast enough and would become extinct. Even in widely distributed species that survive, many populations would presumably become extinct because of a narrowly defined ecological amplitude and the limits of microevolutionary responses (Bradshaw and McNeilly, 1991a). Extinction of entire populations would result in far greater losses of biodiversity than would microevolution within populations.

A SUMMARY FROM A BIODIVERSITY PERSPECTIVE

The following conclusions can be offered:

1. Air pollution has and will continue to influence the genetic structure of sensitive plant species; ecotypic differentiation is likely to be more common than has been documented;
2. The most probable response is a subtle, progressive shift in gene pool that is not readily amenable to investigation using conventional methodologies;
3. From the perspective of evolution at the population level, studies of air pollution as a stress offer unique research opportunities because of the variable exposure dynamics (selection coefficient), distribution (local/regional/global), and potential for interaction with other stresses of natural and anthropogenic origin;
4. The probability of evolving resistance to air pollution stress is both constrained (e.g., low selection coefficients) and enhanced (e.g., habitat fragmentation); relative to that observed for many other classes of environmental stress, it is hypothesized that the role of the latter significantly increases the probability that any given population's genetic

structure will shift in response to air pollution;

5. One of the most unresolved issues is the role of adaptations based on phenotypic plasticity in response to air pollution stress, whereby the genetic basis underlying an individual's capacity to rapidly adjust its physiology or biochemistry is subject to selection; based on first principles in ecology, physiology, and evolution, it is hypothesized that this phenomenon plays a major role;
6. When air pollution causes population extinction, significant losses of genetic diversity may occur, but there is little evidence that microevolution results in permanent losses of variability that reduce the evolutionary potential of the species; and
7. In the context of biodiversity, it is concluded that changes in genetic diversity due to air pollution stress do not rival the corresponding changes in biodiversity due to the other major environmental stresses, including global climate change, tropical deforestation, and landscape modification.

ACKNOWLEDGMENT

The senior author acknowledges partial support during manuscript preparation from the Electric Power Research Institute's Program "Response of Plants to Interacting Stresses" (RP-2799-2).

REFERENCES

1. Amthor, J. 1988. Growth and maintenance respiration in leaves of bean (*Phaseolus vulgaris* L.) exposed to ozone in open-top chambers in the field. *New Phytol.* 110:319-25.
2. Antonovics, J., A.D. Bradshaw, and R. Turner. 1971. Heavy metal tolerance in plants. *Adv. Ecol Res.* 7:1-85.
3. Barrett, S.C.H., and E.J. Bush. 1991. Population processes in plants and the evolution of resistance to gaseous air pollutants. In *Ecological genetics and air pollution*, ed. G.E. Taylor, Jr., L.F. Pitelka, and M.T. Clegg, pp. 137-65. New York: Springer-Verlag.
4. Bell, J.N.B., M.R. Ashmore, and G.B. Wilson. 1991. Ecological genetics and chemical modifications of the atmosphere. In *Ecological genetics and air pollution*, ed. G.E. Taylor, Jr., L.F. Pitelka, and M.T. Clegg, pp. 33-59. New York: Springer-Verlag.
5. Bergmann, F., and F. Scholz. 1985. Effects of selection pressure by SO_2 pollution on genetic structures of Norway Spruce (*Picea abies*). In *Population genetics in forestry*, ed. H.-R. Gregorius, pp. 267-75. New York: Springer.
6. Bergmann, F., and F. Scholz. 1989. Selection effects of air pollution in Norway Spruce (*Picea abies*) populations. In *Genetic effects of air pollutants in forest tree populations*, ed. F. Scholz, H.-R. Gregorius, and D. Rudin, pp. 143-60. New York: Springer-Verlag.
7. Bishop, J.A., and L.M. Cook. 1981. *Genetic consequences of man made changes*. New York: Academic Press.
8. Bradshaw, A.D., 1972. Some of the evolutionary consequences of being a plant. *Evol. Biol.* 5: 25-47.
9. Bradshaw, A.D., and K. Hardwick. 1989. Evolution and stress—genotypic and phenotypic components. *Biol. J. Linn. Soc.* 37:137-55.
10. Bradshaw, A.D. and T. McNeilly. 1981. *Evolution and pollution*. London: Edward Arnold.
11. Bradshaw, A.D., and T. McNeilly. 1991a. Evolutionary response to global climate change. *Ann. Bot.* 67:5-14.

12. Bradshaw, A.D., and T. McNeilly. 1991b. Evolution in relation to environmental stress. In *Ecological genetics and air pollution*, ed. G.E. Taylor, Jr., L.F. Pitelka, and M.T. Clegg, pp. 11-31. New York: Springer-Verlag.
13. Caldwell, M. 1979. Plant life and ultraviolet radiation: Some perspective in the history of the earth's UV climate. *Bioscience* 29:520-25.
14. Carson, H.L. 1990. Increased genetic variance after a population bottleneck. *Trends Ecol. & Evol.* 5:228-30.
15. Clausen, J., D.D. Keck, and W.M. Hiesy. 1940. Experimental studies on the nature of species. 1. Effect of varied environments on western North American plants. *Carnegie Inst. Wash. Publ.* 520:1-442.
16. Davis, M.B. 1989. Lags in vegetation response to greenhouse warming. *Clim. Change* 15:75-82.
17. Dunn, D.B. 1959. Some effects of air pollution on *Lupinus* in the Los Angeles area. *Ecology* 40:621-25.
18. Emanuel, W.R., H.H. Shugart, and M.P. Stevenson. 1985. Climate change and the broad-scale distribution of terrestrial ecosystem complexes. *Clim. Change* 7:29-43.
19. Farrow, S., T. McNeilly, and P.D. Putwain. 1981. The dynamics of natural selection for tolerance in *Agrostis canina* L. subs. montana Hartm. In *International conference, heavy metals in the environment*, pp. 289-95. Edinburgh: C.E.P. Consultants.
20. Fowler, D.J.N., J.N. Cape, J.D. Deans, I.D. Leith, M.B. Murray, R.I. Smith, L.J. Sheppard, and M.H. Unsworth. 1990. Effects of acid mist on the frost hardiness of red spruce seedlings. *New Phytol.* 113:321-55.
21. Gold, W.G., and M.M. Caldwell. 1983. The effects of ultraviolet-B radiation on plant competition in terrestrial ecosystems. *Physiol. Plant.* 58:435-44.
22. Gordon, A.G., and E. Gorham. 1960. Ecological aspects of air pollution from an iron-sintering plant at Wawa, Ontario. *Can. J. Bot.* 41:1063-78.
23. Graber, D.M. 1983. Rationalizing management of natural areas in national parks. *The George Wright Forum* 3:48-56.
24. Guderian, R. 1977. *Air pollution.* New York: Springer-Verlag.
25. Gunderson, C.A., and G.E. Taylor, Jr. 1988. Kinetics of inhibition of foliar gas exchange by exogenous ethylene: An ultrasensitive response. *New Phytol.* 110:517-24.
26. Hogsett, W.E., R.M. Raba, and D.T. Tingey. 1981. Biosynthesis of stress ethylene in soybean seedlings: Similarities to endogenous ethylene biosynthesis. *Physiol. Plant.* 53:307-14.
27. Holt, R.D. 1990. The microevolutionary consequences of climate change. *Trends Ecol. & Evol.* 5:311-15.
28. Johnson, D.W., and G.E. Taylor, Jr. 1989. Role of air pollution in forest decline in eastern North America. *Water Air Soil Pollut.* 48:21-43.
29. Karnosky, D.F. 1991. Ecological genetics and changes in atmospheric chemistry: The application of knowledge. In *Ecological genetics and air pollution*, ed. G.E. Taylor, Jr., L.F. Pitelka, and M.T. Clegg, pp. 321-36. New York: Springer-Verlag.
30. Kettlewell, H.B.D. 1955. Selection experiments on industrial melanism in *Lepidoptera*. *Heredity* 9:323-42.
31. Koshland, D.E., Jr., A. Goldbeter, and J.B. Stock. 1982. Amplification and adaptation in regulatory and sensory systems. *Science* 217:220-25.
32. Mansfield, T.A., and P. Freer-Smith. 1984. The role of stomata in resistance mechanisms. In *Gaseous air pollutants and plant metabolism*, ed. M.J. Koziol and F.R. Whatley, pp. 131-46. London: Butterworth.
33. McLaughlin, S.B., and R.J. Norby. 1991. Atmospheric pollution and terrestrial vegetation: Evidence for changes, linkages, and significance to selection process. In *Ecological genetics and*

air pollution, ed. G.E. Taylor, Jr., L.F. Pitelka, and M.T. Clegg, pp. 61-101. New York: Springer-Verlag.

34. Neighbour, E.A., D.A. Cottam, and T.A. Mansfield. 1988. Effects of sulfur dioxide and nitrogen dioxide on the control of water loss by birch (*Betula* spp.) *New Phytol.* 108:149-57.
35. Parsons, P.A. 1988. Conservation and global warming: A problem in biological adaptation to stress. *Ambio* 18:323-25.
36. Parsons, D.J., and L.F. Pitelka. 1991. Plant ecological genetics and air pollution stress: A commentary on implications for natural populations. In *Ecological genetics and air pollution,* ed. G.E. Taylor, Jr., L.F. Pitelka, and M.T. Clegg, pp. 337-43. New York: Springer-Verlag.
37. Pitelka, L.F. 1988. Evolutionary responses of plants to anthropogenic pollutants. *Trends Ecol. & Evol.* 3:233-36.
38. Roose, M.L., A.D. Bradshaw, and T.M. Roberts. 1982. Evolution of resistance to gaseous air pollutants. In *Effects of gaseous air pollution on agriculture and horticulture*, ed. M.H. Unsworth and D.P. Ormrod, pp. 379-409. London: Butterworth.
39. Roush, R.T., and J.A. McKenzie. 1987. Ecological genetics of insecticide and acaricide resistance. *Annu. Rev. Entomol.* 32:361-80.
40. Scholz, F., H.-R. Gregorius, and D. Rudin. 1988. *Genetic effects of air pollutants in forest tree populations.* New York: Springer-Verlag.
41. Strain, B.R. 1991. Possible genetic effects of continually increasing atmospheric CO_2. In *Ecological genetics and air pollution*, ed. G.E. Taylor, Jr., L.F. Pitelka, and M.T. Clegg, pp. 235-43. New York: Springer-Verlag.
42. Sultan, S.E. 1987. Evolutionary implications of phenotypic plasticity in plants. *Evol. Biol.* 21:127-78.
43. Taylor, G.E., Jr. 1978. Plant and leaf resistance to gaseous air pollution stress. *New Phytol.* 80:523-34.
44. Taylor, G.E., Jr., and W.H. Murdy. 1975. Population differentiation of an annual plant species, *Geranium carolinianum* L., in response to sulfur dioxide. *Bot. Gaz.* 136:212-15.
45. Taylor, G.E., Jr., D.T. Tingey, and C.A. Gunderson. 1986. Photosynthesis, carbon allocation, and growth of sulfur dioxide ecotypes of *Geranium carolinianum* L. *Oecologia* 68:350-57.
46. Taylor, G.E., Jr, L.F. Pitelka, and M.T. Clegg. 1991. *Ecological genetics and air pollution.* New York: Springer-Verlag.
47. Tingey, D.T., and C.P. Andersen. 1991. The physiological basis of differential plant sensitivity to changes in atmospheric quality. In *Ecological genetics and air pollution*, ed. G.E. Taylor, Jr., L.F. Pitelka, and M.T. Clegg, pp. 209-34. New York: Springer-Verlag.
48. Tingey, D.T., C. Standley, and R.W. Field. 1976. Stress ethylene evolution: A measure of ozone effects on plants. *Atmos. Environ.* 10:969-74.
49. Turesson, G. 1922. The genotypical response of the plant species to the habitat. *Hereditas* 3: 211-350.
50. Turner, M.G. 1987. *Landscape heterogeneity and disturbance.* New York: Springer-Verlag.
51. Whitehead, C.W., and C.M. Switzer. 1963. The differential response of strains of wild carrots to 2,4-D and related herbicides. *Can. J. Plant Sci.* 43:255-62.
52. Winner, W.E., J.S. Coleman, C. Gillespie, H.A. Mooney, and E.J. Pell. 1991. Consequences of evolving resistance to air pollutants. In *Ecological genetics and air pollution,* ed. G.E. Taylor, Jr., L.F. Pitelka, and M.T. Clegg, pp. 177-201. New York: Springer-Verlag.
53. Woodwell, G.M. 1970. Effects of pollution on the structure and physiology of ecosystems. *Science* 168:429-33.
54. Wulff, R.D., and H.M. Alexander. 1985. Intraspecific variation in the response to CO_2 enrichment in seeds and seedlings of *Plantago lanceolata* L. *Oecologia* 66:458-60.

8

Air Pollution Effects on Plant Reproductive Processes and Possible Consequences to Their Population Biology

Roger M. Cox

INTRODUCTION

The important measure of Darwinian fitness is the number of offspring an individual contributes to the next generation. In higher plants, sexual reproduction provides seeds that offer the ecological advantages of dormancy and dispersal. Seeds also provide an array of individuals carrying recombinations of parental genetic material on which natural selection can operate. This chapter is concerned with the action of environmental factors, such as air pollution, on fertility through effects on pollen and stigma and their interactions at the time of pollination. Variation in the response of fertility to air pollutants within populations may change the levels of participation of individuals in the breeding systems. The effects on population genetics and population biology of such changes in fertility could have profound influences on the balance of natural ecosystems and their biological diversity.

Since industrialization, there has been a rapid increase in primary air pollutants, defined as those that have the greatest potential for affecting vegetation over the largest areas. Such pollutants are carbon dioxide (CO_2), and the phytotoxic gases sulfur dioxide (SO_2), hydrogen fluoride (HF), and nitrogen oxides (NO_x). Furthermore, the photochemical reactions between NO_x and hydrocarbons (HCs) produce secondary oxidant pollutants such as ozone (O_3) and peroxyacetyl nitrate (PAN). The contribution of these secondary pollutants poses a greater economic hazard to agriculture in some regions than do climate or insect pathogens (Treshow 1968). Transformations of these pollutants during atmospheric transport lead to formation of wet acidic deposition at sites remote from pollution sources.

Pfahler and Linskins (1983) described the various stages in the life cycle of higher plants that can be affected by chemical pollution. They also outlined the complexity in developing adequate testing methods to determine effects. These difficulties are much greater with tree species due to their size and long generation time.

This chapter describes how plant reproductive structures are exposed to air pollutants. Field observations of some suspected effects on plant reproduction will be described. In addition, approaches aimed at answering some basic questions concerned with both cause-effect and dose-response relationships will be described. Such basic questions are:

1. What are the major constraints in developing meaningful experimental protocols?
2. Can air pollutants be implicated in the reduction of reproductive potential of higher plants in the field by examination of effects in vitro?
3. What are the possible population consequences of air pollution effect on reproductive processes?
4. Can effects on reproductive processes of plants be manipulated to affect the quality or quantity of seeds and increase frequencies of desired traits in the progeny?

PLANT REPRODUCTIVE CYCLES AND AIR POLLUTION

Points of interaction of air pollutants with the reproductive cycles of plants must be understood before describing their effects on sexual processes. Such potential interactions with a generalized higher plant reproductive cycle were outlined by Smith (1981) and by Wolters and Martens (1987). These authors suggested that air pollutants would reduce pollen production, distribution, germination and tube growth, through influences on the flowers or strobili. Other reported influences were reduced seed germination and restricted seedling growth, while mature plants may suffer reduced flower and cone initiation.

Direct exposure must imply contact with the atmosphere where pollutant deposition can take place. It is the process of pollination that is most vulnerable to air pollutants, as it requires transport of the dehydrated pollen to the stigmatic surface which must be well exposed to the pollination vectors, either wind or insects, and therefore exposed to air pollution. During transport, the pollen may be directly exposed to gaseous pollutants, which may affect membrane integrity. Such membrane integrity is important in the pollen's ability to withstand the water stresses on further plasmolysis during transport or on hydration on the stigma. Exposure to wet-deposited pollutants may occur in transport or on the stigma. Pollutant deposition on the stigma may accumulate and affect the pollen directly or interfere with pollen stigma interactions.

In conifers such as pines, the germination of the pollen is delayed, the pollen being floated off the micropylar arms (stigmatic surface) by the pollination droplet, which is exuded from the micropyle like a guttation droplet about 10 days after cones emerge. This droplet occurs at night when the root pressure is high and is withdrawn back into the micropyle along with the incorporated pollen grains as stomates open and xylem tension increases. This process may be repeated with pollen grains being deposited in the pollen chamber behind the micropyle close to the surface of the female gametophyte. Within this chamber, pollen germination occurs with initial penetration of the female gametophyte by the pollen tubes prior to cessation of growth for the winter. In such conifers, pollen may be exposed to air pollutants while on the micropylar arms and when incorporated in the pollination droplet. This droplet may dissolve the accumulated deposition around the micropyle or dissolve gaseous pollutants directly from the atmosphere. In pines, the finite space within the pollen chamber allows only limited numbers (4-5) of pollen grains access to the female gametophyte, which commonly have three to five archegonia (Willson and

Burley 1983). This limitation in the numbers of pollen grains per female gametophyte means that any reduction in the viability or vigor of the pollen will result in reduced fertilization rates of the available ovules and therefore reduce the level of embryo competition. This polyembryony may permit competitive elimination of inbred zygotes or at least those with homozygous leathals without loss of the ovule (Willson and Burley 1983). In conifers, this reduction in fertilization may lead to reduced seed quality due to their characteristically high genetic loads. In addition, at markedly reduced pollen viabilities, ovule abortion will also increase while effects that reduce pollen vigor may result in production of empty seeds (Matthews and Bramlett 1986).

FACTORS AFFECTING DIRECT EXPOSURE OF REPRODUCTIVE STRUCTURES TO AIR POLLUTION

Location

The location of a plant population may affect the level of pollution deposition apart from that associated with gradient effects from local point sources. Altitude can affect both qualitative and quantitative aspects of pollution deposition, i.e., orographic cloud can enhance pollutant concentrations in precipitation 4 to 5 times over levels in rain in nearby valleys (Unsworth and Wilshaw 1989). Cloud acidity can be 4 times that of bulk precipitation and may contribute most acidity to ridgetop sites of the Blue Ridge Mountains (Sigmon, Gilliam, and Partin 1989). Marine fog may have similarly high levels of acidic pollutants (Weathers et al. 1986; Kimball et al. 1988). Deposition of acidic pollution to coastal forests by means of marine fog interception may be as high or higher than that of rain during the growing season (Cox, Spavold-Tims, and Hughes 1990).

Ozone has also been reported to increase in concentration with elevation (Lefohn, Shadwick, and Mohnen 1990); however, the magnitude of this effect is an open question due to the necessary corrections for pressure and temperature on deposition (Larson and Vong 1990).

Position in the Plant Community Structure

Unsworth and Crossley (1987) stated that as a consequence of deposition mechanisms, dominant trees that stood above the general level of the canopy and trees at the edges of a stand would be better collectors of air pollution deposition than the lower shrubs present. Tallest trees also experience highest wind speeds and unimpeded radiation. These canopy conditions will increase concentrations of wet-deposited pollutants from rain and especially cloud and fog by evaporation. This concentration of deposited pollutants increases the potential damage to the plant surfaces (Unsworth 1984) and presumably exposed reproductive structures. The tallest trees may also contribute the highest proportion of reproductive structures within a population, as was demonstrated in red pine and loblolly pine (Kozlowski and Constantinidou 1986 a, b). It is possible that some of these tall trees represent the fittest or most productive individuals and include the most valuable genotypes from a commercial point of view. A reduction in their contribution to future generations may seriously affect future productivity of the stand or population.

The physical positions of plants within the plant community or stand, and thus its exposure to air pollutants, can be due to its life form. Hutchinson and Harwell (1985) have suggested that life form classification such as that described by Raunkiaer (1934) could be used to evaluate the potential susceptibility of plants to abiotic stresses. Applying this life form analogy to the position of the reproductive structures, not just buds, it is certain that this would affect their exposure levels and the type of pollution input. For example, one should consider the differential sensitivity of two alpine plants, *Acomastylis rossii* and *Bistoria vivipora* to acid mist. The greater sensitivity of the former species was attributed to the basal position of its flower buds, which allowed water to pool around them. These buds, formed the previous season, were damaged due to a greater exposure to the wet-deposited acidity (Funk and Bonde 1986). Another example of the importance of position within the ecosystem is the location where significant amounts of throughfall or fog drip are input to the understorey and herbaceous layer. This input may be very different in composition from incident precipitation. Pollution concentrations in throughfall may vary widely between sites, depending on canopy composition (Abrahamsen, Horntveldt, and Tveite 1976; Richter and Granat 1987; and Sigmon, Gilliam, and Partin 1989). Mahendrappa (1983) compared the differential abilities of various eastern Canadian species to modify acidity of rain as it interacts with the canopy and becomes throughfall. These types of interactions must be assessed to determine impacts on understorey or ground flora, which tend to have pollen more resistant to acidity than deciduous canopy species (Cox 1983).

Morphology of Reproductive Structures

The morphology and orientation of flowering parts can influence the penetration of airborne chemical pollutants to sensitive reproductive structures (Cox 1983; Barrett and Bush 1991).

The unique vulnerability of wind-pollinated species is discussed below. However, pollination strategy may also play a part in exposure of pollen, i.e., cleistogamy, where the sensitive pollen is completely protected by floral parts. Flowers that are pendulous would be protected from wet deposition. In addition, diurnal patterns of flower opening and closing would offer protection to pollen and stigma during closure.

Phenology and Pollution Episodicity

The probability of direct exposure of reproductive parts to air pollution rests with the relationship of flowering phenology and the episodisity or seasonality of pollution. This probability must be considered when assessing potential effects of air pollution on plant reproduction in natural conditions. For example, species that have determinate flowering patterns, such as *Betula cordifolia* near the Bay of Fundy coast of New Brunswick, have individuals that have all their flowers exposed to acid marine fogs at the same time. These fog events vary in their acidity, i.e., pH 3.3, 3.75, and 4.7, which represent the 10th, 50th, and 90th percentiles of events during the period of birch stigmatic receptivity (Cox and Hughes 1989, unpublished). This would be in contrast with Newfoundland rain, which was recorded at pH 4.0 or greater 99% of the time (Sidhu 1983). In cases of species with indeterminate flowering patterns, at least a proportion of their flowers may avoid pollution events.

Aerodynamic Considerations in Wind-Pollinated Plants

The reproductive structures of wind-pollinated plants may be particularly vulnerable to air pollution since they have evolved mechanisms to increase pollen capture that incidentally increase pollution deposition. One such mechanism is the reduced diameter of collecting surfaces (stigma) that reduce boundary layer resistance for surface deposition (Whitehead 1983).

Niklas (1985) suggested that ovulate strobili have evolved shapes that act like wind turbines, directing pollen to the ovule, thus increasing pollen capture. The young needles around reproductive branches act like a snow fence, showering the receptive cone with pollen. In addition, the harmonic movement of branches in the wind together with the orientation of receptive ovulate cones maximize the intercepted volume of air space carrying pollen or pollutants.

The same structural characteristics of the coniferous canopy that increase interception of pollen may have incidentally increased nutrient input into conifer stands in the past from cloud, fog, or dust. The ability of conifers to intercept aerosol was demonstrated with the use of pollen as a surrogate in particulate deposition studies—deposition rates in conifer forests were higher than in deciduous forests. A dense conifer stand removed 80% of the pollen from the air within 100 m (Neuberger, Hosler, and Koemond 1967). With the rise in industrial pollution, this enhanced aerosol capture may now prove less beneficial and even threaten the sensitive pollen and pollination processes. This apparent targeting of the pollination system by air pollutants may be a generality with wind-pollinated plants.

FIELD OBSERVATIONS OF REDUCED REPRODUCTIVE POTENTIAL IN POLLUTED AREAS

Reductions in seed set, cone size, and pollen viability in conifers in areas high in acidifying pollution have been recorded (Podzorov 1970; Antipov 1970) and for red and white pine in particular by Houston and Dochinger (1977). Losses in reproductive potential also have been observed in polluted areas in relation to growth in Scotch pine provenances (Oleksyn et al., in press). Reduced seed production or cone dimensions linked to air pollution (SO_2 or O_3) have been observed in ponderosa pine, Scotch pine, white pine, and red pine, reviewed by Smith (1981). Luck (1980) also reported reduced seed production over a 6-year period in mature ponderosa and jeffrey pines that had been damaged by oxidant air pollution compared to undamaged trees. Other observations by Pospisil and Alferi (1987a, b) in the Beskids Mountains of Czechoslovakia indicate that air pollution stress negatively influences reproductive processes of Norway spruce. Pollen germination in this species, as well as teratological changes and unusual pollen tube branching, was related to air pollution at these mountain sites.

Another example of a field observation that led to laboratory investigations is the work of Kratky et al. (1974). They noted reduced fruit set and size of tomatoes grown at a site in the lee of an active volcano receiving acid rain at <pH 4.0. These rain acidities were later shown to inhibit tomato pollen germination.

The impacts on plant reproduction of other air pollutants such as fluorides have also been observed. Stanforth and Sidhu (1984) noted a 21- and 10-fold reduction of reproductive potential (seed production) of wild blueberry and raspberry, respectively, growing in sites receiving highest fluoridation levels close to the source. These losses were due to flower mortalities of 89% and 78% near the source, compared to 27% and 26%, respectively, at a distance of 18.7 km.

By their nature, field observations are at best correlative, indicating only possible cause-effect relationships with emissions. Such emissions are often mixtures of pollutant compounds and are subject to transformations during atmospheric transport. More rigorous experimental approaches are required to substantiate cause-effect relationships with reproductive processes. Effects of known concentrations and mixtures can be measured and dose responses computed. These responses can then be related to exposure and estimated doses in the field.

POTENTIAL OF POLLUTANTS TO AFFECT POLLEN FUNCTION

One aspect of a more rigorous experimental approach would be the determination of the direct in vitro effects of pollutant concentrations on pollen function. This would identify potential risks at ambient concentrations of pollutants that could be followed up with in vivo experiments.

Simulated Wet Deposition on Pollen

Early indications of pollen sensitivity to acidity were recorded by Cooper (1939) in a study of media requirements for fungal spores and pollen germination and by Brewbaker and Majumder (1961) in an in vitro study of the "population effect" and self-incompatibility inhibition. The effects of some principal air pollutants on pollen germination in various species together with response thresholds are reviewed in Table 8-1.

The examination of pollen sensitivity of 13 forest species to acidity (Cox 1983) revealed that, while little inhibition of in vitro germination occurred at pH 4.6 and above, many species were significantly inhibited ($p<0.05$) at pH 3.0 and below. It was further noted that all pollen tested was inhibited at acidities now occurring in wet acid deposition. Similar results were found for additional forest species by Van Ryn et al. (1986) (Table 8-1) and for *Picea glauca* (Moench) Voss (Sidhu 1983) (Table 8-1). Paoletti and Bellani (1990) supported the view that pollen of gymnosperms were more tolerant than those of angiosperms and that sulfuric acid in a 2:1 molar ratio with nitric acid was most harmful to the pollen of gymnosperm and angiosperms, respectively.

Kratky et al. (1974) observed reduced tomato crop yield in the lee of an Hawaiian volcano and showed that pollen viability was affected by culture of pollen in collected rain water of low pH. In addition to the acidity effects, temperature effects were also noted. Prewashing pollen with rain solutions before assay increased inhibitions at 13.3° C compared to incubations at 18.7° C. Hughes (1990) also found significant effects of temperature on pollen germination with two birch species, both with optima between 17 and 22° C, with a significant temperature, pH interaction apparent for *Betula papyrifera* but not for *B. cordifolia*. Pollen from two apple cultivars, Empire and Richard Red, were assayed for their response to acidity by Waldron and Craker (1987). They found that in vitro germination was completely inhibited at pH 3.0 and reduced by 60% and 40% respectively at pH 4.0. Karnosky and Stairs (1974) suggested that acidification of the agar growth media during a 4-hour fumigation of moist *Populus deltoides* pollen with 0.3-0.7 ppm SO_2 may have accounted for much of their inhibition.

Masaru et al. (1980) conducted a study of the response of *Camellia* pollen to inorganic components of acid rain. Their work verified the importance of H^+ ion inhibition of pollen tube growth by assaying pollen in three inorganic acids (HNO_3, HCl, and H_2SO_4) and their corresponding ammonium salts. The pollen tubes were markedly inhibited by pH 3.2. They also studied the tube growth responses to Mg, Mn, and Pb as nitrate salts. Although inhibitions were noted at 5.0, 13.0, and 17.0 μM for the metals, respectively, stimulations were noted for Pb and Mn at lower concentrations. Stimulation of pollen germination by low concentrations of metal ions was also noted for *Trillium grandiflorum* in response to 1.0 μM cadmium (Cox 1986a) and for copper (Cox 1988). Furthermore, copper (0.05-0.40 mg l^{-1}) acted antagonistically towards acidity effects on *Pinus resinosa* and *Oenothera parviflora* pollen but was shown to be synergistic to the effects of acidity in *Populus tremuloides* pollen (Cox 1988).

Strickland and Chaney (1979) were also able to demonstrate a significant inhibition of tube growth and stimulation of respiration in *Pinus resinosa* pollen at 0.045 μM Cd after

TABLE 8-1. In Vitro Pollen Germination Assays and Response Threshold Values for Acidity

Species	Treatment	No. of Popula-tion	No. of Indiv-iduals	pH at response threshold $^{\$}$	Reference
Acer saccharum	pH	1	1	3.9^{+}	Cox 1983
Populus tremuloides	pH	3	3	$3.95\text{-}3.61^{+}$	"
Oenothera parviflora	pH	2	3	$3.7\text{-}3.5^{+}$	"
Betula papyrifera	pH	4	29	$4.6\text{-}3.04^{+}$	Cox 1983, 1989
Pinus strobus	pH	4	25	$4.7\text{-}2.6^{+}$	" "
Maianthemum canadense	pH	1	1	3.55^{+}	Cox 1983
Trillium grandiflorum	pH	2	2	$3.5\text{-}3.2^{+}$	"
Prunus pensylvanica	pH	1	2	$3.4\text{-}3.3^{+}$	"
Diervilla lonicera	pH	1	2	$3.5\text{-}3.16^{+}$	"
Pinus banksiana	pH	1	1	3.2^{+}	Cox 1983, 1989
Pinus resinosa	pH	3	7	$2.95\text{-}2.75^{+}$	" "
Tsuga canadensis	pH	1	2	$2.96\text{-}2.92^{+}$	Cox 1983
Betula alleghaniensis	pH	8(provs)	31	$5.3\text{-}3.12^{+}$	"
Betula lenta	pH	1	2	3.8^{x}	Van Ryn, Lassoie, and Jacobsen 1986
Betula alleghaniensis	pH	1	2	3.8^{x}	" "
Cornus florida	pH	1	2	3.4^{x}	" "
Acer saccharum	pH	1	2	3.8^{x}	" "
Sambacus nigra	pH	1	1	4.5^{*}	Paoletti and Bellani 1990
Pinus cembra	pH	1	1	3.5^{*}	" "
Picea glauca	pH	1	1	3.6^{*}	Sidhu 1983
Petunia spp.	pH	1	1	4.5*	Brewbaker and Majumder 1960
Betula papyrifera	pH	1	12	4.30^{+}	Hughes 1990
Betula cordifolia	pH	1	12	3.76^{+}	"

$^{\$}$ = point on the dose response curve that first gives a significant reduction in germination
$^{+}$ = effective dose, probit
^{x} = estimated threshold from actual values
* = significant diff. from pH 5.6 control $P<0.05$

3.5 h, whereas the sensitivity of *Petunia alba* pollen to 0.2 μM Cd was demonstrated by Kapur and Malik (1976). Although these concentrations are in excess of those in rain water (Galloway et al. 1982), they may approach those levels expected to accumulate on plant surfaces from combined dry and wet deposition (Lindberg et al. 1981; Unsworth 1984), especially near industrial or urban areas intercepting cloud or fog.

These results indicate that pollen from a variety of species responds to the chemical composition of wet deposition within its environment. This is not unexpected, as hydrated pollen is also receptive to many chemical recognition signals that take place on the stigmatic surface (Heslop-Harrison 1975). If exposure of the pollen to deposited pollutants occurs on the stigma for long enough to affect the pollen before significant buffering or alteration occurs, there is a clear potential for effects on the reproductive system. In addition, any chemical interference by wet-deposited pollutants with the array of interactions between pollen and stigma may have more subtle effects such as interference with chemical recognition signals and adherence that can also influence the breeding system.

SO_2 Effects on Pollen

The effects of sulfur dioxide (SO_2) on pollen germination in vitro are reviewed in Table 8-2. The inhibitory effects of SO_2 on pollen viability have been known since the pioneering work of Sabachnikoff (1912). Dopp (1931) assayed SO_2 sensitivity of germination in pollen from many species under different humidities. Some threshold concentrations, beyond which a significant inhibition of pollen germination is caused, obtained from Dopp (1931) are shown in Table 8-2. Clear demonstrations of the increased toxicity of SO_2 under high relative humidity (RH) conditions were given by Karnosky and Stairs (1974) and by Varshney and Varshney (1981) (Table 8-2).

Short fumigations of SO_2 up to 4 hours required response thresholds of 0.3 ppm and above. Longer fumigations of three pine pollens, for example (Table 8-2), were substantially reduced in viability by concentrations ranging from 0.02 to 0.075 ppm of the gas (Keller and Hedwig 1984). Also noted was that stored pollen of *Pinus mugo* and *P. sylvestris* seemed more sensitive to 0.075 ppm than fresh pollen. Beda (1982) also observed that fumigated pollen lost its viability sooner in storage than unfumigated pollen. Ma et al. (1973) determined chromatid aberrations in *Tradescantia* pollen fumigated with concentrations from 0.05 to 0.75 ppm—the damage was attributed to disruption of DNA metabolism. The pollen sensitivity of this species to hydrated forms of SO_2 or the gas was demonstrated by Ma and Khan (1976). Mitotic activity of the generative nucleus was inhibited by as little as 0.075 ppm SO_2, which was far more sensitive than pollen tube growth, with dose response between 50 and 750 ppm.

Both in vitro and in vivo fumigations of pollen of *Lepidium virginicum* under high RH conditions with 0.6 ppm SO_2 for 4 hours produced significant reductions in pollen germination (DuBay and Murdy 1983b). Also noted was the reduction of pollen grains retained on the stigma of fumigated plants. Fumigation of 11 clones of *P. sylvestris* with 170 and 270 μg M^{-3} SO_2 (von Scholz, Vornweg, and Stephan 1985) demonstrated much variability in the pollen response of the clones. This variability suggests short-term SO_2 fumigations can exert differential effects on the pollen from different individuals, thus altering their genetic contribution to future generations (fertility selection).

Masaru, Syozo, and Saburo (1976) fumigated *Lilium longiflorum* pollen with various combinations of SO_2, NO_2, O_3, H-CHO (formaldehyde), and CH_2=CH-CHO (acrolein). All

TABLE 8-2. In Vitro Pollen Germination Assays and Response Threshold Values for SO_2

Species	Pollutant	No. of Popula-tion	No. of Indiv-iduals	Concentration at response threshold $	Reference
Populus tremuloides	SO_2 (moist) (4h)	1	1	0.3 ppm*	Karnosky and Stairs, 1974
Pinus resinosa	"	1	1	"	"
Pinus nigra	"	1	1 pooled	"	"
Picea pungens	"	1	1	1.4 ppm*	"
Abies alba	SO_2 (16h)	1	1	0.075 ppm*	Keller and Bada 1984
Pinus sylvestris	SO_2 (16h)	1	1	0.075 ppm**	"
Pinus nigra	"	1	1	0.025 ppm*	"
Pinus mugo	"	1	1	0.075 ppm+	Von Scholz, Vornweg, and Stephan 1985
Pinus sylvestris	SO_2 (moist)	Clones	11	170-270 $\mu g\ m^{-3}$ x	"
Cicer arietemum	SO_2 (moist) (3-5h)	?	?	131 $\mu g\ m^{-3}$ x	Varshney and Varshney 1981
Nasturtium indicum	"	"	"	"	"
Petunia alba	"	"	"	"	"
Tradescanita axillaris	"	"	"	"	"
Lepidium virginicum	SO_2 (wet) 4h	1	1	0.6 ppm x	DuBay and Murdy 1983
Lupinus luteus	SO_2 (damp) (0.25h)	1	1	10.0 ppm x	Dopp 1931
Pinus malus	SO_2 (damp) (2.75h)	1	1	1.0 ppm x	"
Quercus rubra	SO_2 (damp) (0.25h)	1	1	10.0 ppm x	"

$ = concentration on a dose response curve that first gives a significant reduction in germination
* = significant diff from control treatment $P<0.05$
** = significant from control treatment $P<0.01$
\+ = effective dose, probit
x = estimated threshold from actual values

gases individually reduced pollen tube growth with the exception of O_3. The threshold of response to SO_2, where further increase in concentration caused a significant inhibition in tube growth compared to the control, was 0.71 and 0.40 ppm for 1 to 2-h and 5-h exposures, respectively. The NO_2 threshold was 1.70 ppm for 1 to 5-h, whereas the thresholds for formaldehyde and acrolein were both 1.40 ppm for a 2-h exposure. Synergistic interaction occurred only between SO_2-NO_2, O_3-NO_2, and NO_2-H-CHO.

Ozone Effects on Pollen

Feder (1968) was the first to report significant inhibition of in vitro pollen germination and tube growth in the cigar-wrap tobacco varieties caused by O_3 fumigation. Pollen germination was significantly reduced (40-50%) by fumigation by 0.1 ppm for 5.5 to 24 hours. This exposure level also reduced pollen tube elongation by 50%. It was also noted that the effects were similar whether the pollen was exposed in vitro on agar disks or on the plant. Differential susceptibility of pollen from two varieties of tobacco and two varieties of *Petunia* was correlated to foliar relative sensitivity of the varieties (results in Table 8-3). This technique was suggested as a way of screening for O_3 resistance without the use of whole plants. As little as 0.03 ppm O_3 was also shown to induce a 50% increase in free amino acids in corn pollen (Mumford et al. 1972), while pollen tube growth was inhibited by 0.06 ppm O_3; this also increased peptide accumulation, which further increased with ozone dose. Changes in sugars and amino acid accumulations suggested autolysis of structural glycoproteins and stimulation of amino acid synthesis. Ultrastructural changes (lack of organelles except ribosomes) in peripheral cytoplasm of pollen of ozone-sensitive cultures of *Petunia* were noted after exposure to 0.50 ppm for 3 h at 25° C. Few pollen grains showed such changes in an ozone-tolerant variety. It was suggested by Mumford et al. (1972) that movement of organelles away from the plasma membrane would reduce cell wall development and hence tube growth in sensitive pollen. Feder (1981) outlined an ozone biomonitoring system based on pollen tube length distributions in O_3-sensitive and -resistant varieties of tobacco and *Petunia*. He also suggested that the system could be used for biomonitoring other gases and their mixtures in the atmosphere.

Pollen from ozone-sensitive varieties of tobacco and *Petunia* exposed to ozone (0.120 ppm for 3 h) and UV-B (300 UM cm^{-2} for 30 min.) separately and in combination demonstrated that the combination inhibited pollen tube growth more than the two individual treatments (Feder and Shrier 1990).

Exposure of pollen of white pines in situ until anthesis did not affect their germination in vitro (Benoit et al. 1983). However, field-shed pollen exposed when wet demonstrated reduced germination in vitro. There was no relationship between foliar and pollen sensitivity, based on symptom evaluation of the trees, and the response of their pollen.

EXPERIMENTAL EVIDENCE OF DIRECT EFFECTS ON REPRODUCTION

The sensitivity of the reproductive systems of higher plants to air pollutants is, in part, due to the relatively high metabolic rate of germinating pollen and porous membranes necessary for the many pollen stigma interactions that characterize pollination (Pfahler and Linskens 1983; Sari-Gorla, Frova, and Redaelli 1986).

Effects of SO_2 on Stigma Receptivity

It was Dopp (1931) who first carried out SO_2 fumigation of pollen in vivo, demonstrating the effects of high concentrations on both germination and tube growth in vivo. He also reported the importance of humidity in enhancing the toxicity of the gas to pollen. It was also implicit in his report that the reduced toxicity of SO_2 noted for pollen on the stigma of a number of species was due to a buffering effect of the acidity of SO_2 in solution. Another study, under lower, more realistic concentrations, by Murdy and Ragsdale (1980) also suggested the increased inhibitory effects of the gas under high RH. High RH conditions were thought to emulate conditions on the stigma and this may be responsible for the

TABLE 8-3. In Vitro Pollen Germination Assays and Response Threshold Values for O_3 and Some Trace Metals

Species	Pollutant	No. of Population	No. of Individuals	Concentration at response threshold $	Reference
Pinus strobus	O_3 (moist)	1	2	0.15 ppm	Benoit et al. 1983
Tobacco (Bel-W3)	O_3 (moist) (5h)	variety	1	0.1 ppm[x]	Feder 1968 Feder and Sullivan 1969
Petunia (white Cascade)	O_3 (moist) (5h)	variety	1	0.1 ppm[x]	Feder and Sullivan 1969
Petunia (Blue lagoon)	"	variety	1	1.0 ppm[x]	"
Tobacco (Bel-B)	"	variety	1	1.0 ppm[x]	"
Zea maize	O_3 (moist) (2h)	variety	1	0.06 ppm[x]	Mumford et al. 1972
Pinus resinosa	Cu	1	1	0.0356 ppm*	Chaney and Strickland 1984
Acer saccharum	Cu	1	1	0.104 ppm[+]	Cox 1988
Betula alleghaniensis	Cu	1	1	0.252 ppm[+]	"
Diervilla lonicera	Cu	1	1	0.382 ppm[+]	"
Tsuga canadensis	Cu	1	1	0.440 ppm[+]	"
Pinus resinosa	Cu	1	1	0.496 ppm[+]	"
Populus tremuloides	Cu	1	1	0.530 ppm[+]	"
Betula papyrifera	Cu	1	1	0.579 ppm[+]	"
Pinus banksiana	Cu	1	1	1.449 ppm[+]	"
Pinus resinosa	Pb	1	1	1.808 ppm*	Chaney and Strickland 1984
Pinus strobus	Pb	1	1	0.05 ppm*	Cox 1988
Pinus resinosa	Cd	1	1	0.063 ppm*	Chaney and Strickland 1984
Trillium grandiflorum	Cd	1	1	0.061 ppm[x]	Cox 1986b
Pinus resinosa	Hg	1	1	0.223 ppm*	Chaney and Strickland 1984
Pinus resinosa	Zn	1	1	4.653 ppm*	"
Pinus strobus	Al	2	24	50µm/pH4.6	Cox 1986a

$ = concentration on a dose response curve that first gives a significant reduction in germination
x = estimated threshold from actual values
* = significant diff. from 5.6 control $P<0.05$
\+ = effective dose (probit)

inhibition of germination and initial pollen tube growth observed on exposed stigmas. Such inhibition of germination was demonstrated by DuBay and Murdy (1983a, b) in *Geranium arolinianum* at 90% RH and *Lepidium virginicum*. This is in contrast to the report of Facteau and Rowe (1981) who found no effect on cherry and apricot pollen in vivo using concentrations up to 6.0 ppm SO_2; however, there was no accounting of RH conditions or buffering by the stigma. Von Scholz, Vornweg, and Stephan (1985) demonstrated inhibition of poplar pollen function on the stigma by 900 µg SO_2 m^{-3}. However, a long-term exposure of *Abies alba* to 0.027-0.2 ppm SO_2 up to anthesis was a stimulation to subsequent pollen germination in vitro. Visible injury to generative organs in this case was only evident after leaves were affected (Beda 1982).

Effects of Ozone on Stigmatic Receptivity

The sensitivity of pollen to ozone was demonstrated by Feder (1968) using tobacco as well as other species, described above. In vivo inhibition of germination in *Lycopersicum esculentum* and *Petunia hybrida* pollen by 0.8 ppm O_3 was demonstrated after only 2 h of incubation in vitro but not with fresh pollen (Feder et al. 1982; Krause, Riley, and Feder 1975). This may indicate that pollen is sensitive only at a certain stage of germination. Ozone fumigation of reproductive branchlets of white pine with 0.1 ppm of the gas until anthesis did not influence pollen production or in vitro germination (Benoit et al. 1983).

Effects of Fluoride on Stigma Receptivity

Fumigation of tomato plants ("Michigan State Forcing") with HF gas at concentrations as low as 4.2 μg F m^{-3} reduced number of pollen grains retained on the stigma, reduced pollen germination on stigma, and reduced percentages of germinated pollen that produced tubes reaching the ovules. Sulzbach and Pack (1972) also noted that increased calcium nutrition reduced HF sensitivity and that exposure of both parents increased reproductive sensitivity. Similar results with reduced pollen germination and tube growth were found after in vivo HF fumigation of *Prunus avium* (Facteau, Wang, and Rowe 1973) and Tilton apricot (Facteau and Rowe 1977).

Effects of Wet-Deposited Pollution

In vivo effects of wet acid deposition on pollen function and stigmatic receptivity have been investigated. Cox (1984) demonstrated that a single exposure of the receptive stigmas of *Oenothera parviflora* to simulated rain (26 mm), at or below pH 3.6, prior to hand pollination, significantly reduced both pollen germination and the initial tube growth required to penetrate the stigmatic surface. Computed effective dosages of pH, that would reduce pollen germination and subsequent stigmatic penetration by 50% were pH 3.45 and 4.55, respectively. This indicates that initial pollen tube growth in this species is more sensitive to the acidified stigmatic environment than pollen germination. A similar study was carried out by Van Ryn, Lassoie, and Jacobsen (1988), using excised flowering branches of red maple (*Acer rubrum* L.). Here, both the number of pollen grains germinated and number of tubes that reached the bottom of the style decreased as acidity increased. They also reported that some pollen grains germinated on the stigma when treated with pH 2.6 rain simulant, indicating that some buffering had occurred, as in vitro assays had shown complete inhibition at pH 4.4 (Van Ryn, Jacobsen, and Lassoie 1986).

In corn (*Zea mays*), Wertheim and Craker (1987) noted that percent pollen germination on silks previously treated with simulated acid rain solution was directly related to acidity of the solution, while inhibitions were recorded for rain simulants of pH 4.6 and below. The reduced stigmatic receptivity produced by the pH 2.6 simulant was not reversible by rinsing silks in pH 5.6 simulant after 1.5 minutes.

Fruit crops such as grape and apple have also been investigated. Treatment of various cultivars of grape with simulated acid rain (pH 2.75) during anthesis reduced pollen germination in four cultivars. However, fruit set was only reduced in one cultivar (Forsline et al. 1983a). Similar experiments carried out by the same researchers (Forsline et al. 1983b) on apple cultivars found that the pH 2.5 rain simulant reduced pollen germination in the Empire cultivar while the McIntosh showed slightly reduced fruit set.

Such in vivo studies have also been carried out on native plants, such as *Populus tremuloides* L., using simulated wet acid deposition varying in trace element composition (Cox 1988). Rain simulants varying in pH and copper concentration were applied to catkins 1 h after pollination. The responses of pollen germination, fruit abortion, seed to fruit ratio, and the number of placenta with seeds were recorded. Results indicated an overall significant ($p<0.05$) reduction of in vivo pollen germination between pH 4.6 and 3.6 of about 25%, whereas the effects of copper were not significant, although inhibition by pH was more severe in the presence of copper. Pooled data indicated a highly significant ($p<0.01$) negative correlation between pollen germination and pH of rain simulant. Effects on seed set and fruit abortion are described below in the section on abortion.

In Vivo Effects on Seed Set, Fruit Set, and Fruit and Seed Abortion

Yield losses in the ozone-sensitive tomato cultivar (Tiny Tim) exposed to 80-100 ppm O_3 were related to interference with fruit set, which was ascribed to direct effects on pollen or pollination in the species (Manning and Feder 1976).

Craker and Waldron (1986) showed that acidity of simulated wet acid deposition prior to pollination with a low number of grains (9) would reduce seed set in corn. However, Dubay (1989) found that acidity of simulant applied before pollination with large numbers of grains (85 per silk) had no effect on seed set, while acidity of rain applied 1 hour after pollination resulted in a significant negative correlation with seed set. This negative effect of acidity occurred even though the pollen tubes had already penetrated the silks. These results suggest that wet-deposited acidity may affect corn stigma receptivity prior to pollination and tube growth after pollination.

In the study by Cox (1988) on effects of pH and copper on aspen reproduction described above, simulant pH when applied after controlled pollination of catkins was significantly correlated ($r=-0.325$, $p<0.05$) with fruit abortion. A 30% increase in fruit abortion rate was achieved with the acidities used, whereas copper concentration was not correlated with abortion. Both the seeds:fruit ratio and percentage placenta with developed seeds were highly correlated with fruit abortion, which in turn showed weakly significant correlations ($p=0.054$) with in vivo pollen germination. Also noted was a slight reduction of fruit abortion rates only in the absence of copper at the lowest pH (2.6). This may be due to the buffering effect on the stigmatic surface or induced change in stigmatic resistance to pollen tube penetration. It was concluded that the results support the hypothesis that inhibition of the pollen on the stigma by acidic wet deposition may reduce seed set beyond the minimum number required to prevent fruit abortion.

A similar experiment was conducted in a clonal seed orchard (Cox, in press) to study the effects of wet deposition chemistry on seed set in clones of a white pine (*Pinus strobus*) and a red pine (*Pinus resinosa*). Wet deposition simulants varying in acidity, lead, and copper concentrations were applied to receptive female cones 45 minutes after controlled pollinations. Significant ($p<0.05$) effects of pH treatments on seed set were found in both species but the trends were in opposite directions. The acidity (5.6-2.6) stimulated seed production (full-sized seeds) in red pine but inhibited seed set in white pine (Figure 8-1). This difference also relates to the in vitro sensitivity of pollen from the two species determined in previous experiments (described above), in which red pine was shown to have substantially lower pH optima for pollen germination and lower pH thresholds of response than those of white pine (Cox, in press).

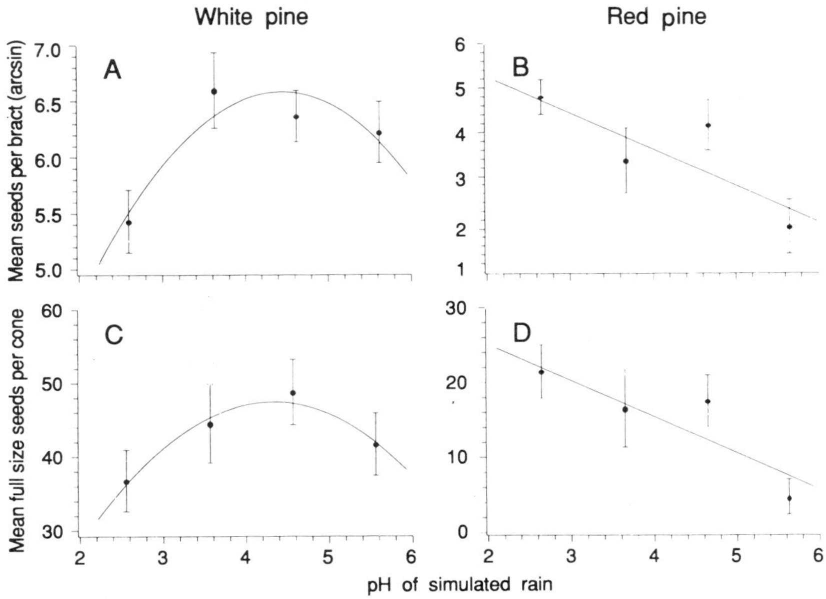

FIGURE 8-1. The effects of wet-deposited acidity after controlled pollination on seed set in clones of a white and a red pine. (Values are means of seven replicates with no more than three cones for each treatment ±1SE). A) & B) are responses of mean seeds per bract in white and red pine, respectively. C) & D) are responses of mean full-size seeds per cone in the respective clones. (Cox 1991)

Wet deposition of acidity at the time of pollination can directly stimulate or inhibit apparent pollination efficiency (APE) as defined by production of full-sized seeds (Matthews and Bramlett 1986). In addition, the continued stimulation of red pine APE after treatments with pHs (2.6) that cause substantial inhibition in vitro indicate a "buffering" effect in vivo. Pollination effectiveness in white pine showed a pH optimum that was obviously higher than that of red pine and similar to in vitro responses to pH (Figure 8-2). The differences in pollen response may be due to differences in pollination droplet pH, which has been shown to vary among white pine populations on acid and calcareous soils (Cox 1989). However, further examination of pollination droplet pH is required before final explanations of these differences can be made.

Controlled pollinations in conifers rarely give levels of seed set as high or higher than those achieved naturally (Willson and Burley 1983). Application of acidified spray treatment of specific pHs after pollination may be a method of increasing seed yield in pines following controlled pollinations or augmented mass pollination.

In an examination of the effects of simulated acid marine fogs on reproduction in two birch species under field conditions, Hughes (1990) demonstrated significant reductions of seed set (fertilized seeds) in *Betula cordifolia* using simulated fog of pH 3.0-5.6.

INDIRECT EFFECTS OF AIR POLLUTION STRESS ON REPRODUCTION

As early as the late 1960s, ozone (0.05-0.10 ppm) was known to retard floral initiation resulting in reduced flower production in cultivars of geranium, carnation, and petunia. In

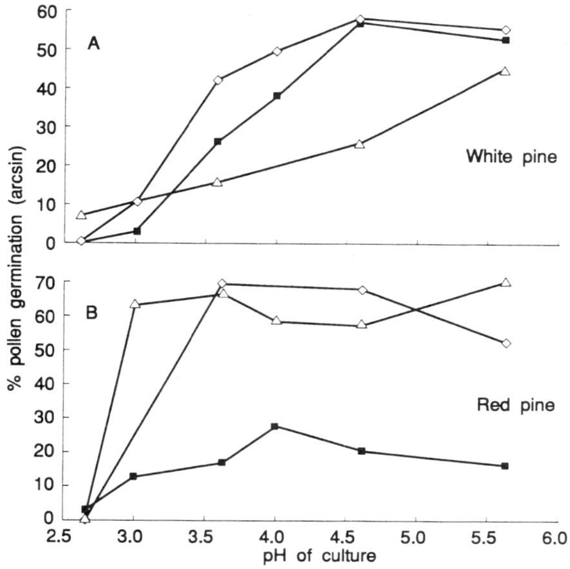

FIGURE 8-2. The in vitro germination response to acidity of A) pollen from three white pines from different populations and B) pollen from three red pines from different populations. (Values are means of seven replicates.) (Cox 1991)

geranium and carnations, the reduced flowering was attributed to reduced side branching (Feder 1970). In a similar experiment with *Lemna perpusilla*, reduction in frond production was accompanied by delayed floral development. These effects may be due to changes in carbon availability and or allocation.

Ernst, Tonneijk, and Pasman (1985), in an investigation of tolerance of native edaphic ecotypes of the herb *Silene cucubalus* to SO_2 and O_3, found complete inhibition of flower production by O_3 (70 μg m^{-3}) irrespective of gas combination or environmental conditions used. Multiple edaphic and atmospheric stresses, such as specific heavy metals in the soil, and air pollutant conditions (O_3, SO_2, and NO_2) similar to concentrations found at their sites of origin, were applied to ecotypes of *Silene cucubalus* (Dueck et al. 1987). Treatments resulted in significant effects on seed number, individual seed weight, and flower production, with the latter contributing most to the loss of reproductive capacity. Results generally suggested that plants in soil conditions most closely resembling their site of origin obtained optimal reproductive capacity, which exceeded that of those in control soil. The effects of the gaseous pollutant mixtures generally further decreased reproductive capacity except in the SO_2-tolerant ecotypes. In this case, flower production increased over those in carbon filtered air.

Yield loss in crops due to air pollution varies according to pollutant and species (Reinert and Heck 1982; Reich 1987; Cooly and Manning 1987) and with cultivar (Heggested et al. 1986). The effects of ozone on reproductive components of crop yield have been investigated and reviewed by Manning and Feder (1976) and Cooly and Manning (1987). They concluded that reproductive sinks of carbon become stronger as the plants mature and, while reduction in the number of fruits or seeds occurred from either direct effects on the reproductive system or indirectly from a lack of carbon availability, the remaining fruits and seeds were comparable in weight with those of unstressed plants. Possible exceptions include pepper (*Capsicum anuum*) fruit weight and cotton (*Gossypium hirsutum*) boll weight; however, precise information on seed set was not available. In addition, the linear decline in soybean yield caused by low levels of O_3 and/or SO_2 exposure was due both to reduction of pods per plant and reduction of seeds per pod (Reich and Amundson 1984). The impact of ozone on yield of corn hybrids consisted of both reduction of seed weight and numbers (Kress and Miller 1985). Seed sterility in panicles of rice cultivars and reduced seed weights have also been reported in response to ozone (Kats et al. 1985).

The indirect effects of air pollution, such as carbon limitation or induced reallocation of carbon, are often confounded with direct effects on specific reproductive processes in experiments designed to assess crop losses. This suggests that evaluation of such factors as flower initiation and abortion of reproductive structures, as well as effects of pollutants on reproductive processes, should be included in experimental designs if knowledge of specific mechanisms of crop loss is required. This would further our ability to predict the genetic consequences of air pollution stress not only in our selection for fertility in inbred lines for hybrid seed production, but also on the consequences of natural selection for fertility in mass-pollinated crops and their native relatives.

Most experimentation on the effects of air pollutants on trees have been conducted to date on seedlings and thus did not involve reproductive individuals. However, work with larger trees under ozone stress has indicated that carbon is reallocated from the older leaves to the younger leaves, while the older leaves also show early senescence. This occurs while carbon is reduced in the older tissues such as the bole and the roots due to reduced export from the foliage (McLaughlin et al. 1982; Cooley and Manning 1987). These researchers also suggested that changes in partitioning would be more damaging to the health of the trees than photosynthesis reduction. The relative strength of seed carbon sinks in forest trees under air pollution stress is little studied, as are many processes that are potentially affected in the canopy. The constraints in developing in situ methods of investigation of canopy processes including reproduction are mainly logistic in nature. Access to the canopy of mature trees is difficult and expensive, while the necessity for replicating treatments precludes or limits population level studies. Other difficulties being addressed with the use of branch chambers are the precise control over gaseous concentrations delivered to the canopy to allow duplication of experiments, reduction of chamber effects, and adequate protection of fruits and cones from pests. However, the knowledge gained by such canopy studies would be of considerable value in determining the potential for reproductive effects of air pollutants and the mechanisms of such effects in organisms that comprise a major component of terrestrial ecosystems.

POLLUTION AND POLLINATOR INTERACTIONS

One reported injury that may reduce pollinator attraction is "dry sepal" of orchids caused by ethylene at concentrations as low as 2.0 ppb. Carnations are also similarly affected at higher concentrations of ethylene (10 ppb). This pollutant is released in the production of certain plastics and from car exhausts; it can also cause failure of flower buds to open or to abscise (Heggested 1968).

Artificial pollinations increase seed set in many boreal forest herbs, suggesting that these are pollinator limited (Barrett and Helendrum 1987). Any reduction in pollinator activity or population levels could reduce reproductive potential and dispersal of these plants. Ostler and Harper (1978) noted that forested ecosystems with a predominance of animal-pollinated species are our most diverse ecosystems. Smith (1981) suggested that adverse impacts on pollinators might result in competitive weakness and reduction in species diversity. Few studies have been carried out on the effects of air pollution on pollinators; however, Dewey (1973) noted increased body burdens of fluoride in pollinators within 1 km of an aluminum plant. Another way pollinators may be affected is by the ingestion of contaminated pollen or nectar. Ernst and Bast-Cramer (1980) found elevated heavy metals in pollen and honey from two Compositae species growing on contaminated soil or in contaminating environments such as near heavy road traffic.

CONSEQUENCES OF POLLUTION EFFECTS ON PLANT REPRODUCTION

Fertility Selection

Reproduction is a decisive bottleneck in the transmission of genetic information to the next generation. Genetic variation in reproductive response to air pollution can lead to differential fertility selection. The importance of environmentally dependent genetic selection of sex-specific fitness components was stressed by Muller-Starck and Ziehe (1984). Such fitness components could be pollen tolerance to low or high temperatures or to specific pollutants. These authors also noted the probability that such selection would affect ecological and economic characteristics of the offspring via linkage and epistatic effects. By testing the effects of specific air pollutants on certain steps in the reproductive processes, we may begin to understand how fertility responds to air pollution. Fertility selection may have other indirect genetic effects through reduced effective population size (breeding members of the population) at reproductive events. This may be important in small populations if individuals exhibit sexual asymmetry or if flowering is irregular (Muller-Starck and Ziehe 1984; Muona 1990), but may be less important in larger populations during mast years where many individuals are reproductive at the same time. There are examples where effective population size has been estimated from breeding individuals in the population by determination of variance in allozyme frequencies at neutral loci (Crow and Denniston 1988). Estimates of effective population size (N_e) thus obtained have also been compared with actual size (N) in subpopulations of *Avena fatua* (Jain and Rai 1974). The average effective sizes of the *Avena* subpopulations were in every case smaller than the actual population sizes. Barrett and Husband (1990) have suggested that the magnitude and the direction of the differential between these measures of population size will vary, both within and among species, depending on their life histories, reproductive systems, and colonization patterns. Estimates of N_e/N ratios for some irregularly flowering forest trees are low (0.15-0.35) (Muona 1990). Populations of such species may be at greater risk of the genetic consequences of small effective population size due to additional variation in fertility induced by air pollutants. Reduction in the effective population size may lead to loss of rare alleles and reduced heterozygosity through effects of inbreeding and genetic drift.

Loss in Pollen Quality or Quantity

The literature on actual seed yield resulting from controlled pollination of trees is sparse in part due to the complications of insect damage (Bramlett and O'Gwynn 1981). The development of intensive insect protection (DeBarr, Bramlett, and Squillace 1975) enabled the effect of pollen limitation to be investigated (Bramlett 1977). Results with slash and loblolly pines indicate greater seed yields and larger cones with increased quality and quantities of pollen applied to the cones (Bramlett and O'Gwynn 1981; Matthews and Bramlett 1986; Moody and Jett 1990).

The effect of reduced pollen viability on seed set in loblolly pine was investigated by Matthews and Bramlett (1986), who concluded that yield of developed seed and apparent pollination effectiveness were more strongly influenced by increased viability than by increased pollen quantity. In addition, Moody and Jett (1990), after testing pollen of different ages and viabilities, found loblolly pine pollen viability and vigor were good predictors of total seed set. These authors suggested that pollen quality can affect seed set in two ways: first, viability, which indicated the ability of the pollen to germinate on the nucellus and initiate seed formation; second, vigor, which is the ability of the germinated grain to complete fertilization. It was further suggested that viable pollen of low vigor could produce large quantities of empty seeds.

Reductions in the ability of pollen to germinate and to penetrate the stigma caused by air pollutants occur on stigmatic surfaces with only limited indications of buffering from within the stigma (Cox 1984, 1988). These acid deposition effects reduce the level of pollen competition by decreasing effective pollen loadings, which can result in reductions in progeny fitness components such as vigor (Mulcahy 1979; McKenna 1986; Stephenson, Winsor, and Davis 1986; Lee and Hartgerink 1986).

Seed Production

Willson and Burley (1983) reported that only one out of five pines investigated gave more seeds under controlled pollinations than under open pollination. This low success with controlled pollinations was assigned to technical difficulties with wind pollination systems. One such difficulty with pollination bags is the role of wet deposition as an aid and as an augmentation to the pollination droplet mechanism in pines as suggested by the study of the effects of simulated wet deposition on seed set in red and white pine (Cox, in press). This augmentation of the mechanism in red pine may occur under more acid conditions than in white pine and may relate to the pH of the pollination droplet (Cox 1989). Wet deposition as an aid to the pollination droplet mechanism was also mentioned by Greenwood (1986) after investigations with the pollination mechanism of loblolly pine. He suggested this mechanism to explain the "first come, first served" effect reported by Franklin (cited in Greenwood 1986) with regard to the effectiveness of early pollination by long distance-transported pollen. These results suggest that simulated wet deposition of suitable pH applied to conelets after controlled or mass pollination may be a way of improving seed yields in pine breeding orchards or breeding halls.

Genetic Overlap and Gametic Selection

Correlation between sporophyte and microgametophyte tolerances has been demonstrated (Feder and Sullivan 1969; Zamir, Tanksley, and Jones 1982; Bowman 1986; Searcy and Mulcahy 1986). In addition, genetic variation in air pollutant sensitivity of eastern white pine and other tree species (Berry 1973; Houston and Stairs 1973; reviewed by Gerhold 1977; Karnosky 1989) is apparent. This information implies the presence of ample variation to allow natural selection within populations of pollen grains (microgametophytes). Such direct selection by air pollution, should it occur at the time of pollination, may result in changes of gene frequencies in the progeny that would be adaptive if gametophytic and sporophyte environments correlate but would reduce fitness where the inverse correlation exists, i.e., selection for acid tolerance in a calcareous environment (Cox 1989).

Moran and Griffin (1985) provided evidence of nonrandom contribution of pollen to progeny of polycrosses of *Pinus radiata*, and suggested gamete competition as a possible cause. This competitive element may emphasize the advantages of any differences in tolerances of individual pollen grains to air pollution and may change existing male fitness components within populations.

Age Structure

Effects of quantity of seed produced in a population may have demographic effects. As a rule, trees do not have significant seed longevity, with the exception of some specialists (Harper 1977). Following a disaster, such as fire, age structure of the colonizing population may depend on proximity of seed parents and availability of dispersed seed. If parents are distant or are inhibited in their reproduction by air pollution, founding individuals will have to grow to reproductive age before full stocking by the particular species occurs. This will result in mixed age classes. Here founding individuals recruited in the presence of air pollution may be better adapted to pollutants due to fertility selection of parents and cohort selection of seedlings. However, such founding populations will probably have less genotypic variation. This assumes, of course, that the stocking rates are sufficient for replacement of the original populations as described for loblolly pine by McQuilkin (1940) reviewed by Harper (1977).

In shade-tolerant tree species, regeneration from large-scale disturbances (with the exception of fire), such as hurricanes, spruce budworm, or other pest outbreaks, is from a reserve of established seedlings. These seedlings can be recruited over a relatively long period. In this case, there would be little effect on regenerative capacity due to reduced reproductive potential caused by air pollution unless effects on fertility were constant over time. Regeneration from an aging residue of formerly suppressed seedlings offers less opportunity for an evolutionary response to environmental change unless fire or similar disturbance removed the seedling/sapling stock. With the frequency of natural disturbances, it is open to question what sort of age structure is expected at "equilibrium" if indeed this is ever achieved. Even in old-growth forest, phases of recruitment and mortality are evident (Harper 1977). Departures from stable distributions of age and size classes appear typical of many populations (Platt, Evans, and Rathburn 1988).

Age distribution certainly affects reproduction within a population due to age-specific fecundities. Longleaf pine, for example, increases cone production as it ages. In addition, year-to-year variation in cone production declines with age (Platt, Evans, and Rathburn 1988). This development in cone production was in contrast to other species investigated (Enright and Ogden 1979; Pinero, Martinez-Ramos, and Sarukhn 1984). Platt, Evans, and Rathburn (1988) also suggested that any single age or size distribution is unlikely to characterize stable populations of long-lived conifers and that small effects on growth rates can produce large shifts in stable age structures. These factors lead to the suggestion that the effective size of the breeding population may be affected by subtle changes in growth rate due to air pollution.

CONCLUSIONS

Many factors are involved in the exposure of plant reproductive structures to effects of air pollution. Factors such as altitude and latitude influence extent and concentration of pollutant deposition. In addition, ecological factors governing position of plants in the plant community and the position of reproductive structures on the plant are important in determining exposure to air pollutants. Morphological structures that have evolved in association with breeding systems are also of importance. These attributes of plants and their populations, together with the timing of reproduction in relation to pollution events, must be considered when assessing air pollution effects on plant reproduction in natural conditions.

Field observations of natural populations suggest that reproductive potential of some species is being affected by levels of pollutants in the atmosphere derived from both local point sources and regional sources. However, these observations must be backed up by sound experimentation to determine cause-effect and dose-response relationships.

There are logistic constraints in designing experiments to treat reproductive structures of forest trees in situ. However, with trees that take months or even years to complete their reproductive processes, there is little choice other than in situ experiments. Access to the canopy of mature trees is expensive, as is the delivery of pollutant gases to the reproductive structures. The use of branch chambers with replicate treatments allows some hypotheses to be tested if chamber effects can be minimized. However, investigations at the population level are limited and have to be conducted over different years, using reproducible pollutant exposure regimes. These experiments require precise control of concentrations of delivered gases and the use of air previously filtered to remove pollutants. Another constraint is the often irregular flowering of forest trees. These difficulties require that these studies be long term, and combined with other studies, such as pollutant effects on photosynthesis or carbon allocation. These other studies are also necessary to determine indirect effects of the pollutants on reproduction. There are fewer constraints in carrying out such experiments with agricultural crops, except the number of crops and cultivars involved. The open-topped field chamber and different chamberless techniques have been put to use, as well as various laboratory chambers. However, these experiments, designed to investigate yield effects, often confound the direct effects on reproductive processes.

One way to attempt to implicate specific pollutants in the observed decreases in reproductive potential was to conduct in vitro pollen assays. These assays of pollen germination and tube growth indicated a potential risk to plant reproduction in a variety of species. These risks may result from deposition of pollutants in the pollen environment expected from both local sources and long-range transport. In vitro sensitivity of pollen to pollutants was also reflected in vivo with only limited buffering from the stigma; hence, variation in pollen sensitivity, screened in vitro, can indicate sensitivities in natural conditions. In addition, variation that exists within species and between species in pollen sensitivity to specific pollutants may be used as sensitive air pollution bioindicators.

Variation in pollen response to air pollution has implications for gamete selection and fertility selection, which in turn affects genetic structure of populations. Failure of seed set is a symptom of interference with fertility of individuals in the population. The subtle action on fertility selection by environmental stressors such as air pollutants will reduce effective size of breeding populations leading to increases in inbreeding and genetic drift. Critical effective sizes are largely unknown for natural plant populations. However, there are some indications that some conifer species have low effective population size to actual population size ratios and would be most at risk if further decreases in the number of interfertile individuals were induced by air pollutants. These additional stresses placed on the breeding system of trees by air pollution, together with restrictions of actual population sizes by changes in land usage and forestry practices, increase the chance of inbreeding and the risk of eventually reducing genetic multiplicity through increased homozygosity. This multiplicity seems essential for the adaptiveness required of long-lived forest tree species. The maintenance of genetic multiplicity would be increasingly important to trees if they are to cope with the accelerating atmospheric and climatic changes.

In mass-pollinated agricultural crop cultivars, changes in fertility induced by air pollutants over generations may gradually change genetic structure and affect yields. This must lead to consideration of genetic conservation of agricultural crops and forest tree gene pools in polluted breeding areas, as well as genetic conservation of important native populations thus threatened. Expression of genetic traits in the pollen (gametophyte generation) and the plant (sporophyte generation) concerned with tolerance of certain environmental conditions has been demonstrated. This genetic overlap suggests capability for manipulation of the pollen environment at the time of pollination to select for certain traits in the pollen that can increase the proportion of these traits in the progeny. Potentially useful genetic traits showing genetic overlap are heat shock resistance, cold tolerance, vigor, herbicide tolerance, and tolerance to trace metals. This would represent the applied use of gamete selection in biotechnology and provides much potential for plant breeding if the required traits can be found and their genes or expression selected for in the pollen. Forestry has much to gain from gametic selection due to the long generation times involved in tree breeding. The study of gametic selection by environmental factors during plant reproduction in natural populations will also increase our understanding of the importance of this process in the evolutionary response of plants to changes in the physical and chemical climate.

Another possible manipulation of the reproductive processes of pines is the use of sprays of suitable pH which may be used in concert with controlled crosses or augmented mass pollination to increase seed yield in seed orchards or breeding halls. The importance of pH in the reproductive processes of conifers was demonstrated by the increased seed yields following controlled pollination in the presence of certain wet deposition regimes.

Differential effects of air pollutants on reproduction in different species, be it direct or indirect, has long-term implications to species dispersal, competitive ability and survival, with potential effects on the biodiversity of ecosystems intercepting air pollutants.

If natural selection occurs in the pollen population at the time of pollination, changes in the frequency of genotypes in the progeny may be expected. These changes may be adaptive if the micro-gametophyte environment correlates with that of the sporophyte. The importance of natural selection in gametic populations should be investigated as it may contribute significantly to the evolutionary adaptation of plants to their environment. Fertility selection by air pollutants, however, can reduce the effective population size, thus increasing the risks of inbreeding and reducing genetic multiplicity. This may reduce genetic fitness of long-lived forest trees and increase the risk of declines, which strongly impact biodiversity. This risk increases with the additional stresses of the predicted accelerated change in climate. Tree declines are not well understood and much work needs to be done to relate them to genetic parameters such as levels of individual multiplicity. Such a model will give insight into not only the causes of declines in tree populations but also the importance of biodiversity and multiplicity at higher levels of biological organization such as communities and ecosystems.

ACKNOWLEDGMENTS

I thank S.C.H. Barrett, J.M. Bonga, J.A. Loo-Dinkins, and K.E. Percy for their comments on the manuscript, and Forestry Canada—Maritimes Region for their support.

REFERENCES

1. Abrahamsen, G., R. Horntveldt, and B. Tveite. 1976. Impacts of acid precipitation on coniferous forest ecosystems. In *Proc. 1st Int. Symp. Acid Precipitation and Forest Ecosystem*, pp. 991-1009. USDA Forest Service, Gen. Tech. Rep. No. NE-23, Upper Darby, Pennsylvania.
2. Antipov, B.G. 1970. The effects of SO_2 on the reproductive organs of woody plants. Ohrana priody naturale. *Sverdlovak* 7:31-35 (*For. Abstr.* 32 (4): 752).
3. Barrett, S.C.H., and E.J. Bush. 1991. Population processes in plants and the evolution of resistance of gaseous air pollution. In *Ecological genetics, terrestrial vegetation and anthropogenic changes in the atmosphere*, ed. M.T. Clegg, L.F. Petelka, and G.E. Taylor, Jr. New York: Springer-Verlag.
4. Barrett, S.C.H., and K. Helendrum. 1987. The reproductive biology of boreal forest herbs. I. Breeding systems and pollination. *Can. J. Bot.* 65:2036-46.
5. Barrett, S.C.H., and B.C. Husband. 1990. The genetics of plant migration and colonization. In *Plant population genetics, breeding, and genetic resources*, ed. A.H.D. Brown, M.T. Clegg, A.L. Kahler, and B.S. Weir, pp. 254-77. Sunderland, MA: Sinauer Assoc.
6. Beda, H. 1982. Der Einfluss einer SO_2 Begasung auf Bildung und Keimkraft des Pollens von Weisstanne, *Abies alba* (mill.). *Mitt. Eidg. Anst. Forstl. Versuchswes.* 58:163-223.
7. Benoit, L.F., J.M. Skelly, L.D. Moore, and L.S. Dochinger. 1983. The influence of ozone on *Pinus strobus* L. pollen germination. *Can. J. For. Res.* 13:184-87.
8. Berry, C.R. 1973. The differential sensitivity of eastern white pine to three types of air pollution. *Can. J. For. Res.* 3:543-47.
9. Bowman, R.N. 1986. Factors influencing artificial gametophytic selection using synthetic stigmas. In *Biotechnology and ecology of pollen*, ed. D.L. Mulcahy, G.B. Mulcahy, and E. Ottaviano, pp. 113-18. New York: Springer-Verlag.
10. Bramlett, D.L. 1977. Pollen quantity affects cone and seed yields in controlled slash pine pollinations. 14th South. For. Tree Improv. Conf. Proc., pp. 28-34.
11. Bramlett, D.L., and C.H. O'Gwynn. 1981. Controlled pollination. In *Pollen management handbook*, ed. E.C. Franklin, pp. 44-51. USDA Forest Serv. Agric. Handbook No. 587.
12. Brewbaker, J.L., and S.K. Majumder. 1961. Cultural studies of pollen population effects and self-incompatibility inhibition. *Am. J. Bot.* 48:457-62.
13. Chaney, W.R., and R.C. Strickland. 1984. Relative toxicity of heavy metals to red pine pollen germination and germ tube elongation. *J. Environ. Qual.* 13 (3): 193-94.
14. Cooly, D.R., and W.J. Manning. 1987. The impact of ozone on assimilate partitioning in plants: A review. *Environ. Pollut.* 47:94-113.
15. Cooper, W.C. 1939. Vitamins and the germination of pollen grains and fungal spores. *Bot. Gaz.* 100:844-52.
16. Cox, R.M. 1983. Sensitivity of forest plant reproduction to long-range transported air pollutants: In vitro sensitivity of pollen to simulated acid rain. *New Phytol.* 95:269-76.
17. Cox, R.M. 1984. Sensitivity of forest plant reproduction to long-range transported air pollutants: In vitro and in vivo sensitivity of *Oenothera parviflora* pollen to simulated acid rain. *New Phytol.* 87:67-70.

18. Cox, R.M. 1986a. In vitro and in vivo effects of acidity and trace elements on pollen function. In *Biotechnology and ecology of pollen,* ed. D.L. Mulcahy, G.B. Mulcahy, and E. Ottaviano, pp. 95-100. New York: Springer-Verlag.
19. Cox, R.M. 1986b. Contamination and effects of cadmium in native plants. *Experientia suppl.* 50:101-09.
20. Cox, R.M. 1988. Sensitivity of forest plant reproduction to long-range transported air pollutants: the effects of wet deposited acidity and copper on the reproduction of *Populus tremuloides. New Phytol.* 110:33-38.
21. Cox, R.M. 1989. Natural variation in sensitivity of reproductive processes in some boreal trees to acidity. In *Genetic effects of air pollutants in forest tree populations,* ed. F. Schulz, H.R. Gregorius, and D. Rudins, pp. 77-88. Berlin: Springer-Verlag.
22. Cox, R.M. In press. The effects of wet deposition chemistry on reproductive processes in two pine species: Pollination effectiveness in relation to species pollen sensitivity. *Water Air Soil Pollut.*
23. Cox, R.M., J. Spavold-Tims, and R.N. Hughes. 1990. Acid fog and ozone: Their possible role in birch deterioration around the Bay of Fundy, Canada. *Water Air Soil Pollut.* 48:263-76.
24. Craker, L.E., and P.F. Waldron. 1986. Pollen number as a contributing factor to an acid rain/seed set effect in corn. In *Agron Astracts,* p. 93. Madison, WI: ASA.
25. Crow, J.F., C. Dennison. 1988. Inbreeding and variance: Effective population numbers. *Evolution* 42:482-95.
26. DeBarr, G.L., D.L. Bramlett, and A.E. Squillace. 1975. Impact of seed insects on controlled-pollinated slash pine clones, pp. 177-81. 13th South. For. Tree Improv. Conf. Proc.
27. Dewey, J.E. 1973. Accumulation of fluorides by insects near an emission source in western Montana. *Environ. Entomol.* 2:179-82.
28. Dopp, W. 1931. Uber die Wirkung der Schwefligen Saure auf Blutenorgane. *Ber. Dtsch. Bot. Ges.* 49:173-221.
29. DuBay, D.T. 1989. Direct effects of simulated acid rain on sexual reproduction in corn. *J. Environ. Qual.* 18:217-21.
30. DuBay, D.T., and W.H. Murdy. 1983a. Direct adverse effects of SO_2 on seed set in *Geranium carolinianum* L.: A consequence of reduced pollen germination on the stigma. *Bot. Gaz.* 144:376-81.
31. DuBay, D.T., and W.H. Murdy. 1983b. The impact of sulphur dioxide on plant sexual reproduction: in vivo and in vitro effects compared. *J. Environ. Qual.* 12:147-49.
32. Dueck, Th. A., H.G. Walting, D.R. Moet, and F.J.M. Pasman. 1987. Growth and reproduction of *Silene cucubalus* Wib. intermittently exposed to low concentrations of air pollutants, zinc, and copper. *New Phytol.* 105:633-45.
33. Enright, W., and J. Ogden. 1979. Application of transition matrix models in forest dynamics: *Araucaria* in Papua New Guinea and *Nothofagus* in New Zealand. *Aust. J. Ecol.* 4:3-23.
34. Ernst, W.H.O., and W.B. Bast-Cramer. 1980. The effects of lead contamination of soils and air on its accumulation in pollen. *Plant Soil* 57:491-96.
35. Ernst, W.H.O., A.E.C. Tonneijk, and F.J.M. Pasman. 1985. Ecotypic response of *Silene cucubalus* to air pollutants (SO_2, O_3). *J. Plant Physiol.* 118:439-50.
36. Facteau, T.J., and K.E. Rowe. 1977. Effect of hydrogen fluoride and hydrogen chloride on pollen tube growth and sodium fluoride on pollen germination in Tilton apricot. *J. Am. Soc. Hortic. Sci.* 102:95.
37. Facteau, T.J., and K.E. Rowe. 1981. Response of sweet cherry and apricot pollen tube growth to higher levels of sulfur dioxide. *J. Am. Soc. Hortic. Sci.* 106:77-79.

38. Facteau, T.J., S.W. Wang, and K.E. Rowe. 1973. The effect of hydrogen fluoride on pollen germination and pollen tube growth in *Prunus avium* L. Royal. Ann. *J. Am. Soc. Hortic. Sci.* 98:234.
39. Feder, W.A. 1968. Reduction in tobacco pollen germination and tube elongation induced by low levels of ozone. *Science* 160:1122.
40. Feder, W.A. 1970. Plant response to chronic exposure to low levels of oxidant type air pollution. *Environ. Pollut.* 1:73-79.
41. Feder, W.A. 1981. Bioassaying for ozone with pollen systems. *Environ. Health Perspec.* 37:117-23.
42. Feder, W.A., G.H.M. Krause, B.H. Harrison, and W.D. Riley. 1982. Ozone effects on pollen tube growth in vivo and in vitro. In *Effects of gaseous air pollution in agriculture and horticulture*, ed. M.H. Unsworth and D.P. Ormrod. London: Butterworth Scientific.
43. Feder, W.A., and R. Shrier. 1990. Combination of U.V.-B and ozone reduces pollen tube growth more than stress alone. *Environ. Exp. Bot.* 30 (4): 451-54.
44. Feder, W.A., and F. Sullivan. 1969. Differential susceptibility of pollen grains to ozone injury. *Phytopath. Abs.* 59:399.
45. Forsline, P.L., R.C. Musselman, R.J. Lee, and W.J. Kinder. 1983a. Effects of acid rain on grapevines. *Am. J. Enol. Vitic.* 34:17-22.
46. Forsline, P.L., R.C. Musselman, W.J. Kinder, and R.J. Lee. 1983b. Effects of acid rain on apple (*Malus domestica*): Tree productivity and fruit quality. *J. Am. Soc. Hortic. Sci.* 108:70-7.
47. Funk, D.W., and E.K. Bonde. 1986. Effects of artificial acid mist on growth and reproduction of two alpine species in the field. *Am. J. Bot.* 73:534-38.
48. Galloway, J.N., J.D. Thorton, S.A. Norton, H.L. Volchok, and R.N.A. McLean. 1982. Trace elements in atmospheric deposition. A review and assessment. *Atmos. Environ.* 16:1677-1700.
49. Gerhold, H.D. 1977. Effect of air pollution on *Pinus strobus* L. and genetic resistance—literature review. US EPA Ecol. Res. Series, EPA Report No. 600/3-77-002.
50. Greenwood, M.S. 1986. Gene exchange in loblolly pine: The relation between pollination mechanisms, female receptivity, and pollen availability. *Am. J. Bot.* 73 (10): 1443-51.
51. Harper, J.L. 1977. *Population biology of plants*, pp. 634-36. New York: Academic Press.
52. Heggested, H.E. 1968. Diseases of crops and ornamental plants incited by air pollutants. *Phytopathology* 58:1089-97.
53. Heggested, H.E., J.H. Bennett, E.H. Lee, and L.W. Douglass. 1986. Effects of increasing doses of sulfur dioxide and ambient ozone on tomatoes: Plant growth, leaf injury, elemental composition, fruit yield and quality. *Phytopathology* 76:2047-56.
54. Heslop-Harrison, J. 1975. Incompatibility and the pollen-stigma interaction. *Ann. Rev. Plant Physiol.* 26:403-25.
55. Houston, D.B., and L.S. Dochinger. 1977. Effects of ambient air pollution on cone, seed, and pollen characteristics in eastern white and red pines. *Environ. Pollut.* (Series 14) 12:1-5.
56. Houston, D.B., and G.R. Stairs. 1973. Genetic control of sulfur dioxide and ozone tolerance in eastern white pine. *For. Sci.* 19 (4): 267-71.
57. Hughes, R.N. 1990. Effects of acid fog on reproduction in *Betula papyrifera* and *Betula cordifolia* (Betulaceae). M Sc. Thesis in Forestry. Faculty of Forestry, University of New Brunswick.
58. Hutchinson, T.C., and M.A. Harwell. 1985. *Environmental consequences of nuclear war. Vol. II. Ecological and agricultural effects*. Chichester, England: John Wiley & Sons.
59. Jain, S.K., and K.N. Rai. 1974. Population biology of Avena. IV. Polymorphism in small populations of *Avena fatua*. *Theor. Appl. Genet.* 44:7-11.

60. Kapur, A., and C.P. Malik. 1976. Effects of metabolic inhibitors on pollen germination and pollen tube growth of *Petunia alba*. *Plant Sci.* (Lucknow) 8:26-27.
61. Karnosky, D.F., and G.R. Stairs. 1974. The effects of SO_2 on in vitro forest tree pollen germination and tube elongation. *J. Environ. Qual.* 3:406-9.
62. Karnosky, D.F., F. Scholz, Th. Geburek, and D. Rudin. 1989. Implications of genetic effects of air pollution on forest ecosystems-knowledge gaps. In *Genetic effects of air pollutants in forest tree populations*, ed. F. Scholz, H.-D. Gregorius, and D. Rudin, pp. 199-201. Berlin: Springer-Verlag.
63. Kats, G., P.J. Dawson, A. Bytnerowicz, J.W. Wolf, C.R. Thompson, and D.M. Olszyk. 1985. Effects of ozone or sulfur dioxide on growth and yield of rice. *Agric. Ecosyst. & Environ.*14:103-18.
64. Keller, T., and H. Bada. 1984. Effects of SO_2 on the germination of conifer pollen. *Environ. Pollut.* (Series A) 33:237-43.
65. Kimball, K.D., R. Jagels, G.A. Gordon, K.C. Weathers, and J. Carlisle. 1988. Differences between New England coastal fog and mountain cloud water chemistry. *Water Air Soil Pollut.* 39:383-93.
66. Kozlowski, T.T., and H.A. Constantinidou. 1986a. Responses of woody plants to environmental pollution. Part I. Sources and types of pollutants and plant responses. *For. Abs.* 47:5-51.
67. Kozlowski, T.T., and H.A. Constantinidou. 1986b. Environmental pollution and tree growth. Part II. Factors affecting responses to pollution and alleviation of pollution effects. *For. Abs.* 47:105-32.
68. Kratky, B.A., E.T. Fukunaga, J.Q. Hylin, and R.T. Nakano. 1974. Volcanic air pollution: Deleterious effects on tomatoes. *J. Environ. Qual.* 3:138-40.
69. Krause, G.H.M., W.D. Riley, and W.A. Feder. 1975. Effects of ozone on petunia and tomato pollen tube elongation in vivo. *Proc. Am. Phytopathol. Soc.* 2:100.
70. Kress, L.W., and J.E. Miller. 1985. Impact of ozone on field corn (*Zea mays*) yield. *Can. J. Bot.* 63:2408-15.
71. Larson, T.V., and R.J. Vong. 1990. The theoretical investigation of the pressure and temperature dependence of atmospheric ozone deposition to trees. *Environ. Pollut.* 67:179-89.
72. Lee, T.D., and A.P. Hartgerink. 1986. Pollination intensity, fruit maturation pattern, and offspring quality in *Cassia fasciculate* (Leguminosae). In *Biotechnology and ecology of pollen*, ed. D.L. Mulcahy, G.B. Mulcahy, and E. Ottaviano, pp. 417-422. New York: Springer-Verlag.
73. Lefohn, A.S., D.S. Shadwick, and V.A. Mohnen. 1990. The characterization of ozone concentrations at a select set of high-elevation sites in the eastern United States. *Environ. Pollut.* 67:147-78.
74. Lindberg, S.E., R.R. Turner, D.S. Shriner, and D.D. Huff. 1981. Atmospheric deposition of heavy metals and their interaction with acid precipitation in a North American deciduous forest. In *Proc. Int. Conf. Heavy Metal in the Environment*. Amsterdam, Sept. 1981, pp. 306-9. Edinburgh, UK: CEP Consultants Ltd.
75. Luck, R.F. 1980. Impact of oxidant air pollution on ponderosa and Jeffrey pine cone production. In *Proc. Symp. Effects of Air Pollutants on Mediterranean and Temporate Forest Ecosystems*, ed. P.R. Miller. Gen. Tech. Rep. PSW-43 USDA, Pacific Southwest Forest and Range Experimental Station.
76. Ma, T.H., D. Isbandi, S.H. Khan, and Y. Tseng. 1973. Low level of SO_2 enhanced chromatid aberrations in Tradescantia pollen tubes and seasonal variation of the aberration rates. *Mutat. Res.* 12:93-100.
77. Ma, T.H., and H. Khan. 1976. Pollen mitosis and pollen tube growth inhibition by SO_2 in cultured pollen tubes of *Tradescantia*. *Environ. Res.* 12:144-49.

78. Mahendrappa, M.K. 1983. Chemical characteristics of precipitation and hydrogen input in throughfall and stemflow under some eastern Canadian forest stands. *Can. J. For. Res.* 13:948-55.

79. Manning, W.J., and W.A. Feder. 1976. Effects of ozone on economic plants. In *Effects of airpollutants on plants,* ed. T.A. Mansfield, pp. 47-60. Cambridge: Cambridge Univ. Press.

80. Masaru, N., F. Syozo, and K. Saburo. 1976. Effects of exposure to various injurious gases on germination of lily pollen. *Environ. Pollut.* 11:181-87

81. Masaru, N.F., F. Katsuhisa, T. Sankichi, and W. Yukata. 1980. Effects of inorganic components in acid rain on tube elongation of *Camellia* pollen. *Environ. Pollut.* 21:51-57.

82. Matthews, F.R., and D.L. Bramlett. 1986. Pollen quantity and viability affect seed yields from controlled pollinations of loblolly pine. *South. J. Appl. For.* 10 (2): 78-80.

83. McKenna, M. 1986. Heterostyly and microgametophytic selection: The effect of pollen competition on sporophytic vigor in two distylous species. In *Biotechnology and ecology of pollen,* ed. D.L. Mulcahy and E. Ottaviano, pp. 443-448. New York: Springer-Verlag.

84. McLaughlin, S.B., R.K. McConathy, D. Durick, and L.K. Mann. 1982. Effects of chronic air pollution stress on photosynthesis, carbon allocation, and growth of white pine trees. *For. Sci.* 28:60-70.

85. McQuilkin, W.E. 1940. The natural establishment of pine in abandoned fields in the Piedmont Plateau region. *Ecology* 21:135-47.

86. Moody, W.R., and J.B. Jett. 1990. Effects of pollen viability and vigour on seed production of loblolly pine. *South. J. Appl. For.* 14 (1): 33-38.

87. Moran, G.F., and A.R. Griffin 1985. Non-random contribution of pollen in polycrosses of *Pinus radiata* D. Don. Silvae *Genet.* 34:117-21.

88. Mulcahy, D.L. 1979. The rise of angiosperms: A genecological factor. *Science* 206:20-23.

89. Muller-Starck, G., and M. Ziehe. 1984. Reproductive systems in conifer seed orchards: 3 Female and male fitness of individual clones realized in seeds of *Pinus sylvestris* L. *Theor. Appl. Genet.* 69:173-77.

90. Mumford, R.A., H. Lipke, D.A. Loufer, and W.A. Feder. 1972. Ozone-induced changes in corn pollen. *Environ. Sci. & Technol.* 6:427-30.

91. Muona, O. 1990. Population genetics in forest tree improvement. In *Plant population genetics, breeding, and genetic resources,* ed. H.D. Brown, M.T. Clegg, A.L. Kahle, and B.S. Gudileir, pp. 282-98. Sauderland, MA: Sinauer Assoc.

92. Murdy, W.H., and H.L. Ragsdale. 1980. The influence of relative humidity on direct sulfur dioxide damage to plant reproduction. *J. Environ. Qual.* 9:493-96.

93. Neuberger, H., C.C. Hosler, and C. Koemond. 1967. Vegetation as an aerosol filter. In *Biometeorology* 2, ed. S.W. Tromp, and W.H. Weihe, pp. 693-702. New York: Pergamon Press.

94. Niklas, K.J. 1985. The aerodynamics of wind pollination. *Bot. Rev.* 51:328-86.

95. Oleksyn, J., R. Chalupka, M. Tjoelker, and P.B. Reich. In press. Geographic origin of *Pinus sylvestris* populations influence flowering and growth response to air pollution. *Water Air Soil Pollut.*

96. Ostler, W.K., and K.T. Harper. 1978. Floral ecology in relation to plant species diversity in the Wasatch Mountains of Utah and Idaho. *Ecology* 59:848-430.

97. Paoletti, E., and L.M. Bellani. 1990. The in vitro response of pollen germination and tube length to different types of acidity. *Environ. Pollut.* 67:279-86.

98. Pfahler, P.L., and H.F. Linskens. 1983. Methods for assessing the effects of chemicals on reproduction in higher plants. In *Methods for assessing the effects of chemicals on reproductive function,* ed. V.B. Vouk and P.J. Sheenan, pp. 499-514. London: SCOPE.

99. Pinero, D., M. Martinez-Ramos, and J. Sarukhn. 1984. A population model of *Astrocaryum mexicanum* and a sensitivity analysis of its finite rate of increase. *J. Ecol.* 72:977-91.
100. Platt, W.J., G.W. Evans, and S.L. Rathburn. 1988. The population dynamics of a long-lived conifer (*Pinus palustris*). *Am. Nat.* 131:491-525.
101. Pospisil, J., and L. Alferi. 1987a. Investigation of airborne pollen on the female generative organs of Norway spruce (*Picea abies* L. Karst.) in the Beskids Mountains at localities influenced by air pollution stress. *Lesnictvi* (Prague) 33:193-210.
102. Pospisil, J., and L. Alferi. 1987b. The influence of air pollution on pollen quality in Norway spruce (*Picea abies* L. Karst.) in the Beskids Mountains Czechoslovakia. *Lesnictvi* (Prague) 33:15-32.
103. Raunkiaer, C. 1934. *The life forms of plants and statistical plant geography*. Oxford: Oxford Univ. Press.
104. Reich, P.B. 1987. Quantifying plant response to ozone: A unifying theory. *Tree Physiol.* 3:63-69.
105. Reich, P.B., and R.G. Amundson. 1984. Low level O_3 and/or SO_2 exposure causes a linear decline in soybean yield. *Environ. Pollut.* 34:345-55.
106. Reinert, R.A., and W.W. Heck. 1982. Effects of nitrogen dioxide in combination with sulphur dioxide and ozone on selected crops. In *Air pollution by nitrogen oxides*, ed. T. Schneider and L. Grant, pp. 533-46. New York: Elsevier Scientific.
107. Richter, A., and L. Granat. 1987. Pine forest throughfall measurements. Report AC-43, Dept. Meterology, Univ. Stockholm. Stockholm: Inter. Meterology Inst.
108. Sabachnikoff, V. 1912. Action de l'acide sulfureux sur le pollen. *Comptes rendus de la Soc. de Biol. Paris.* 72s:191-93.
109. Sari-Gorla, M., C. Frova, and E. Redaelli. 1986. Extent of gene expression at the gametophytic phase in maize. In *Biotechnology and ecology of pollen*, ed. D.L. Mulcahy, G.B. Mulcahy, and E. Ottaviano, p. 32. New York: Springer-Verlag.
110. Searcy, K.B., and D.L. Mulcahy. 1986. Gametophytic expression of heavy metal tolerance. In *Biotechnology and ecology of pollen*, ed. D.L. Mulcahy, G.B. Mulcahy, and E. Ottaviano, pp. 159-64. New York: Springer-Verlag.
111. Sidhu, S.S. 1983. Effects of simulated acid rain on pollen germination and pollen tube growth of white spruce (*Picea glauca*). *Can. J. Bot.* 61:3095-99.
112. Sigmon, J.T., F.S. Gilliam, and M.E. Partin. 1989. Precipitation and throughfall chemistry for a montane hardwood forest ecosystem: Potential contribution from cloud water. *Can. J. For. Res.* 19:1240-47.
113. Smith, W.H. 1981. *Air pollution and forests: Interactions between air contaminants and forest ecosystems.* New York: Springer-Verlag.
114. Stanforth, R.J., and S.S. Sidhu. 1984. Effects of atmospheric fluorides on foliage, flower, fruit, and seed production in wild raspberry and blueberry. *Can. J. Bot.* 62:2827-34.
115. Stephenson, A.G., J.A. Winsor, and L.E. Davis. 1986. Effects of pollen load size on fruit maturation and sporophyte quality in zucchini. In *Biotechnology and ecology of pollen*, ed. D.L. Mulcahy, G.B. Mulcahy, and E. Ottaviano, pp. 429-34. New York: Springer-Verlag.
116. Strickland, R.C., and W.R. Chaney. 1979. Cadmium influence on respiratory gas exchange of *Pinus resinosa* pollen. *Physiol. Plant.* 47:129-33.
117. Sulzbach, C.W., and M.R. Pack. 1972. Effects of fluoride on pollen germination, pollen tube growth, and fruit development in tomato and cucumber. *Phytopathology* 62:1247-53.
118. Treshow, M. 1968. The impact of air pollutants on plant populations. *Phytopathology* 58:1108-13.

119. Unsworth, M.H. 1984. Evaporation from forests in cloud enhances the effects of acid deposition. *Nature* 312:262-64.
120. Unsworth, M.H., and A. Crossley. 1987. Consequences of cloudwater deposition on vegetation at high elevations. In *Effects of atmospheric pollutants on forests, wetlands and agricultural ecosystems*, ed. T.C. Hutchinson, and K.M. Meema, pp. 171-88. Berlin: Springer-Verlag.
121. Unsworth, M.H., and J.C. Wilshaw. 1989. Wet occult and dry deposition of pollutants on forests. *Agric. For. Meteorol.* 47:221-38.
122. Van Ryn, D.M., J.S. Jacobsen, and J.P. Lassoie. 1986. Effects of acidity on in vitro pollen germination and tube elongation in four hardwood species. *Can. J. For. Res.* 16:397-400.
123. Van Ryn, D.M., J.P. Lassoie, and J.S. Jacobsen. 1988. Effects of acid mist on in vivo pollen tube growth in red maple. *Can. J. For. Res.* 18:1049-52.
124. Varshney, S.R.K., and C.I. Varshney. 1981. Effect of sulphur dioxide on pollen germination and pollen tube growth. *Environ. Pollut.* 24:87-92.
125. von Scholz, F., A. Vornweg, and B.R. Stephan. 1985. Wirkungen von luftverunreinigungen auf die pollenkeimung von waldbaumen. *Forstarchiv.* 56 Jahrgang. 121-24.
126. Waldron, P.F., and L.E. Craker. 1987. Sensitivity of apple pollen to acid rain. *Hortic. Sci.* 22:1075.
127. Weathers, K.C., G.E. Likens, F.H. Bormann, J.S. Easton, W.B. Bowden, J.C. Anderson, D.A. Cass, J.N. Galloway, W.C. Keene, K.D. Kimball, P. Huth, and D. Smiley. 1986. A regional acidic cloudfog water event in the eastern United States. *Nature* 319:657-58.
128. Wertheim, L.S., and L.E. Craker. 1987. Acid rain and pollen germination in corn. *Environ. Pollut.* 48:165-72.
129. Whitehead, D.R. 1983. Wind pollination: Some ecological and evolutionary perspectives. In *Pollination biology*, ed. L. Real, pp. 97-107. Orlando, FL: Academic Press.
130. Willson, M.F., and N. Burley. 1983. *Mate choice in plants: Tactics, mechanisms, and consequences*. Princeton: Princeton Univ. Press.
131. Wolters, J.H.B., and M.J.M. Martens. 1987. Effects of air pollution on pollen. *Bot. Rev.* 53 (3): 372-414.
132. Zamir, D., S.D. Tanksley, and R.D. Jones. 1982. Haploid selection for low temperature tolerance of tomato pollen. *Genetics* 101:129-37.

9

Air Pollution Effects on the Diversity and Structure of Communities

Thomas V. Armentano and James P. Bennett

INTRODUCTION

As ecological units, biotic communities consist of aggregations of populations, interacting with other biotic and abiotic components of the ecosystem. Communities thus possess a set of emergent properties not understandable solely from inferences derived from the study of their constituent populations (O'Neill et al. 1986). Ideally then, delineating air pollution effects upon communities would involve measuring community attributes rather than attempting to infer community responses from individual plant or population measurements. However, the great body of air pollution effects literature is primarily based on individual organism responses, which provide little basis for inferring community response. Data limitations are especially acute when considering perennial plant communities, the focus of this chapter.

Evidence that communities possess attributes distinct from their constituent species is available from multivariate studies of many kinds of communities (Gauch 1982). In terrestrial ecosystems, species assemblages often are functionally linked in relation to nutrients or microenvironmental requirements (Tilman 1988). As a result, distinct species groupings are associated temporally with stages of succession or spatially within distinct microenvironments or patches within the larger ecosystem. Thus, it is conceptually logical that alteration of ecosystems by external stresses like air pollution, acting either directly upon susceptible populations or indirectly on essential resources, can affect community properties. Mechanisms causing the community change could include alteration of availability and quality of resources and shifts in competitive relationships of species within the community. Review of the literature, however, shows that unambiguous measurements of community change have seldom been made except in cases of severe pollutant stress.

Smith (1981) suggested dividing ecosystem responses to air pollution exposure into three classes ranging from profound alteration, as seen close to ore smelters, to subtle effects varying from stimulatory to slightly adverse. The well-documented restructuring of ecosystems near large point sources of air pollutants (e.g., Gordon and Gorham 1963) graphically reveals how air pollution stress can radically alter communities and ecosystems. The most challenging questions of pollutant-induced community effects, however, lie in regions exposed to chronic pollutant levels well below the levels found near large point sources. Although effects in this larger area may be small or difficult to measure, the fact that extensive areas of the world are exposed to such levels means that collective impacts could be great. As one measure of the adequacy of the current status of knowledge, this paper summarizes the current understanding of community responses to air pollution, with emphasis on problems associated with measuring subtle and moderate stress effects upon higher plant communities. Selected case studies are emphasized rather than attempting an exhaustive review (see Sigal and Suter 1987). Topics requiring further study are noted and research strategies suggested.

RESEARCH APPROACHES TO ANALYZING POLLUTION EFFECTS ON COMMUNITIES

A variety of research approaches reported in the literature provide some insight into mechanisms of community response to air pollution (Table 9-1). However, most of the methods, reviewed briefly below, provide no direct inferences into community responses.

Controlled environment growth chamber and greenhouse studies form the bulk of the reports in the air pollution literature. In recent years, there has been a trend toward more realistic fumigation regimes and recognition that artificial conditions modify plant susceptibility to pollutant effects. For example, some recent studies have used native soils and evaluated stomatal conductance as components of "effective dose" (e.g., Foster, Loats,

TABLE 9-1. Methods for Investigating Pollution Effects and Their Applicability to Plant Community Responses

Method	Value	Problems
Controlled Environment	Species comparisons; symptomology; dose-response curves; replication; populations.	No extrapolation to field conditions.
Field Symptom Scoring	Rapid assessment of many individuals and species in natural environment; good for mature trees.	Uncertain relationship to other responses; only short-term response is assessed.
Field Chamber Fumigation	Community response analysis; dose-response curves; replication in simple communities.	Chamber microclimate effects; replication problems in diverse communities; limited to small-statured communities.
Field Fumigation—Open System	Community response analysis in field setting; dose-response curves.	Replication and sample size problems; expensive; use limited to small-statured communities.
Computer Simulation Models	Project long-term ecosystem responses; evaluate systems-level mechanisms.	Limited validation; requires much data.
Quasi-experimental Approaches	Directly measures ecosystem responses; applicable to different spatial scales.	Control extraneous variable incomplete; dose must be reconstructed.

and Jensen 1990). Although this approach permits only limited extrapolation to field conditions, understanding of probable levels of response and associated mechanisms can be gained.

Open-top chambers situated in field environments reduce but do not minimize the confounding influences of microclimate effects on plant response. Another benefit is the opportunity to administer ambient pollutant doses or filter out extraneous pollutants. This method has been used primarily to evaluate agricultural crops, but work with individual trees has provided valuable insight into growth responses of immature stages of native species. In relatively few cases, field chambers have been used to investigate responses of small samples of herbaceous communities (e.g., Duchelle et al. 1983). However, problems of replication generally limit this approach to relatively uniform vegetation strata consisting of only a few species.

Open-air fumigation is a relatively expensive approach but provides direct measures of community response (Lauenroth and Milchunas 1985). Results of the multiyear zonal air pollution study (ZAPS), for example, and related linear gradient exposure systems (Reich, Amundsen, and Lassoie 1982) have elucidated the hierarchical nature of community response and underscored the need to determine long-term behavior of the community so that phenological, life-cycle, and interannual environmental fluctuations can be properly evaluated. The approach thus satisfies many of the criticisms of enclosure systems, but has been used only with small-statured herbaceous communities.

Field symptom diagnosis depends on the constancy of diagnostic expression of air pollution injury such as ozone-induced leaf stipple. Field symptoms often closely resemble symptoms induced in controlled studies, thus providing verification of the causative agent. Field diagnosis allows for relatively rapid evaluation of populations of mature trees or other plants under native conditions. With statistical approaches, symptoms can be quantitatively related to presumptive causal or influencing factors (e.g., Armentano and Menges 1987; Muir and McCune 1987). However, the method requires information about pollutant dose and environment conditions that often is lacking. A greater drawback is the unreliability of visible injury as an indicator of physiological effects.

Mathematical models of ecosystem response to air pollutants serve as tools for evaluating the long-term behavior of communities subject to chronic pollution exposure. They are especially useful for projecting future responses and for exploring differing assumptions of pollutant exposure, environmental conditions, and forest type (Shugart and McLaughlin 1985). Forest gap models simulate the growth of individual trees in defined areas and can estimate the long-term response of mixed-species forests to low-level pollution exposures, one of the few approaches to this difficult topic. Realistic models are given parameters based on species-specific and site-specific data, including species-specific pollution sensitivity information. The models cannot be fully validated because of the near absence of long-term pollution response information but have been shown to be valid for simulating shorter term responses to pollution or to nonpollutant stresses such as the chestnut blight (Weinstein and Shugart 1983). The validity of extrapolating model results beyond specific test sites needs further evaluation as it may be limited by structural and site differences and by site history influences (Dale and Doyle 1987), which are difficult to quantify.

Because the models are based on species sensitivity developed largely from fumigation of tree seedlings in controlled environments, their success in simulating ecological responses is tied to the state of the art of pollution effects research. Thus, improved understanding of species sensitivity under field conditions may alter previous sensitivity rankings and change modelled outcomes. Use of the models for understanding chronic air pollution effects also is limited by quantitative information on effects and mechanisms of interaction at the system level (Dale and Gardner 1987).

Quasi-experimental approaches (Cook and Campbell 1979), somewhat analogous to epidemiological studies of human health, seek to determine community or ecosystem responses with partial control of factors such as soil chemistry and stand age that can influence community response (e.g., McClenahen 1978; Westman 1979). Site selection procedures and prior information on study sites is essential for obtaining study sites that are "ecologically analogous" and representative of ecosystems of interest. One difficulty of the method involves the necessity for determining long-term pollution exposures using indirect methods.

Ideally, quasi-experimental approaches rely on the presence of a gradient in the pollutant of interest within an area relatively uniform in other potentially confounding factors. However, because this ideal is perhaps unattainable, results may be interpretable primarily in the context of correlative relationships and sometimes lack strong statistical power. The strength of the approach is its promise for inferring pollutant responses of complex communities such as forests under native conditions. The extent to which an analogous situation is achieved will influence the detection limit of the approach although the availability of a well-developed database on ecosystem properties allows for some statistical control of influencing factors. Field gradient studies require an intensive level of onsite data collection and are best conducted where the extent of spatial and temporal variation in the response and in the influencing factors is understood through previous study.

As a general axiom, the more subtle the effects or deviations from preexisting patterns, the harder they are to detect. The validity of the axiom has been discussed in relation to environmental stress in general (Auerbach 1981) and to air pollution (Sigal and Suter 1987). As a corollary, the more subtle the effects, the greater the need for long-term studies to elucidate the effects. For example, an important database is that of Falkengren-Grerup (1986), who described forest understory patterns and related them to a half-century of soil acidification in southern Sweden. Included among the responses was an increase in species assemblages defined as nitrophilous because of their association with high levels of available soil nitrogen. The chronic deposition of pollution-derived atmospheric nitrogen was regarded as a probable cause for the community trends. Such results support the recognition that subtle effects are not necessarily acceptably small or nondetectable except over a limited time scale and within the context of a specific experimental design. Likens (1989) showed how 18 years of time series data were required to demonstrate a statistically significant trend in the acidity of precipitation. Such findings are dependent upon the maintenance of long-term plots, documentation of methods and locations, and preservation of data integrity.

INSIGHTS FROM SELECTED PAPERS ON COMMUNITY RESPONSES TO AIR POLLUTION

Although insights into community responses can be derived from any of the methods, field studies of natural communities are perhaps most useful. However, basic questions remain as to the applicability of the findings from specific sites to other sites, to other pollution stress regimes, and to other ecosystem types. To illustrate the current understanding of community responses, selected papers are briefly reviewed (Table 9-2).

TABLE 9-2. Representative Studies of Plant Community Responses to Air Pollution

Community	Stress	Responses	References
Oak Forest	25 years of SO_2 ozone?	R* & D* declined hyperbolically; growth less sensitive; herbs & shrubs more sensitive than canopy layer.	Rosenberg, Hutnik, and Davis (1979)
Oak Forest	SO_2, NO_x, Cl, F, ozone?	R & D declined near sources except shrub layer; canopy more sensitive than other layer; tree density declined; species shifts in shrubs & herb layer.	McClenahen (1978)
Deciduous bottom-land Forest	50+ year of acidic drainage; ozone? other?	Canopy species D inversely related to soil H^+, Al^{3+}, and directly to soil bases.	Cribben and Scacchetti (1977)
Great Lakes pine-hardwood-spruce	SO_2, heavy metals (point source)	Understory R, cover more sensitive than D; overstory basal area more sensitive than R and D; overstory more sensitive than understory.	Freedman and Hutchinson (1980)
Boreal Forest (Canada)	7 year of SO_2 (point source)	Total understory D decreases within 3 km of source while cover decreases out to 15 km; mosses more sensitive than herbs; tree layer least sensitive.	Winner and Bewley (1978)
Boreal Forest (Sweden)	50+ years of acid deposition	Shifts in herb layer species composition related to nitrogen and hydrogen inputs; increase in herb layer D.	Falkengren-Grerup (1986)
Short grass prairie	5 year SO_2 (field fumigation)	Lowered biomass in dominant grass; no effect on D and on cover.	Lauenroth and Milchunas (1985)
Sonoran Desert grassland	60 years of SO_2, heavy metals (point source)	R declined 4.2 km from source; sensitivity of vegetation layers proportional to rooting depth.	Dawson and Nash (1977)
Southern California coastal sage scrub	Decades of high summer oxidants.	Reduced cover and R, increased species dominance along oxidant gradient.	Westman (1979)
Southern California coastal sage scrub	27 years of returning SO_2	Replacement of shrubs by annuals; decrease in cover; increase in D.	Preston (1988)

*R=species richness; D=diversity index (e.g., Shannon-Wiener index)

Community Response Patterns Around Point Sources

Two papers provide examples of community responses to strong point sources in forested environments. Freedman and Hutchinson (1980) describe the vegetation pattern around the Sudbury, Ontario, smelter complex, which had been operating for nearly a century. A central devastated zone was bordered by areas of heavily damaged forest, with diminishing injury extending outward to a distance of about 40 km where no effects were measurable. Both understory and overstory of the mixed pine-spruce-northern hardwood forests were analyzed for diversity and cover or basal area, respectively (Fig. 9-1). Variability in the vegetation response was high, particularly near the source and beyond 30 km. Understory cover and overstory basal area correlated better with pollution exposure than did the Shannon-Weiner diversity index. In both strata, the number of species per area also served as a more sensitive indicator of effects than the diversity index. Species per area paralleled percentage of cover well, particularly in the understory. However, the authors concluded that the understory was a poorer indicator of community effects than the overstory, conforming with concepts of ecosystem behavior under stress (Rapport, Regier, and Hutchinson 1985).

Winner and Bewley (1978) examined community responses in a boreal forest exposed to oil refinery emissions (Fig. 9-2). Understory canopy cover was more sensitive to emissions than was species diversity, which showed local increases in response to colonization of openings by nonforest "weed" species. Otherwise, diversity did not decline markedly until beyond 1 km from the source, while cover decreased to beyond 6 km from the source. Relative cover of the moss stratum unexpectedly exceeded vascular plant understory cover, although the effect reversed within 3 km of the source. In terms of species diversity, however, the moss stratum responded more sensitively at relatively low

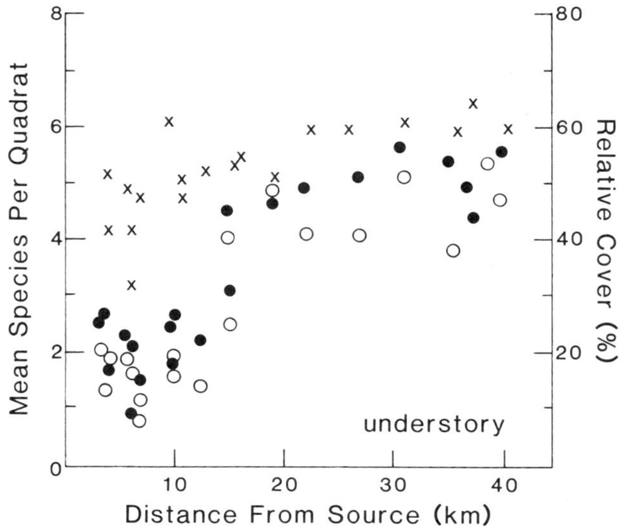

FIGURE 9-1. Understory cover (solid circles), Shannon-Weiner index (x's), and species per quadrate (open circles) around the Sudbury smelter. Redrawn from Freedman and Hutchinson (1980).

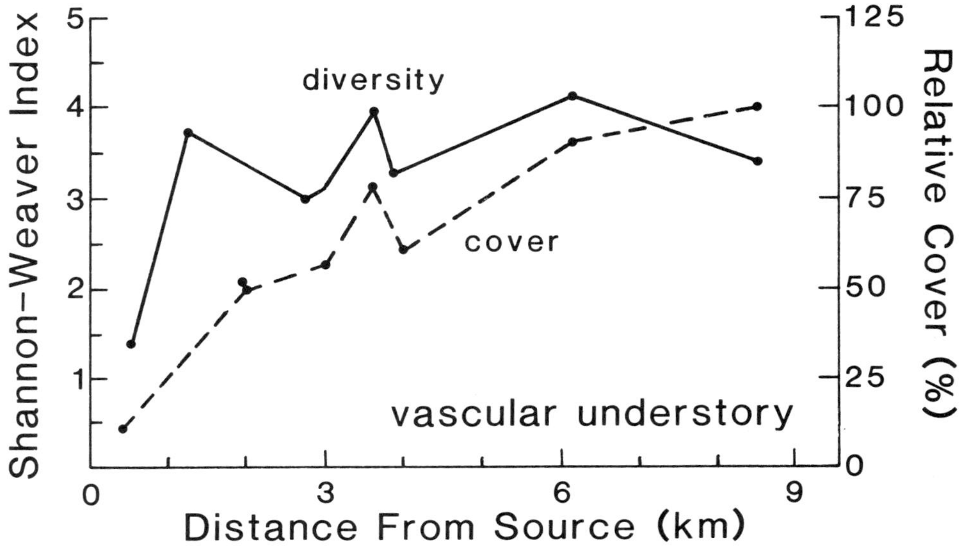

FIGURE 9-2. Vascular understory species diversity and cover in the vicinity of an oil refinery. Redrawn from Winner and Bewley (1978).

exposures, but increased within 3 km of the source, exceeding diversity at background exposures.

Both of these studies show that community structure (i.e., plant cover, biomass, species diversity) decreased along the exposure gradient but that beyond the severe damage zone, variability and "signal to noise ratio" increased. Results suggest that varying local conditions and choice of sampling approaches as well as intrinsic biological factors influence findings.

These factors probably are involved in any study but become more critical when pollution stress is moderate or subtle, as in the sense of Smith (1981). As Freedman and Hutchinson (1980) note, reductions in emissions in recent years may have led to different recovery rates between forest strata, reflecting the differing life cycles of the understory and overstory. If so, maximum effects on understory diversity may have been masked by partial recovery. Also, because the potential species overstory pool was much lower than the understory pool (the forest at 39 km distance consisted of 28 understory species and 6 tree species, of which 3 were rare), the absence of a single overstory species (due to chance or to microsite effects where plots were established) could have caused a relatively large decrease in the evenness component of diversity and a large increase in species dominance. For example, the community pattern at Sudbury was influenced by quaking aspen, often classified as a pollution-sensitive species in fumigation studies (Davis and Wilhour 1976). Aspen was dominant between 3 and 8 km from the source, perhaps due to its root-sprouting ability. Thus, a species-specific adaptation determined a key aspect of the community pattern. Winner and Bewley (1978) noted that synecological effects were complex at their study site and concluded that patterns of community recovery from stress generally were not simple or predictable.

Community Responses to Ambient Air Pollution Gradients

Factors other than intrinsic ecological forces would be expected to influence assessments of low-level chronic pollutant exposures, which inevitably encounter natural variability and multiple-factor interaction. Several reports of community responses to chronic but nonsevere pollution regimes, summarized below, are used as examples (Table 9-2).

McClenahen (1978) studied seven deciduous forest stands on similar sites located along 50 km of the Ohio River Valley, an area found within a larger region long exposed to industrially derived air pollutants. The work represents one of the few efforts to detect pollution responses by stratum in a forest. The sites were selected to represent points along an apparent gradient of chronic exposure to chlorine, sulfur dioxide, and fluorine gases. Exposures were not measured but assumed to decline with distance down the valley from the most industrialized area.

Species diversity of overstory, subcanopy, and herb strata, but not of the shrub layer, declined near the pollutant sources. The percentage of similarity of total stand composition decreased with increasing air pollution exposure when compared with composition of the least polluted stand. Overstory tree density decreased along the gradient, but density of other strata increased, suggesting a release from competition. Among canopy species, sugar maple's (*Acer saccharum*) relative importance was reduced in all strata while buckeye (*Aesculus octandra*) increased in importance with increasing pollution exposure. Also, the species-distribution pattern of the subcanopy suggested that sugar maple was partially replaced by more tolerant species.

The community pattern suggested that air pollution had reduced overstory density more than subordinate woody strata, thus partially releasing the understory from competitive suppression (McClenahen 1978). Results of the study thus conformed with the concepts of community response: namely, that greater effects in canopy trees could be explained by their relatively low photosynthesis to respiration ratio, that the magnitude of change in community structure correlates with exposure, and that resistant species expand at the expense of sensitive species through a shift in competitive ability (Rapport, Regier, and Hutchinson 1985).

Rosenberg, Hutnik, and Davis (1979), and Cribben and Scacchetti (1977) describe somewhat similar results for, respectively, oak forests in a Pennsylvania valley exposed to power plant emissions and deciduous bottomland forests along Ohio River tributaries receiving highly acidic mine drainage.

In the Pennsylvania study, Rosenberg, Hutnik, and Davis found that species richness and the Shannon-Weiner diversity index varied inversely with distance from the power plant. Other findings differed from the previous studies reviewed. Diversity measures were more sensitive indicators of pollution effects than were measures of growth. Overstory trees were least affected, shrub layer next, and the ground vegetation most affected by apparent SO_2 exposure.

In Ohio, Cribben and Scacchetti (1977) found that tree species diversity and equitability were inversely related to high concentrations of exchangeable soil aluminum and hydrogen derived from mine drainage waters and directly related to high concentrations of base cations. River birch (*Betula nigra*) increasingly dominated acidic sites, suggesting resistance to acid soil stress and the ability to gain competitive advantage over a group of apparently sensitive species that declined as acidity increased. A second species group appeared to be tolerant and able to maintain its importance in the community.

Unfortunately, because Cribben and Scacchetti considered only canopy trees, no inferences could be made about understory layers, a response of interest because the pollutant stress was mediated through soils. Strata and species with shallow root systems might have been affected more quickly if root access was confined to shallow, more contaminated soils, as found by Dawson and Nash (1980) for Sonoran Desert communities exposed to smelter effluent. Whether this response holds across greatly differing ecosystem types is of considerable interest. If greater effects had occurred in subordinate layers, community response would have been ordered in reverse of McClenahen's (1978) results but similarly to those of Rosenberg, Hutnik, and Davis (1979).

In Mediterranean ecosystem types of southern California, chronically high photochemical oxidant loadings have demonstrably altered plant community composition and in certain areas, such as the San Bernardino Mountains, are suspected of affecting successional patterns (McBride et al. 1985; O'Leary and Westman 1988). Species are clearly differentiated based on pollutant sensitivity, with clear evidence of altered competitive relationships. Effects on plant cover and species diversity would thus be expected to be well developed.

Westman (1979) reports that regional declines in coastal sage foliar cover correlates better with oxidant loadings than with nonpollutant factors. Species diversity, however, increases in the more polluted areas, reflecting the replacement of native shrubs by a mixture of mostly nonnative grasses and forbs. Given the life form of the better adapted species, the coastal sage community appears to be undergoing a physiognomic shift, thus making it one of the few communities outside of those exposed to strong point sources to be undergoing so profound a change. Sulfur dioxide may have similar effects based on Preston's (1988) report that in sage scrub surrounding an oil refinery releasing sulfur dioxide, greater species richness and reduced dominance were found in high-exposure areas. The processes involved resembled those described by O'Leary and Westman (1988): native shrubs were replaced by annual, nonnative grasses.

In the previously discussed studies, the presence of a well-defined response pattern, although differing from study to study, suggests that the ecosystems involved had been under stress sufficient to elicit clearly measurable community effects. At the three deciduous forest sites where pollutant exposure extended over 30+, 60+, and 25 years, respectively, measurable changes in cover and diversity occurred. In the coastal sage, where exposure durations were of the same order, alteration progressed further to include physiognomic changes similar to high-exposure sites around smelters. A third class of response is seen in the short-grass study (Table 9-2), where 5 years of SO_2 fumigation reduced biomass but did not change species diversity (Lauenroth and Milchunas 1985).

The relative importance of intrinsic differences in the ecosystems, in the mode of action of the various pollutants, or in sampling design choices as determinants of community responses cannot be readily discerned. Only in the prairie study was sufficient pollutant-dose information available to permit analysis of exposure:response relationships and possible cumulative dose effects. There clearly is a need for a common basis by which to compare effects of chronic exposures to the major pollutants. Comparison of ecosystem types, in addition to requiring some standardization of measurement protocols, should involve selection of a common set of community response measures. The existing literature, however, does not make clear which measures are most promising.

ECOLOGICAL FACTORS AFFECTING INTERPRETATION OF COMMUNITY RESPONSE

The representative studies from the literature just reviewed indicate some disagreement as to the relative sensitivity of various community strata and of response measures. Intrinsic differences in the ecosystems or behavior of the pollutants probably at least partially explains the disagreements in the literature. One or more ecological factors probably are involved to some extent at every site and therefore need to be understood when designing a community effects study. Some of the more important ecological factors are discussed below.

Temporal Aspects of Exposure and Expression of Effects

The response of perennial communities to chronic pollution often lags behind exposure to the pollutant, temporally separating cause and effect. Stress response lags at higher levels of organization are widely recognized (Auerbach 1981; Sigal and Suter 1987), but their dynamics have not been well worked out. Lag times may increase in some broadly predictable relationship to effective dose. Lag times may be longer where the primary effect is indirect, such as through soils. Whether the pollutant provides a nutrient (e.g., NO_x vs. O_3), thus delaying deleterious effects, as well as the presence of concomitant stresses (Manion 1981) are other considerations. Table 9-3 presents temporal scales that might be associated with pollution effects. The suggested scales are rough approximations that might apply to a hypothetical forest subject to ambient gaseous pollution exposure. Note that community effects are expressed over relatively long time scales. The importance of considering lag times in designing a study can be seen. For example, if a study is conducted at a site already exposed to pollution for decades, it may be difficult to detect shorter term responses at almost any level including the community because sensitive species or populations in which effects are most easily detected may have been already eliminated from the site. At the time a study is conducted, responses to pollution may be in

TABLE 9-3. Simplified Conceptual Model of the Apparent Time Scales of Physiological and Ecological Processes Associated with Plant Community Responses to Chronic Air Pollution*

Response Variable	Time Scale Interval
Pollutant uptake	10^{-1} to 10^{3} minutes
Reduced photosynthesis; altered membrane permeability	10^{1} to 10^{3} minutes
Reduced labile carbohydrate pool	10^{0} to 10^{1} days
Reduced growth of root tips and new leaves	10^{1} to 10^{2} days
Decreased leaf area	10^{2} to $10^{2.5}$ days
Differences in species growth performance	10^{2} to $10^{2.5}$ days
Reduced community canopy cover	10^{2} to 10^{3} days
Reduced reproductive capacity	10^{2} to 10^{3} days
Shifts in interspecific competitive advantage	10^{2} to $10^{3.5}$ days
Alteration of community composition	$10^{2.5}$ to 10^{4} days
Change in species diversity	10^{3} to 10^{4} days
Change in community structure (physiognomy)	$10^{3.5}$ to $10^{4.5}$ days
Functional ecosystem changes (e.g., decline in nutrient cycling efficiency, net productivity)	$10^{3.5}$ to $10^{4.5}$ days

* Time scales are suggested for a hypothesized forest community exposed to chronic ozone levels comparable to much of the eastern United States of the 1980's. The time-scale intervals, which are not verified empirically and are not intended to be associated with a specific site, are suggested as the ranges within which response symptoms would be clearly detected given current capabilities in pollution effects research.

any stage, ranging from too soon for an effect to be measured to too late for detecting differences in responses of system components.

Because of natural and sampling variability and the lack of controls, all studies of ecosystems will have lower limits of detectability, somewhat analogous to the detection limit of an analytical instrument. Although researchers seldom determine the sensitivity of field methods in the study of ambient pollution effects, three possible "no effects outcomes" must be considered: 1) no effects are detected because there are none; 2) the study does not detect effects because methods are not sensitive enough; and 3) no effects are detected at higher organizational levels such as communities because they are not yet expressed. Statistical aspects of detectability or statistical power, which can be critical to testing hypotheses of community effects, usually are discussed in relation to spatial variability and sample size (e.g., Peterman 1990) but temporal aspects also are critical where chronic pollution is of concern. Repeated measures analysis (e.g., Moser, Saxton, and Perzeshki 1990) can address statistical problems associated with temporally linked data, but pollution effects that scientists also must consider in analyzing community responses are generation times, time lags, and process rates.

Preadaptation of the Community to Pollutant Stress

Preadaptation can occur in many ways. Generally it can be assumed that plants are not directly preadapted to specific pollutants except in special cases where plants have evolved in the vicinity of sulfur-emitting volcanos (Winner and Mooney 1985) or have adapted to metalliferous substrates. But adaptations to other environmental stresses may provide the capacity to adjust to, or tolerate, pollutant exposures. For example, physiological adaptations to drought, such as high stomatal and diffusive resistances, are thought to confer some resistance to gaseous air pollution stress by limiting pollutant uptake (Reich 1987; Winner, Koch, and Mooney 1982). Such adaptations ordinarily are associated with low photosynthetic capacity and thus lower growth rates so that air pollution-resistant species communities may consist of relatively slower growing species. Conversely, rapidly growing early successional species like quaking aspen, white pine, and tulip poplar have lower stomatal resistances and also appear to be air pollution sensitive (e.g., Federer and Gee 1976). Thus, a comparison of responses even within the same community type may need to take into account differences in the physiological status of species on sample plots. Although site selection ordinarily may control for this automatically, difficulties are presented by species like tulip poplar that may persist in forests for centuries. Because gap-phase replacement, the dominant mechanism for community maintenance in many forests (Runkle 1985), creates opportunities for seral species, studies involving permanent plots may encounter increasing variation in community properties over time. Where sample sizes are small, gap formation in or near a few plots can confound results.

Climatic Conditions

Interannual climate fluctuations affected most ecosystem responses in the 5-year Montana prairie study (Lauenroth and Milchunas 1985). Climate appeared to determine pollution sensitivity of the grasses and constrained growth in some years, thus damping biomass accumulation of less polluted stands. LeBlanc (in press) reports a second type of climatic interaction. Dendroclimatological aspects of oak growth showed that climate since 1960 was more favorable for white and black oak growth in the lower midwest. Because the past three decades also has been the period of highest ozone and acid deposition, deleterious effects could have been masked by increased growth. But in other circumstances, climatic fluctuations in the form of drought can predispose trees to pest and pathogen injury and perhaps to pollution injury as well (Manion 1981).

Competitive Relationships Between and Within Species

Some evidence in the literature indicates that air pollutant effects on plant competition must be considered when interpreting community effects. Competitive ability is equated with growth, height, leaf area index, rooting patterns, and allelopathy. Competitive exclusion (Grime 1973) has been observed in nature from measurements of plant spacing and density. Several authors have compared the growth of pollution-injured plants with the growth of uninjured plants and concluded that the reduced growth of injured plants affects their competitive ability. Miller (1983) suggested this for ozone-injured *Pinus ponderosa*, as did Keller (1988) for ozone-injured *Populus tremuloides* sensitive and tolerant clones. Berrang, Karnosky, and Bennett (1989) concluded that intraspecific competition was probably a major factor determining relative abundances of sensitive and tolerant aspen clones in clean and polluted areas of the country. These studies constitute indirect evidence of competition effects, not of the effects of pollution on competition itself.

Direct evidence of air pollution effects on competition can be found in only a few studies. Steubing and Fangmeier (1987) enclosed portions of the understory of a European beech forest in 1.5-m diameter chambers and exposed plants to 4 hours per week of SO_2 over 2 years at levels similar to ambient peak concentrations. Four species of forbs showed decreases in growth and leaf area while two grass species and the vine *Hedera helix* were resistant. Reductions in reproductive capacity also were observed in the sensitive species. Although the authors did not evaluate diversity or other community changes, they surmised that structural alterations of the understory were underway and would become apparent in the future.

Experimental observations of plant performance in mixtures exposed to pollutants were made by Heil et al. (1988) for ammonium exposures and by Heagle et al. (1989) for ozone. The former were made on grassland canopies and the latter on clover-fescue plots. In both studies, the slower growing and/or more pollutant-sensitive species were outperformed by the more tolerant species as exposure increased, leading to substitution of one species by the other. In general, the mixture yield was lower in the pollution treatments compared to the mixture yield in the clean air controls. Similar results have been reported for annual weeds and deciduous tree saplings in mixtures exposed to increased levels of CO_2 (Bazzaz and Garbutt 1988; Williams et al. 1986).

However, these studies failed to compare performance of the mixtures relative to monocultures in response to increased pollution. Observations of mixtures alone without the appropriate monocultures only allow inferences about the pollutant effects on interspecific competition. Studies using mixtures and monocultures or increasing levels of density are needed to compare interspecific with intraspecific competition. The effects of ozone on *Lolium multiflorum* and *Trifolium incarnatum* using 50:50 mixtures and monocultures found mixtures to be less sensitive to the pollutant than the respective monocultures (Bennett and Runeckles 1977). The proportion of ryegrass in the higher ozone treatments was greater than in the clean air controls.

In another study, equal numbers of *Artemisia vulgaris* and *Solidago canadensis* in mixtures exposed to high ozone levels once a week had higher yields than mixtures not exposed to ozone (Cornelius 1982). Similar mixtures of *Aster pilosus* and *Andropogon virginicus* yielded more than monocultures when exposed to higher levels of CO_2 (Marks and Strain 1989). The effects of density and ozone were also studied for crimson clover and annual ryegrass (Bennett 1975). Increased density and ozone stimulated the competitive ability of ryegrass and depressed the competitive ability of clover due to increased tillering by the ryegrass. Kochhar (1974) found evidence for an altered allelopathic mechanism in ozone-treated fescue-clover mixtures.

Although the number of pollution-competition interaction studies is small and limited to O_3 and CO_2, results appear to conform to expectations. In mixtures and monocultures of increasing density, the stronger competitor and/or tolerant species or clones outperform the weaker competitors. In short-term fumigation experiments, mixtures appear to yield better than the respective monocultures. Results suggest that intraspecific competition is more sensitive to pollution than interspecific competition but also that pollution can change the balance of species in communities and can alter species' innate competitive abilities. The literature suggests that community-level effects of pollutant-altered competition is initially clonal (genetic), followed by species substitution.

Whether or not patterns observed in experimental studies, which most often involve only two species, apply to the long term or hold under natural conditions across species mixtures and community types is a major question. Chapin and Waring (1987) note that species-rich communities ordinarily consist of species differing slightly in their resource needs and responsiveness to changes in resource availability. Fluctuations in resources cause changes in direct species responses and in competitive interactions, leading to shifts in importance or growth rate of the constituent species. However, the aggregate effect on the community is dampened by compensatory behavior of some species, leading to smaller changes in the community than observed in the sensitive species.

Allen and Forman (1976) found that removing individual species from old-field communities did not cause a predictable response in the remaining community. Responses depended on interactions of remaining species. Defining recovery in terms of the cover of the remaining vegetation before and after removals, Allen and Forman found that in 9 of the 17 removals recovery was high, but in 8 it was low. Recovery was due principally to a single species, which differed between treatments. Austin and Austin's (1980) study of experimental grass communities along an artificial nutrient gradient supports the view that community responses are specific because of the influence of species interactions. Responses differ between species and along the stress gradient and do not always follow predictable Gaussian curves, although some consistency is seen with Grime's competitive exclusion hypothesis.

Interactions with Nutrient Status

Chapin and Waring (1987) argue that plants with high C:N ratios in tissues are more resistant to herbivores but lose some growth potential because of the metabolic expenditure required to synthesize high levels of carbon-defensive compounds. Nitrogen acquisition also requires carbon expenditure, thus competing with synthesis of defensive compounds for energy. Since acid precipitation adds nitrogen to ecosystems, the C:N ratio of plants may decrease, hypothetically increasing susceptibility to pests and pathogens. If so, sites with high cumulative nitrogen inputs or retention may respond differently to air pollution stress than sites receiving low nitrogen inputs. Screening of sites by determining the C:N ratio of standardized tissue samples may be a worthwhile check to use in site selection for studies of future trends.

Interactions with Other Pollutants

Most of the community studies in the literature considered sulfur dioxide effects, alone or in the presence of toxic metals. Despite widespread agreement that oxidants are by far the source of the greatest losses to plant resources among air pollutants, ambient ozone effects have seldom been considered except in southern California (Miller 1983; Westman 1979). In fact, the studies in Ohio, and perhaps most of the reported studies in the eastern United States, probably were conducted in the presence of phytotoxic ozone concentrations. Thus, sites designated as control or reference sites relative to the toxicant of interest actually may have been exposed to ozone levels sufficient to affect comparisons to treatment sites. Ozone-induced changes in stomatal resistance may reduce or alter uptake of other pollutant gases, thus modifying the effective dose (Jensen and Roberts 1986), but synergistic effects also are possible (Houston 1974).

Where metals such as cadmium, zinc, or nickel are present, plants may be predisposed to greater ozone injury (Czuba and Ormrod 1974). Ecosystems near urban areas commonly carry increased concentrations of trace metals that sometimes far exceed rural levels (e.g., Parker, McFee, and Kelly 1978; Friedland et al. 1984). Although current evidence does not support the presence of deleterious effects at the observed concentrations, the evidence derives mainly from experimental studies of limited duration, controlled conditions, and high doses. A stronger basis is needed for concluding that in ecosystems, over decades, no functional alterations are occurring.

Seed Source and Site History

Both seed source and site history are difficult to quantify, but may leave residual influences on species diversity and other aspects of community structure (Dale and Doyle 1987). In the midwest and south, even where forests were never cleared, selective logging and stock grazing were widely practiced. These practices selectively removed timber and palatable species, thus reducing seed sources, and little is known of how long their effects on composition persist. Archbold (1978) concluded that regrowth around a heavy metal smelter following emission controls was principally limited by the shortage of viable propagules.

CONCLUSIONS AND RECOMMENDATIONS FOR FUTURE RESEARCH

A thorough understanding of plant community responses to air pollution will require much additional research, particularly where pollution levels are chronic but relatively low. Limited evidence available so far suggests that some responses common to communities occur but that natural variation, adaptation potential of individual species, and intrinsic community differences complicate any interpretation. Comparisons of results from the literature also are complicated by varying methods and objectives. A thorough, perhaps formalized, approach employing techniques such as meta-analysis may help fully interpret the literature and ascertain the likelihood of trends common to ecosystems and pollutant regimes.

For future work, the strengths and weaknesses of available approaches argue for including as many approaches as possible in designing studies of community effects. Perhaps the greatest need remains for data from field studies of intact communities, which would capture the spatial and temporal aspects of community behavior and provide a basis for distinguishing variation attributable to pollution and nonpollution sources. Ideally, a determination of air pollution effects on communities under ambient conditions would focus on integrating the various aspects of ecosystems, ranging from detailed mechanistic studies ordinarily conducted under controlled conditions to field analysis of pollutant pathways and repeated inventory of permanent plots. Although advances can be made from more limited approaches, piecemeal studies become increasingly ineffective as exposures and presumptive effects become more subtle. Given the wide diversity of community types (in the United States alone, 156 forest types are recognized), all potentially worthy of separate study, limited resources will necessarily reduce the number of comprehensive field studies to fewer than needed. Thus, to the extent possible, community studies should be conducted at already well-studied sites, where documented, accessible data sets on ecological and other properties can be integrated into the design of the study.

Analysis of spatial and temporal patterns in communities is a fundamental research approach in ecology. It is perhaps most commonly seen in plant community ecology, which has employed multivariate statistical analysis to derive trends and relationships in complex communities. In a related approach, pollution effects in communities often are sought a posteriori by measuring selected structural components and attempting to relate the observed patterns to known stressors by regression methods. Understanding pollution effects on natural systems, however, requires ancillary data on ecosystem processes, effective dose, species-specific pollutant responses, and other process-related properties of the community and its environment. As Cale, Heneby, and Yeakley (1989) point out, however, processes often are not deducible from patterns. In the case of pollution studies, community patterns consistent with preconceptions of pollution responses (e.g., decreased diversity or productivity) may also be consistent with nonpollution factors that were not measured. Also, the absence or low abundance of a species at a site exposed to air pollution does not necessarily imply any relationship to the pollutant or any other factor (Grieg-Smith 1983; Cairns 1974). Patterns in perennial plant communities are influenced markedly by adequacy of seed source and site history. Often these factors are not incorporated into analyses and are not easily quantified. However, at an intensive study site these factors are more likely to have been evaluated than elsewhere. Existing research sites also provide data needed to consider statistical power beforehand (Peterman 1990), which is a particularly critical need in studies addressing, either explicitly or implicitly, a null hypothesis of no effects in complex and variable natural communities.

REFERENCES

1. Allen E.B., and R.T.T. Forman. 1976. Species removal and old-field community structure. *Ecology* 57:1233-43.
2. Archbold, O.W. 1978. Vegetation recovery following pollution control at Trail, British Columbia. *Can. J. Bot.* 56:1625-37.
3. Armentano T.V., and E.S. Menges. 1987. Air pollution-induced foliar injury to natural populations of jack and white pine in a chronically polluted environment. *Water Air Soil Pollut.* 33:395-409.

4. Auerbach, S.I. 1981. Ecosystem response to stress: A review of concepts and approaches. In *Stress effects on natural ecosystems*, ed. G.W. Barrett and R. Rosenberg, pp. 29–42. Chichester, England: John Wiley & Sons.
5. Austin M.P., and B.O. Austin. 1980. Behavior of experimental plant communities along a nutrient gradient. *J. Ecol.* 68:891-918.
6. Bazzaz F.A., and K. Garbutt. 1988. The response of annuals in competitive neighborhoods: Effects of elevated CO_2. *Ecology* 69:937-46.
7. Bennett, J.P. 1975. Effects of low levels of ozone on plant populations. Doctoral Dissertation. Univ. of British Columbia. 94 pp.
8. Bennett, J.P., and V.C. Runeckles. 1977. Effects of low levels of ozone on plant competition. *J. Appl. Ecol.* 14:877-80.
9. Berrang, P., D.F. Karnosky, and J.P. Bennett. Natural selection for gene tolerance in *Populus tremuloides:* Field verification. *Can. J. For. Res.* 19:519-522.
10. Cairns, J., Jr. 1974. Indicator species vs. the concept of community as an index of pollution. *Water Resour. Bull.* 10:338-47.
11. Cale W.G., G.N. Heneby, and J.A. Yeakley. 1989. Inferring process from pattern in natural communities. *BioScience* 39:600-605.
12. Chapin, F.S., and R. Waring. 1987. Plant responses to multiple environmental factors. *BioScience* 37:49-54.
13. Cook, T.D., and D.T. Campbell. 1979. *Quasi-experimentation design and analysis issues for field settings.* Boston: Houghton-Mifflin Co.
14. Cornelius, R. 1982. The influence of ozone on the competition between *Solidago canadensis* L. and *Artemisia vulgaris* L. *Angew. Bot.* 56:243-51.
15. Cribben L.D., and D.D. Scacchetti. 1977. Diversity in tree species in southeastern Ohio *Betula nigra* communities. *Water Air Soil Pollut.* 8:47-55.
16. Czuba, M., and D.P. Ormrod. 1974. Effects of cadmium and zinc on ozone-induced phytotoxicity in cress and lettuce. *Can. J. Bot.* 52:645-49.
17. Dale V.H., and R.H. Gardner. 1987. Assessing regional impacts of growth declines using a forest succession model. *J. Environ. Manage.* 24:83-93.
18. Dale, V.H., and T.W. Doyle. 1987. The role of stand history in assessing forest impacts. *Environ. Manage.* 11:351-57.
19. Davis, D., and R.G. Wilhour. 1976. Susceptibility of woody plants to sulfur dioxide and photochemical oxidants. U.S. EPA Environmental Research Laboratory-Corvallis, Office of R&D. EPA Report No. 600/3-76-102.
20. Dawson, J.L., and T.H. Nash III. 1980. Effects of air pollution from copper smelters on a desert grassland community. *Environ. Exp. Bot.* 20:61-72.
21. Duchelle, S.F., J.M. Skelly, T.L. Sharik, B.I. Chevone, Y.S. Yang, and J.E. Nellessen. 1983. Effects of ozone on the productivity of natural vegetation in a high meadow of the Shenandoah National Park of Virginia. *J. Environ. Manage.* 17:299-308.
22. Falkengren-Grerup, U. 1986. Soil acidification and vegetation changes in deciduous forest in southern Sweden. *Oecologia* 70:339-47.
23. Federer, C.A., and G.W. Gee. 1976. Diffusion resistance and xylem potential in stressed and unstressed hardwood trees. *Ecology* 57:975-84.
24. Foster J.R., K.V. Loats, and K.F. Jensen. 1990. Influence of two growing seasons of experimental zone fumigation on photosynthetic characteristics of white oak seedlings. *Environ. Pollut.* 65:371-80.
25. Freedman B., and T.C. Hutchinson. 1980. Long-term effects of smelter pollution at Sudbury, Ontario on forest community composition. *Can. J. Bot.* 58:2123-40.

26. Friedland, A.J., A.H. Johnson, T.G. Siccama, and D.L. Mader. 1984. Trace metal profiles in the forest floor of New England. *Soil Sci. Soc. Am. J.* 48:422-25.
27. Gauch, H.G., Jr. 1982. *Multivariate analysis in community ecology.* Cambridge: Cambridge Univ. Press.
28. Gordon, A.G., and E. Gorham. 1963. Ecological aspects of air pollution from an iron-sintering plant at Wawa, Ontario. *Can. J. Bot.* 41:1063-78.
29. Grieg-Smith, P. 1983. *Quantitative plant ecology*, 3rd ed. Oxford, England: Blackwell Scientific Publications.
30. Grime, J.P. 1973. Competitive exclusion in herbaceous vegetation. *Nature* 242:344-46.
31. Heagle, A.S., J. Rebbek, S.R. Shafer, U. Blum, and W.W. Heck. 1989. Effects of long-term ozone exposure and soil moisture deficit on growth of ladino clover-tall fescue pasture. *Phytopathology* 79:128-36.
32. Heil, G.W., M.J. Werger, W. de Mol, D. van Dam, and B. Heljne. 1988. Capture of atmospheric ammonium by grassland canopies. *Science* 239:764-65.
33. Houston, D.B. 1974. Response of selected *Pinus strobus* L. clones to fumigations with sulfur dioxide and ozone. *Can. J. For. Res.* 4:65-68.
34. Jensen, K.F., and B.R. Roberts. 1986. Changes in yellow poplar stomatal resistance with SO_2 and O_3 fumigation. *Environ. Pollut.* 41:235-45.
35. Kochhar, M. 1974. Phytotoxic and competitive effects of tall fescue on ladino clover as modified by ozone and/or *Rhizoctonia solani.* Doctoral Dissertation, North Carolina State University, Raleigh, 71 pp.
36. Lauenroth, W.K., and D.G. Milchunas. 1985. SO_2 effects on plant community function. In *Sulfur dioxide and vegetation: Physiology, ecology and policy issues,* ed. W.E. Winner, H.A. Mooney, and R.A. Goldstein. Stanford: Stanford Univ. Press.
37. LeBlanc, D.C. In press. Oak growth-climate relationships along the Ohio River Corridor acidic deposition gradient. II. Temporal and spatial variation of oak growth-climate relationships along a pollution gradient in the midwestern U.S. *Can. J. For. Res.*
38. Likens, G.E. 1989. Some aspects of air pollution effects on terrestrial ecosystems and prospects for the future. *Ambio* 18:172-78.
39. Marks, S., and B.R. Strain. 1989. Effects of drought and CO_2 enrichment on competition between two old-field perennials. *New Phytol.* 111:181-86.
40. Manion, P.D. 1981. *Tree disease concepts.* Englewood Cliffs, NJ: Prentice-Hall.
41. McBride, J.R., P.R. Miller, and R.D. Laven. 1985. Effects of oxidant air pollutants on forest succession in the mixed conifer forest of southern California. In *Air pollution effects on forest ecosystems,* a symposium coordinated by the Acid Rain Foundation, St. Paul, MN.
42. McClenahen, J.R. 1978. Community changes in a deciduous forest exposed to air pollution. *Can. J. For. Res.* 8:432-38.
43. Miller, P.R. 1983. Ozone effects in the San Bernardino National Forest. In *Proc. of the symp. air pollution and the productivity of the forest,* pp. 161–97. Published under a grant from the Isaac Walton League of America Endowment, with cooperation of Pennsylvania State University.
44. Moser, E.B., A.M. Saxton, and S.R. Perzeshki. 1990. Repeated measures analysis of variance: Application to tree research. *Can. J. For. Res.* 20:524-35.
45. Muir, P.S., and B. McCune. 1987. Index construction for foliar symptoms of air pollution injury. *Plant Dis.* 71:558-65.
46. O'Leary, J.F., and W.E. Westman. 1988. Regional disturbance effects on herb successional patterns in coastal sage scrub. *J. Biogeogr.* 15:775-86.

47. O'Neill, R.V., D.L. DeAngelis, J.B. Waide, and T.F.H. Allen. 1986. *A hierarchical concept of ecosystems*. Princeton: Princeton Univ. Press. 253 pp.
48. Muir, P.S., and B. McCune. 1987. Index construction for foliar symptoms of air pollution injury. *Plant Dis*. 71:558-65.
49. Parker, G.R., W.W. McFee, and J.M. Kelly. 1978. Metal distribution in urban and rural ecosystems in northwestern Indiana. *J. Environ. Qual*. 7:337-42.
50. Peterman, R.M. 1990. The importance of reporting statistical power: The forest decline and acidic deposition example. *Ecology* 71:2024-28.
51. Preston, K.P. 1988. Effects of sulphur dioxide pollution on a Californian coastal sage scrub community. *Environ. Pollut*. 51:179-95.
52. Rapport, D.J., H.A. Regier, and T.C. Hutchinson. 1985. Ecosystem behavior under stress. *Am. Nat*. 125:617-40.
53. Reich, P.B. 1987. Quantifying plant response to ozone: A unifying theory. *Tree Physiol*. 3:63-91.
54. Reich, P.B., R.G. Amundsen, and J.R. Lassoie. 1982. Reduction in soybean yield after exposure to O_3 and SO_2 using a linear gradient exposure technique. *Water Air Soil Pollut*. 17:29-36.
55. Rosenberg, C.R., R.J. Hutnik, and D.D. Davis. 1979. Forest communities at varying distances from a coal-burning power plant. *Environ. Pollut*. 10:307-17.
56. Runkle, J.R. 1985. Disturbance regimes in temperate forests. In *The ecology of natural disturbance and patch dynamics*, ed. S.T.A. Pickett and P.S. White, pp. 17-33. Orlando, FL: Academic Press, Inc.
57. Shugart, H.H., and S.B. McLaughlin. 1985. Modeling SO_2 effects on forest growth and community dynamics. In *Sulfur dioxide and vegetation: Physiology, ecology and policy issues*, ed. W. Winner, H.A. Mooney, and R.A. Goldstein. Stanford: Stanford Univ. Press.
58. Sigal L.L., and G.W. Suter II. 1987. Evaluation of methods for determining adverse impacts of air pollution on terrestrial ecosystems. *Environ. Manage*. 11:675–94.
59. Smith, W.H. 1981. *Air pollution and forests: Interactions between air pollution and forest ecosystems*. New York: Springer-Verlag.
60. Steubing, L., and A. Fangmeier. 1987. SO_2 sensitivity of plant communities in a beech forest. *Environ. Pollut*. 44:297-306.
61. Tilman, D. 1988. *Plant strategies and the dynamics and structure of plant communities*. Princeton: Princeton Univ. Press.
62. Weinstein, D.A., and H.H. Shugart. 1983. Ecological modeling of landscape dynamics. In *Disturbance and ecosystems*, ed. H.A. Mooney and M. Godron, pp. 29-47. Berlin: Springer-Verlag.
63. Westman, W.E. 1979. Oxidant effects on California coastal sage scrub. *Science* 205:1001-3.
64. Williams, W.E., K. Garbutt, F.A. Bazzaz, and P.M. Vitousek. 1986. The response of plants to elevated CO_2. IV. Two deciduous forest tree communities. *Oecologia* 69:454-59.
65. Winner W.E., and J.B. Bewley. 1978. Contrasts between bryophyte and vascular plant synecological responses in an SO_2-stressed white spruce association. *Oecologia* 35:311-25.
66. Winner, W.E., G.W. Koch, and H.A. Mooney. 1982. Ecology of SO_2 Resistance. IV. Predicting metabolic responses of fumigated shrubs and trees. *Oecologia* 52:16-21.
67. Winner, W.E., and H.A. Mooney. 1985. Ecology of SO_2 resistance. V. Effects of volcanic SO_2 on native Hawaiian plants. *Oecologia* 66:387-93.

10

Air Pollution Effects on Terrestrial and Aquatic Animals

James R. Newman, R. Kent Schreiber, and E. Novakova

The beauty and genius of a work of art may be reconceived, though its first material expression be destroyed; a vanished harmony may yet again inspire the composer; but when the last individual of a race of living things breathes no more, another heaven and another earth must pass before such a one can be again.

William Beebe (1906)

INTRODUCTION

Air pollution has adversely affected animals since the advent of the industrial revolution (Newman 1980). Currently, the greatest threat to animal biodiversity from air pollution occurs in industrial countries where regional impacts (e.g., acid precipitation, ozone) are causing widespread direct and indirect effects to animals and their habitats. In Eastern Europe, local, regional, and transboundary air pollution is severe. Future threats will occur as underdeveloped countries that have minimal air pollution controls industrialize. Of particular concern are those areas, such as the tropical forest of the Amazon Basin, that harbor the world's greatest biodiversity including many species yet to be described (Wilson 1988).

The significance of air pollution as a stressor on animal biodiversity might be considered as inconsequential, given that: (1) the total number of living animals is large, likely over a million identified species (Wilson 1988), (2) relatively few animal species are reported to be affected, (3) most animal species occur in habitats such as saltwater or soil habitats that are not directly exposed to air pollution, and (4) other stressors such as water pollution and habitat loss may have a greater effect on species diversity than air pollution. However, any stressor of animal diversity is significant given that society values species conservation. To date, however, only passing mention has been made of the effects of air pollution on animal biodiversity (e.g., McNeely et al. 1990).

This chapter addresses the effects of air pollution on animal biodiversity, including impacts on habitats and food resources. Future threats at local, regional, and global scales, ways of monitoring the effects of air pollution on animal biodiversity, and related laws and programs for protection are discussed. "Animal biodiversity" in this chapter refers to the variety of animals as reflected in genetic diversity and species diversity. The value of animal diversity to conservation is reflected in legal and policy frameworks (e.g., endangered species policies), in the biological functioning of animals in ecosystems (e.g., energy transfer), in the ethical and intrinsic value that humans attribute to animals (e.g., existence and recreational values), and in the economic value of animals.

HISTORY OF AIR POLLUTION EFFECTS ON TERRESTRIAL AND AQUATIC ANIMALS

Effects to Terrestrial Animals

Harmful effects to birds and mammals in North America, Europe, Africa, and Japan have been observed for a variety of airborne pollutants ranging from gaseous pollutants to heavy metals. These incidents include death, debilitating industrial-related injury and disease, bioaccumulation of air pollutants, physiological changes associated with stress, and population declines (Newman 1980). Injury and death of terrestrial animals from air pollution have been reported since the 1870's (Tables 10-1 and 10-2). Most of the early reports were of injury or death of domestic animals. One of the earliest reports describes the death of fallow deer (*Dama dama*) in 1887 as the result of arsenic emissions from a silver foundry in Germany (Tendron 1964). In England at the end of the last century, air pollution had caused such widespread effects to the environment that genetic changes in moths (i.e., industrial melanism) were commonly reported in many scientific journals. This

TABLE 10-1. Early Incidents Involving the Adverse Effects of Air Pollutants on Animals

Date	Location	Pollutant(s)	Effects
1873	England	Sulfur dioxide	Death of cattle
1878	England	Smoke	Blinding of cattle near copper works
1887	Germany	Arsenic	Death of fallow deer
1908	USA	Arsenic	Widespread sickness and death to cattle and horses
1914	England	Industrial smoke	Respiratory problems in cattle and reduced wool production in sheep
1915	USA	Lead	Widespread respiratory problems in horses near smelter
1927	USA	Hydrogen sulfide	Death of large number of birds
1930	Belgium	Smoke and fog	Death of cattle from respiratory failure
1931	Austria	Iron-containing flue gas	Stomach and intestinal disorders in cattle
1935	Italy	Fluoride	Death of cattle and goats
1936	Germany	Arsenic	Death of 60 to 70 percent of game populations
1939	Germany	Arsenic	Widespread sickness in cattle, sheep, horses, and poultry
1957	USA	Fluoride	Fluorosis in deer

Source: modified from Newman 1979.

TABLE 10-2. Recent Incidents Involving the Adverse Effects of Air Pollutants on Wildlife

Date	Location	Species	Pollutant(s)	Effects	Reference
1963	South Africa	Baboons and rats	Asbestos	Respiratory lesions	Webster 1963
1965	Czechoslovakia	Small birds	Fluoride	Declining populations	Feriancova-Masarova and Kalivodova 1965
1967	Canada	Whitetail deer	Fluoride	Fluorosis	Karstad 1967
1968	Czechoslovakia	House sparrows	Fluoride	Biological concentration	Balazova and Hluchan 1969
1969	Czechoslovakia	Red and roe deer	Arsenic	Sickness and death	Hais and Masek 1969
1970	USA	House sparrows	Photochemical oxidant smog or particulates	Respiratory lesions	Wellings 1970
1971	Canada	Passerine birds	Hydrogen sulfate	Death of hundreds of birds	Harris 1971
1971	Czechoslovakia	Hares	Sulfur dioxide and fly ash	Hypocalcemia and hypoproteinesis	Novakova and Roubal 1971; Novakova, Finkova, and Sova 1973
1973	Japan	Sparrow	Cadmium	Death of birds	Nishino et al. 1973
1973	USA	Bighorn sheep	Oxidants	Blindness in herd	Light 1973
1974	USA	Voles	Lead	Biological concentration	Hirao and Patterson 1974
1975	Japan	Larks	Urban air pollution	Reduced populations	Miyamoto 1975
1975	England	Sparrow hawks and song thrushes	Cadmium	Food chain accumulation	Martin and Coughtrey 1975, 1976
1975	USA	Small mammals	Oxidants	Reduced populations	Kolb and White 1975
1975	USA	Mule and white-tail deer	Fluoride	Fluorosis	Kay, Tourangeau, and Gordon 1975
1976	USA	Black-tail deer	Fluoride	Fluorosis	Newman and Yu 1976
1977	Czechoslovakia	House martins	Sulfur dioxide, particulates, fluoride, and nitrogen oxides	Reduced nesting	Newman 1977; Newman and Novakova 1977
1977	England	Wood mice and voles	Mercury	Biological concentration	Bull et al. 1977
1978	Canada	Black and mallard ducks	Copper and nickel	Biological concentration	Ranta, Tomassini, and Nieboer 1978
1979	USA	Black-tail deer	Fluoride	Browse contamination	Newman and Murphy 1979
1979	USA	Deer mice	Ozone	Genetic change in sensitivity to ozone	Richkind 1979
1979	Czechoslovakia	Hares	Sulfur dioxide, fly ash	Decrease in corneal proteins	Mikova and Novakova 1979
1980	England	Tawny owls, badgers	Cadmium	Biological concentration	Tjell, Christensen, and Bro-Rasmussen 1983
1982	USA	Rats	Lead	Biological concentration	Way and Schroder 1982
1982	London	Pigeons	Lead	Biological concentration	Hutton 1982
1982	Canada	Ruffed grouse	Copper, nickel, iron	Biological concentration	Rose and Parker 1982
1982	USA	Deer	Fluoride	Fluorosis	J. Fleming, personal communication

TABLE 10-2. Recent Incidents Involving the Adverse Effects of Air Pollutants on Wildlife (continued)

Date	Location	Species	Pollutant(s)	Effects	Reference
1983	Poland	Roe deer	Cadmium, lead, nickel, chromium, iron, copper, zinc, manganese, magnesium	Decline in antler quality	Grodzinska, Grodzinski, and Zeveloff 1983
1984	Wales	Foxes	Fluoride	Biological concentration	Walton 1984
1984	USA	Owls, bats, songbirds, mice	Hydrogen sulfide	Death	Bicknell 1984
1984	Czechoslovakia	Ducks	Trace metals	Biological concentration	Newman et al. 1984
1984	Czechoslovakia	House martins	Sulfur dioxide, fly ash	Decline in nesting with increased emissions	Newman et al. 1985
1986	Czechoslovakia	Pheasants and hares	Mercury, lead, cadmium	Bioconcentration	Urbanek 1986
1986	Finland	Night-flying moths	Radioactivity	Bioconcentration	Mikkola and Albrecht 1986
1987	Poland	Invertebrates, including arthropods	Sulfur dioxide	Densities inversely related to pollution	Chlodny et al. 1987
1987	USA	Passerine birds	Sulfur dioxide	Death to approximately 3,000 birds	Bjorge 1987
1987	Sweden	29 species of birds and mammals	Radiocaesium	Bioconcentration	Mascanzoni 1987
1988	Poland	Roe deer	Lead, cadmium, copper, manganese, iron	Bioconcentration in tissues	Babinska-Werka and Czarnowska 1988
1988	Sweden	Red fox	Metals (cadmium, mercury, lead, copper, zinc, chromium, manganese, nickel)	Interuterine losses negatively correlated with pollution source	C. Lindstrom, personal communication
1988	England	Otters	Radiocaesium	Bioconcentration	Mason and McDonald 1988
1988	England	Roe deer	Radiocaesium	Bioconcentration	Lowe and Horrill 1988
1989	Germany	Magpie	Metals, including zinc and cadmium	Bioconcentration in feathers	Hahn, Hahn, and Stoeppler 1989
1989	Czechoslovakia	Birds	Sulfur dioxide, fly ash	Reduced bird species and density in relation to habitat damage	Flousek 1989; Lemberk 1989
1989	Canada	Caribou	Cadmium	Bioconcentration	Crete et al. 1989
1989	Sweden	Moose	Radiocaesium	Bioconcentration	Crete et al. 1989
1990	Italy	House mice	Radiocaesium	Increased mutagenicity from pre-Chernobyl conditions	Cristaldi et al. 1990

Source: modified from Newman 1980.

genetic response was first observed in 1848 in the peppered moth (*Biston betularia*) in Manchester, England. The black melanic form of this moth completely replaced the typical light-colored form in certain parts of England by 1900 (Kettlewell 1973).

In the 1920's, Yont and Sayers (1927) reported the death of large numbers of birds and other animals by hydrogen sulfide fumes near a Texas oil field. Venting of gas in oil fields continues to kill animals in the United States and Canada (O'Gara 1982; Bicknell 1984; Bjorge 1987).

One of the most detailed early reports on the harmful effects of air pollution on wildlife showed widespread death of game animals by acute and chronic exposure to arsenic emissions in the Tharandt forest, Germany (Prell 1930). Between 60 to 70 percent of the red deer (*Cervus elephus*), roe deer (*Capreolus capreolus*), and wild rabbits (*Oryctolagus cuniculus*) died. The deer exhibited defective hair growth and antler formation, cirrhosis of the liver and spleen, and emaciation.

Although reports of large die-offs of wildlife caused by air pollution are rare, reports of industrial-related debilitating diseases are common. Hais and Masek (1969) documented the loss of balance and hair in red deer and roe deer caused by industrial arsenic emissions in Czechoslovakia. These debilities contributed to the freezing to death of deer during the winter. Industrial fluorosis has been found in white-tailed deer (*Odocoileus virginianus* sp.), black-tailed deer (*Odocoileus hemiomus columbianus* sp.), and mule deer (*Odocoileus h. hemionus* sp.) populations of Canada and the United States. The first report in the United States was of mule deer in Utah (Robinette et al. 1957). In black-tailed deer, severe deterioration of the teeth including pitting, chipping, and excessive tooth wear was observed in both young and old animals exposed to high fluoride levels (Newman and Yu 1976). In white-tailed deer, besides dental disfigurement, a high incidence of jaw fracturing was observed (Karstad 1967). In mule deer, lameness was reported (Kay 1975). In the vicinity of asbestos mines in South Africa, asbestosis has been found in free-living baboons (*Papio* sp.) and rats (*Rattus namaquensis*) (Webster 1963).

Due to atmospheric transport, the harmful effects of air pollutants have been observed at great distances from their source. There is a positive correlation between atmospheric deposition of pollutants and uptake in terrestrial animals (Larsson, Okla, and Woin 1990). A high frequency of total and partial blindness has been found in a herd of bighorn sheep (*Ovis canadensis*) in the San Bernadino Mountains of California, a region with the highest known concentration of photochemical oxidants. The source of these known eye irritants is more than 160 km away in Los Angeles (Light 1973). Higher than normal levels of lead have been found in voles (*Microtus montanus*) from other mountainous areas of California. Again, the source of the pollution is from distant urban areas (Hirao and Patterson 1974).

Bioaccumulation or tissue contamination has been one of the most commonly reported air pollution effects. Fluoride has been found in high concentrations not only in the bones of deer, but also in cottontail rabbits (*Sylvilagus floridanus*), hares (*Lepus* sp.), muskrats (*Ondatra zibethicus*), ground squirrels (*Spermophilus columbianus*), woodchucks (*Marmota monax*), deer mice (*Peromyscus* sp.), voles (*Microtus* sp.) (Gordon 1969a; Karstad 1970), and sparrows (Balazova and Hluchan 1969). High fluoride concentrations have also been observed in carnivores such as the red fox (*Vulpes vulpes*) (Karstad 1970) and the barn owl (*Tyto alba*) (J. Newman, personal observation). Cadmium levels four times higher than normal and a high mortality have been found in house sparrows in industrial areas of Japan (Nishino et al. 1973). High levels of cadmium have also been found in the tissues of wild rabbits (*Sylvilagus nuttali*), ground squirrels, and cricitid rodents near a smelter in Montana (Gordon 1969b). American kestrels (*Accipiter nisus*) and song thrushes (*Turdus philomelos*) living near a lead zinc smelter in England accumulated high levels of cadmium in body tissues (Martin and Coughtrey 1975, 1976). The ecological effects of this food-web-transmitted cadmium were not reported. Airborne mercury from a chlor-alkali plant in England concentrated in the tissues of several small mammals (*Apodemus sylvaticus* and *Clethrionomys glareolus*) (Bull et al. 1977). Recently, bioaccumulation of mercury has been reported in the Florida panther (*Felix concolor coryei*), representing a threat to this endangered species (Jordan 1990). Exposure to mercury appears to result from a shift by the panther from its traditional food source, deer and wild pig, to raccoons that are bioaccumulating mercury from consumption of fishes in the Everglades. The source of mercury in the Everglades and other parts of Florida is being investigated. Incineration, fossil fuel burning, burning of sugar cane, as well as past mercurial pesticide use are all suspected causes. With a few exceptions, the effects (i.e., toxicity) of bioaccumulation in animals are not reported.

Effects to Aquatic Animals

The aquatic impacts from airborne pollutants, primarily a consequence of the industrial revolution, have been noted since the latter part of the 19th century (Cowling 1982). Acidic deposition has directly affected poorly buffered aquatic ecosystems throughout the world. Some of the first effects were noted in Norwegian salmon hatcheries at the turn of the century (Leivestad et al. 1976). By the 1970's, damages in Scandinavia had focused world attention on the problem. Surveys indicated about 20 percent of the lakes and approximately 90,000 km of streams and rivers in Sweden were acidified by airborne pollutants (Dickson 1985; Bernes and Thornelof 1990). Early damages also were reported in northern Europe, Canada, and the United States. In the Netherlands, comparison of historical and recent data on fish occurrence indicate fishes formerly inhabited at least 67 percent of the present extremely acidified waters (Leuven and Oyen 1987). It is estimated that 390,000 lakes in eastern Canada are sensitive to acidification (Kelso et al. 1990). In over a third of the secondary watersheds in eastern Canada, Minns et al. (1990) predicted a loss of at least 20 percent of the potential species richness in at least 20 percent of all lakes. Results of models based on chemical survey data and geochemical assumptions combined with information on species acid tolerances, indicate substantial losses of biota in certain parts of the United States (Schindler, Kaslan, and Hesslein 1989). Lakes in the Adirondacks, Poconos-Catskills, and southern New England have lost perhaps 50 percent or more of the species in certain taxonomic groups. In 1980, the United States initiated a 10-year research program to assess the extent and effects of acidification on aquatic and terrestrial ecosystems, which culminated in a series of state-of-science and technology reports (NAPAP 1990) and Federal Agency reports (e.g., Villella 1989).

All trophic levels in susceptible freshwater systems can potentially suffer damage from decreased pH and increased trace metals caused by acidification processes (Haines 1981; Baker et al. 1990). The consequences to a particular aquatic community reflect the sensitivities of the resident species and their interactions. Although the relationship between invertebrate communities, acidity, and trace metals is complex (Smith et al. 1990; Stokes, Howell, and Kratzberg 1989), it has provided an early warning system for acidification (e.g., Raddum and Fjellheim 1984, Fjellheim and Raddum 1990). For example, many species of mayflies and caddisflies are particularly sensitive to low pH, and density, diversity, or richness may decrease in acidic waters. Invertebrate assemblages in acidic waters additionally are influenced by the remaining populations of acid-tolerant fish species (Eriksson et al. 1980). Overall community structure in acidified waters can be affected by the shift in predator-prey relationships resulting from the differences in acid tolerance of the species (Yan et al. 1991).

Fishes differ in their response to acidity by species, life stages, strains, and populations (see Baker et al. 1990 for discussion of mechanisms of effects). Although most fish species tolerate pH levels above 5.5 but not below 4.5, acid sensitivity within that range varies widely among species. Acidity and associated elevated levels of aluminum affect fish species directly through toxicity (disrupting normal ion regulation at the gill) and indirectly through changes in the physicochemical environment and food chain. Loss of fish species and populations in acidic waters generally results in lower species richness.

Aquatic-breeding species of frogs and salamanders can be affected by acidification of their breeding habitats (Freda 1986). The effects of acidity depend on the species, life stage, genetics, temperature, and interactions of calcium, aluminum, and dissolved organic acids. Ephemeral ponds, used by many amphibians for breeding, may be particularly influenced by acidic snowmelt in the spring runoff. Reproductive success is reduced by low pH [e.g., greater incidence of fungi-infested eggs (Leuven et al. 1986)]. Changes in the local distribution and abundance of amphibians can result (e.g., Freda and Dunson 1986; Harte and Hoffman 1989; Beebee et al. 1990). However, the regional losses of some amphibians due to low pH is not proven (Corn and Vertucci, in press).

Birds and mammals that depend on the aquatic environment may be affected by air pollution for some part of their life history. The loss of important acid-sensitive prey (invertebrates, amphibians, and fishes) can alter population distribution and affect reproduction of birds. Haramis and Chu (1987) reported impaired growth and reduced survival of American black duck (*Anas rubripes*) ducklings feeding in acidified wetlands. Reduced survival or breeding success must also be interpreted in relation to other indirect effects from acidification, such as competition with fishes and metal toxicities (Blancher and McAuley 1987). Decreased fish production and availability of invertebrate prey may reduce populations of both piscivorous and nonpiscivorous birds (Graveland 1990; Kerekes 1990). Effects on mammals also are primarily indirect, through loss of prey, bioaccumulation of toxic metals, or habitat degradation (Schreiber and Newman 1988).

AIR QUALITY CONDITIONS CAUSING EFFECTS TO ANIMALS

Types and Sources of Air Pollutants

Air pollutants having the greatest potential for effects on animals can be organized into functional categories based on their physical and chemical properties and common effects, the similarities in atmospheric dispersion and deposition, corresponding pathways of exposure to animals and their habitats, and similar responses by various components of the ecosystem. These categories are:

1. Acidifying air pollutants, e.g., primary gaseous emissions [sulfur oxides (SO_x), nitrogen oxides (NO_x), ammonia (NH_3), and chlorine (Cl^-)] and derivative acids;
2. Particulate matter, e.g., trace elements, metals, nonmetallic ions, organic compounds, and radioactive particulates;
3. Photochemical oxidants, e.g., secondary atmospheric pollutants (ozone, PAN) and their organic precursors;
4. Other organic compounds, e.g., airborne pesticides.

Combustion of fossil fuels (particularly coal) produces the greatest amounts and variety of acidifying air pollutants (Dvorak et al. 1978; Soholt and Wiedenbaum 1981). Sulfur and nitrogen oxides, the primary acidifying air pollutants, comprise about 31 percent of the total gaseous pollutants emitted in the United States. Fuel-combustion facilities contribute 81 percent of sulfur dioxide gaseous emissions (US EPA 1986) and are the primary precursors of the acids that contribute to regional acidity impacts in the United States.

Particulates, defined as dispersed matter existing in the condensed phase (Fennelly 1976), range from molecular clusters (< 0.005 μm) to visible material (> 100 μm). The deposition and transport of particulates and, ultimately, the location of potential effects is inversely related to the size of the particulates. Stationary fuel-combustion sources and industrial processes contribute 64 percent of the total particulates emissions (US EPA 1986). Mobile sources, such as automobiles, contribute 19 percent. Radioactive particulates also are emitted from fossil fuel combustors, but the most dramatic releases have been from nuclear bomb testing and nuclear accidents (e.g., Chernobyl accident in 1986).

Photochemical oxidants are formed as products of atmospheric reactions involving precursor emissions (e.g., organic pollutants and NO_x associated with mobile sources and point-sources such as fossil fuel), oxygen, and sunlight. These pollutants involve complex chemical transformations and both short-range and long-range transport to receptor sites.

The last category of air pollutants affecting animals is other organic compounds, such as volatile organic compounds that are emitted principally by transportation (33 percent) and industrial processes (39 percent) (US EPA 1986), and pesticides, which through dispersion after application are transported from their point of application. Pesticides and derivatives of pesticides such as DDT isomers, alpha- and gamma-BHC (benzene hexachloride), PCBs (polychlorinated biphenyls), and HCB (hexachlorobenzene) have been measured in the precipitation from Lake Superior, New Brunswick, Nova Scotia, Prince Edward Island, and Southern Labrador (Lockerbie and Clair 1988). The presence of these organic compounds in remote areas has been observed since the banning of DDT and the restriction of the use of other compounds. Studies of isolated lakes in southern Labrador revealed long-range transport of long-lived organic compounds into northeastern Canada.

Modes of Exposure of Animals to Air Pollutants

Primary modes of exposure of terrestrial animals to air pollutants are inhalation, adsorption, and ingestion. One of the most startling examples of air pollution effects by inhalation was the die-off of several hundred songbirds near a pulp mill in British Columbia that emitted high concentrations of hydrogen sulfide (H_2S) and other pollutants. The birds showed internal hemorrhaging in the lungs and liver (Harris 1971). Inhalation effects from H_2S to birds (owls) and mammals (bats, moose, and antelope) have occurred in the vicinity of oil and gas wells (Bicknell 1984). Wellings (1970) reported pulmonary anthracosis in urban but not in rural sparrows (*Passer domesticus*) in California. Limited studies suggest secondary effects such as secondary poisoning can also occur via inhalation (DeMent et al. 1986).

Adsorption of air pollutants may involve the adhesion of gases or particulates to the external surfaces or membranes, for example to the cornea of eyes in mammals (Light 1973). Investigators in Czechoslovakia have observed premature aging of the cornea in hares (*Lepus europaeus*) in areas with heavy sulfur dioxide emissions and particulate deposition from power plants and other industries. These investigators postulated that the hares living in these areas are exposed to heavy particulate deposits on the ground and on vegetation; as the hares move through the vegetation, these particulates are adsorbed on the eyes and cause corneal damage (E. Novakova, pers. communication).

Ingestion is likely the most common pathway of contamination for terrestrial animals. For example, some terrestrial animals may swallow contaminants during grooming. Exposure can be through surface contamination of food by air emissions or through contaminated plant or animal tissues. Different pollutants have different bioaccumulation properties (Table 10-3). Jenkins (1980) presents a comprehensive literature review on the bioaccumulation (uptake and storage of pollutants) and bioconcentration (selectively accumulated and concentrated pollutants in certain tissues) of toxic trace elements in fish and wildlife as well as other organisms. Certain trace elements, (e.g., cadmium and copper) have greater propensity to bioaccumulate than other trace elements, such as beryllium and cobalt. Although biomagnification (increasing concentration with increasing trophic levels) is often the reported result of chemical contamination of ecosystems, for some trace elements the opposite condition of biominification (decreasing concentration with increasing trophic levels) can also occur (Jenkins 1980). In general, reported bioaccumulation incidents in animals often lack important information on the atmospheric or food source concentration level and physiological and behavioral symptoms resulting from exposure. Injury and death to wildlife from ingestion of contaminated food and water have been reported for a variety of wildlife species (Tables 10-1 and 10-2).

For aquatic animals, the primary modes of exposure have been in response to the direct deposition of acidifying air pollutants, such as sulfates and particulate matter, and subsequent reaction to the physical and biogeochemical changes created indirectly by these atmospheric pollutants in the aquatic medium. For example, the ability of surface waters to neutralize the acids created by atmospheric transport and transformation of sulfur and nitrogen gases is dependent on watershed soils and weathering processes. The loss of alkalinity and decrease in pH and in other factors due to the acidification process

TABLE 10-3. Relative Importance of Biomagnification and Biominification in Food Chains

	Terrestrial			Aquatic		
Pollutant	Plants	Herbivores	Carnivores	Plants	Herbivores	Carnivores
Antimony	–	+ –	–	B	+	M
Arsenic	M	+ –	M	B	+ O M	+ O M
Beryllium	+ –					
Boron	B –	–				
Cadmium	B M	B O M	+ O –	B M	B	B O M
Chromium	+ M	+ M	O	B		–
Cobalt	M	O		B	M	M
Copper	M	+ –		B –	M	O M
Lead	B O M	B O M	O M	B –	–	M
Mercury	+ M	+ O	B O –	B O	B O	B O
Nickel	B M	–		B	–	O –
Selenium	B M	+ M		B		–
Tin	+ O M	–		+		
Vanadium	+ M	M		B	B*	

Note: B = biomagnification reported two or more times; + = biomagnification reported once; M = minification reported two or more times; – = minification reported once; O = no change in trace metal level; * = tunicates

Source: Jenkins 1980.

subsequently alter the aquatic environment, causing changes in species composition, community structure, energy transfer, and nutrient cycling. Acidified water, passing through the watershed, may leach aluminum and increase the solubility of many other potentially toxic metals, such as cadmium, zinc, lead, and manganese, rendering them biologically accessible.

For any animals or animal habitat subject to air pollution, one or more of the pathways of exposure are possible (Table 10-4). For example, in Czechoslovakia, both adsorption and ingestion of particulate emissions have adversely affected hares. Habitat alteration and subsequent ecological effects to animals are associated with acidifying air pollutants and photochemical oxidants.

Factors Affecting Exposure of Animals to Air Pollutants

Predicting the pathways of exposure and potential effects to animals requires an understanding of the physical and chemical behavior of air pollutants and atmospheric processes. Ground-level and atmospheric concentrations of air pollutants reflect dilution rates and the temporal and spatial variation of atmospheric conditions. Also, concentrations are influenced by physical characteristics of the emission source, e.g., stack height, physical and chemical properties of the emitted effluents, meteorological conditions, and the nature of intercepting vegetation (Dvorak et al. 1978). Effects to animals from gaseous and particulate pollutants can occur from acute (high concentrations and short-term) as well as chronic (lower concentrations and longer term or periodic) exposures to air pollutants (Table 10-4). Acute exposure may occur locally from episodic conditions such as fumigations of acidifying air pollutants. For example, sulfur dioxide created by plume downwash or at some distance can affect aquatic systems, where snowbound acids are released in the spring snowmelt (Dvorak et al. 1978).

Chronic exposure may occur locally from particulate deposition from coal-fired power plants or regionally from ozone resulting from distant urban sources. Particulates that can be transported long distances are of particular biological concern, e.g., as respirable contaminants and as an indirect threat to animals through bioaccumulation. The exposure

TABLE 10-4. Relationship of Air Pollutant Categories and the Media, Pathways, and Conditions for Exposure

		Pathways for Exposure		
Category of Pollutant	Media (Habitat) for Exposure	Terrestrial Animals	Aquatic Animals	Conditions for Exposure
Acidifying air pollutants	Air and water through air	Inhalation, habitat alteration	Ingestion, habitat alteration	Acute and chronic deposition and fumigation
Particulate matter	Air, soil, water, and vegetation through air	Inhalation, ingestion/adsorption	Ingestion, adsorption	Acute and chronic deposition
Photochemical oxidants	Air and vegetation through air	Inhalation, adsorption, habitat alteration	None	Chronic fumigation
Organic compounds	Air through water	Inhalation, ingestion	Ingestion	Acute and chronic fumigation

of animals to radionuclides has been associated with both the movement of contaminated air masses and the migration of animals through contaminated areas. Acute and chronic effects from particulates occur in both primary consumers and other trophic groups (Newman 1979, 1980). In aquatic ecosystems, atmospheric contamination by particulates and subsequent degradation of water quality and aquatic habitat can alter food-chain dynamics and cause sublethal and lethal effects at both organismic and populational levels (Martin 1987). Episodic acidification of aquatic habitats causes significant stress and mortality to sensitive aquatic species and has resulted in fish kills, declines in Atlantic salmon stock in Norwegian rivers, and decreased diversity in aquatic invertebrates in New York and Pennsylvania (Baker 1991).

Photochemical oxidants and other organic compounds such as pesticides also can be transported long distances from their precursor emission source and, besides direct effects to animals, may interact synergistically and intensify the effects of primary gaseous emissions (Dvorak et al. 1978). Ozone is a toxic and biologically active pollutant that can damage vegetation, altering habitat suitability for wildlife (Miller 1980). Reviews of the physical and chemical nature of air emissions are in Stern (1977) and Chapter 5 of this volume.

Industrial accidents can also affect animals. Although episodes of major air pollution have been reduced by air pollution legislation and emission controls in developed countries (excluding Eastern European countries), wildlife is at risk from accidental releases associated with emission sources. Accidents at Chernobyl and Bhopal have demonstrated this. However, accidental releases or episodes may occur because of equipment malfunctions or meteorological conditions, or during start-up operations of any industrial plant. In 1985, during an accident at a chemical plant in northern Florida, sulfur emissions were high enough to damage vegetation and noticeably affect humans and domestic animals 5 miles (8 km) from the plant. Levels of sulfur dioxide exceeded the upper limit of the 1-hour secondary National Ambient Air Quality standard (1,300 $\mu g/m^3$) (J. Brown, Florida Department of Environmental Regulation, pers. communication). Death and injury in waterfowl and vultures were reported by local farmers. The incident lasted only several hours, and no systematic assessment of wildlife losses was made.

TYPES OF EFFECTS INFLUENCING ANIMAL BIODIVERSITY

General Types of Effects

The effects of air pollutants on animals can be either direct or indirect. Direct effects result from exposure to air pollutants, such as inhalation of hydrogen sulfide gases near oil fields, leading to physiological injury or death. In some cases, these conditions have been the result of rare meteorologic conditions—such as fog in the vicinity of a pollutant source (e.g., Harris 1971)—that caused acute level of pollutants from industrial accidents where higher than permitted levels of pollutants have been emitted, or from the exposure of wildlife to normally high levels of pollutants (e.g., birds perching on the stacks of flare gas wells). Such direct effects generally are associated with gaseous pollutants. There is evidence of sublethal effects, generally in urban wildlife, from direct exposure to particulates, e.g., respiratory lesions in birds (Wellings 1970).

Indirect effects result from secondary exposure to air pollutants through food resources and habitats (e.g., heavy metal contamination of animal forage, changes in key habitat elements, such as acidification of streams). These effects are widespread and appear to affect animal biodiversity.

The effects of air pollution on animals can be either acute, resulting in sudden mortality, or chronic, causing injury, debilitating disorders, or reproductive effects. The magnitude of effect is a function of the pollutant, its ambient concentration, mode of exposure, duration of exposure and the age, sex, reproductive condition, nutritional status, and health of the organism(s) at the time of exposure.

Aquatic animals are affected indirectly by air pollution from changes in food resources or habitat and these changes can result in acute and chronic effects. Currently, the best information base on air pollution effects on aquatic animals is from surface water acidification. The effects of acidification on fishes and other aquatic biota have been widely documented (e.g., Haines 1981; Baker and Schofield 1985; Haines and Baker 1986; Baker et al. 1990).

Both ecological and physiological effects to animals from air pollutants have been reported (Table 10-5). The physiological responses of animals to air pollutants often depend on the specific contaminant. For example, photochemical oxidants directly affect the neutrophil and lymphocyte counts, whereas sulfur dioxide affects the blood vitamin C of mammals. Newman and Schreiber (1988) discuss the physiological effects of a number of air pollutants on birds and mammals. Ecological effects are less pollutant-specific.

Animals affected by air pollutants often show abnormal behavior (Figure 10-1), ranging from listlessness and lethargy to violent movements. Abnormal behavior is the first indication of a problem and is often symptomatic of specific poisoning (Newman 1980). Other abnormal behaviors, such as lameness in mammals, are not pollutant-specific. Abnormal behaviors observed in wildlife and others animals in the vicinity of emission sources strongly suggest contamination.

Finally, air pollution can cause phenotypic and genotypic changes in insects (Kettlewell 1973), amphibians (Andren, Marden, and Nilson 1989), birds (Murton, Westwood, and Thearle 1973), and mammals (Richkind 1979).

TABLE 10-5. Ecological and Physiological Effects of Air Pollutants to Animals

Ecological Effects	Physiological Effects
Change in population numbers, e.g., density and diversity	Direct pathological effects, i.e., mortality and morbidity
Change in spatial distribution, e.g., dispersal, migration	Change in blood chemistry or physiology
Change in quality and quantity of habitat	Change in cellular enzymes
Change in quality and quantity of food resources	Change in energy requirement for normal activities
Change in selective pressure and competitive ability	Change in growth rates
Change in appearance, e.g., phenotypic changes	Genetic change, including mutagenic effects
Change in birth rate	Lowered resistance to natural environmental stress, e.g., immunosuppression
Change in death rate	Teratogenic and carcinogenic effects
Behavioral change, including abnormal behavior	Residue accumulation, bioaccumulation, and biomagnification

Source: modified from Newman 1980.

SELECTED AIR POLLUTANTS

RESPONSES	Arsenic	Asbestos	Barium	Beryllium	Boron	Cadmium	Carbon Monoxide	Chromium	Fluoride	Hydrocarbons	Hydrochloric Acid	Hydrogen Sulfide	Iron	Lead	Manganese	Mercury	Molybdenum	Nickel	Nitrogen Dioxide	Particulates [b]	Phosphorus	Photochem. Oxidants	Selenium	Sulfur Dioxide	Vanadium	Zinc
Changes in Population Numbers	■	■				■			■			■							■	■		■		■		
Changes in Blood Chemistry or Physiology	■			■		■	■					■		■			■	■	■	■		■		■	■	
Changes in Cellular Enzymes	■			■					■	■	■		■						■	■	■	■		■		■
Changes in External Appearance	■		■						■			■		■			■		■	■			■			
Change in Population Distribution									■											■				■		
Change in Death Rate (In Free-living Animals)	■	■		■	■	■			■			■		■						■		■	■	■		
Change in Birth Rate	■													■								■				
Change in Growth Rate				■					■		■			■					■					■		
Change in Genetic Resistance																				■		■	■			
Abnormal Behavior	■		■	■	■		■	■	■	■		■		■	■	■	■		■		■	■	■	■	■	■
Physiological Changes Observed in Autopsy and Histological Analysis	■	■				■	■		■			■		■	■			■		■	■		■		■	■
Lowered Resistance to Natural Environmental Stress	■								■										■	■		■		■		
Residue Accumulation In Body Tissues	■	■		■	■	■	■	■	■			■	■	■	■	■	■		■	■	■		■		■	■
Teratogenic, Mutagenic, Or Carcinogenic		■	■	■		■		■										■				■	■			

a Domestic and laboratory animals and wildlife.
b Particulates refer to unidentified pollutants.

FIGURE 10-1. Responses of birds and mammals[a] to acute chronic exposure to selected air emissions.

Source: modified from Newman 1980.

Effects to Terrestrial Animals

Physiological Effects

Extensive studies on the effects of air pollution on the physiology of animals have been conducted in western Czechoslovakia, a region of heavy industry, and widespread air pollution from sulfur dioxide, nitrogen oxides, fly ash, and cement dust. This region also has intensively managed game populations. Wild hares (Lepus europaeus) from areas of high sulfur dioxide (up to 0.35 mg/m3) and fly ash (up to 301 t/km2/yr) have a more acid urine than hares from pollution-free areas. Hares from regions with heavy cement dust of up to 50 t/km2/yr have a more basic urine (Novakova 1970). Also, changes were observed in the calcium/phosphorus ratio of the blood. In areas of heavy sulfur dioxide and fly ash, calcium deficiency in hares approaches a hypocalcemic condition. This difference is greater in younger hares and adult females. The concentration of phosphorus in the blood increases in adult hares, especially males (Novakova and Roubal 1971). Changes in the erythrograms of hares are associated with variations in calcium and phosphorus levels in the blood. Hypoproteinesis, including declines in albumins, a-globins, and °-globins, is found in hares from areas of heavy air pollution. These changes in erythrograms are comparable to those observed in animals with infections or allergic reactions (Novakova, Finkova, and Sova 1973). Along with these reactions are changes in the age structure of these hare populations. In regions of high sulfur dioxide and fly ash, the ratio of 1-year-old hares to adult hares was 30 percent below that observed in pollution-free areas. In areas with heavy cement dust, the ratio of 1-year-old hares to adult hares was 35 percent greater than in control areas (Novakova 1969). Although the particular relationship between the physiological differences, differential survival in age classes of hares, and air emission levels is not clear, there is a strong correlation in hare populations exhibiting signs of physiological stress, differential survival, and levels of air pollution. Overall biodiversity of an area will reflect the differential sensitivity of the species to the pollutants.

Air emissions may not be lethal by themselves but may cause injury or death by lowering the resistance of animals to natural stresses. Although no extensive evaluation has been made of the overall health of deer populations affected by fluoride emissions, deer exhibiting fluorosis are considered to be generally in poor health. In Washington State, several severely affected deer had their molars and premolars worn down to, and even below, the gum line (Newman and Yu 1976; Newman and Murphy 1979). In cattle, this condition leads to impaired food intake and digestion. Besides abnormal tooth wear, Karstad (1967) observed a high incidence of fractured jaws in fluoride-contaminated deer in Montana. Although no information was reported on the survival of these deer, injuries such as abnormal tooth wear and jaw fracturing would likely lead to infections, nutritional problems, and diminished competitive ability. Such conditions likely lead to higher mortality rates during times of environmental stress.

Air pollutants have also had unexpected effects on some wildlife diseases. In Czechoslovakia, for example, nematode infections of pheasants (Phasianus colchicus) and protozoan infections in hares were less prevalent in areas of heavy air pollution. The incidence of syngamosis (gapeworm disease) in pheasants declined by 50 percent when sulfur dioxide emissions exceeded 0.15 mg/m3 and fly ash deposition was between 150 to 300 t/km2/yr. Nematode infections disappeared completely at sulfur dioxide emission levels above 0.35 mg/m3 and fly ash deposition rates above 501 t/km2/yr (Novakova and Temmlova 1973). A similar relationship in protozoan infections (coccidiosis) in hares has been observed. Although the cause for the decline in the incidence of syngamosis (and coccidiosis) is not known, the intermediate hosts of these parasitic nematodes, or the nematodes themselves, may be adversely affected by sulfur dioxide and fly ash, thereby reducing the chance for infection in pheasants (E. Novakova, pers. observation).

Although no direct effects to the physiology of mammals from acid deposition have been reported, mammals may be indirectly affected through the acidic mobilization of metals. Elevated levels of cadmium were reported in the tissues of moose (Alces alces), roe deer, and hares in southern Sweden, where the mobilization of metals by acidification is considered a likely mechanism (Frank, Petersson, and Morner 1981; Mattsson, Albanu, and Frank 1981).

Evidence suggests that the photosynthetic capability and nutritional value of vegetation used by wildlife may be influenced by acid deposition (Grodzinska 1977; Lechowicz 1981; Shaw 1981). A change in the quality of browse, such as nutrient content or metal contamination, may be reflected in the consumer population. For example, in Poland, a population of roe deer living in a pine forest affected by sulfur dioxide and particulate emissions exhibited a significant decline in the length, volume, weight, and trophy value of the antlers (Sawicka-Kapusta 1978, 1979; Jop 1979). In Czechoslovakia, Novakova and Hanzl (1974) also assessed the roe deer populations and found an inverse relation between wildlife quality and air pollution levels. Studies of livestock and their forage show a link between increased sulfate fertilization and reduced selenium content of browse in selenium-deficient soils (Allaway 1970). This relationship might occur with acid deposition. It has been demonstrated that excessive sulfur in a plant community can inhibit the transfer and function of selenium in the soil-plant-animal food chain (Davies and Watkinson 1966; Gissel-Nielsen 1973; Shaw 1981). Excess sulfate in the diet of herbivores may induce excretion of selenium, which is important in enzymatic and other metabolic functions of the animal (Harr 1978). Symptoms of selenium deficiency documented for domestic herbivores include white muscle disease and reproductive failure, including abortions, infertility, and neonatal mortality (Oldfield 1972; Eisler 1985a). Tests of this phenomenon in wild animals are incomplete, but several cases were reported for wild herbivores (Herbert and Cowan 1971; Stoszek, Kessler, and Willmers 1978; Flueck 1990).

An overlooked aspect of air pollution is the combined effects of industrial air emissions and pesticides. Death and injury to wildlife from the indiscriminate use of pesticides are well known and many wildlife populations carry detectable amounts of pesticides and their chemical derivatives in their tissues (Peterle 1991). However, the synergistic effects of air pollutants and pesticides on animal populations are not known.

Ecological Effects

Adverse ecological effects to animals may result from contamination of terrestrial and aquatic food webs and alteration of habitat by air pollutants. Acute and chronic effects have been reported in terrestrial animals exposed to air pollutants bioaccumulated in food webs. Dewey (1973) has reported biomagnification of fluoride in the insect food webs near an aluminum plant in Montana. Newman and Murphy (1979) studied the trophic transfer of fluoride emitted from an aluminum plant in Washington State, observing differential accumulation of fluoride in the primary browse of black-tailed deer. These deer heavily used two browse species that had the highest fluoride concentration of all winter browse. Severe fluorosis included dental disfigurement, abnormal tooth wear, and hyperostosis in long bones. Hirao and Patterson (1974) reported the bioaccumulation of lead from urban areas through the terrestrial food chain of remote watersheds in California. Similar bioaccumulation of metals was observed in small mammals in the northeastern United States, which receives deposition of metals from industrial areas of the Midwest and Middle Atlantic States (Schlesinger and Potter 1974).

Studies of soil that concentrations of heavy metals from vehicular traffic emissions demonstrated that primary and secondary vertebrate consumers are accumulating and biomagnifying these metals (Scanlon 1979). Although no problems with these animals were observed, the measured levels of residues suggest potential problems (Clark 1979). Other studies have shown the interaction between metals may affect their uptake, complicating the interpretation of air pollution effects. For example, low mercury levels were found in mammals, fishes, and invertebrates sampled near the Sudbury smelters in Canada, even though the environment was extremely acidified and contaminated from the emission sources (Wren and Stokes 1988). It was postulated that elevated concentrations of other metals, such as selenium, may inhibit methylmercury production and bioaccumulation in animals.

Soil ecosystems are a sink for many air pollutants. Wildlife that depends upon the soil environment, such as burrowing and litter-inhabiting species, are sensitive to soil contamination caused by air emissions. Air emissions can cause reductions in soil organisms and shifts in trophic structures. Freitag and Hastings (1973) observed an inverse relationship between ground beetle population numbers and sulfur dioxide emissions levels in the vicinity of a pulp mill. Bengtsson, Nordstrom, and Rundgren (1983) found that zinc in soil from brass mill emissions caused declines in earthworm densities. In experimental studies, Beyer and Anderson (1985) demonstrated that toxicity of zinc to woodlice due to concentrations in soils could occur up to 19 km from a zinc smelter. Lime dust emissions have been implicated in changing the soil pH around industrial locations in Germany (Bassus 1968), causing a reduction in the population density of existing nematode populations and changes in the species compositions (trophic structure from a semiparasitic-dominated food web to a saprophage-predatory nematode food web). Also, a reduction or change in decomposers can result in a decrease in litter decomposition and nutrient cycling (Jackson and Watson 1977). Thus, soil animal biodiversity, especially of invertebrate species, is affected by air pollution. But these organisms and their level of activity may be the base of the ecosystem support that is initially harmed and triggers further responses at higher trophic levels. Soil animals such as earthworms and millipedes have lower populations in acid soils (Altshuller and Linthurst 1984), and they are highly important in the energy flow and nutrient cycling of forests. For example, the density of insectivorous bird species has been found to be correlated with changes in arthropod biomass in the New Jersey Pine Barrens (Brush and Stiles 1986).

Habitat loss or alteration occurs as the result of injury or death to vegetation that provides cover, reproductive habitat, and food for wildlife. This loss of habitat can affect numbers and types of animals present and the biodiversity of the affected area. Many examples exist of extensive damage to wildlife habitat from air emissions; however, comprehensive assessments of the effects of this habitat loss to wildlife and the carrying capacity of affected habitats are not common.

Acid precipitation and air pollution have been considered contributing factors in forest decline in Europe and North America (Cowling 1989). Large areas of mixed coniferous forests in southern California (more than 40,000 ha) have been injured from photochemical oxidants. No comprehensive assessment was made of the loss of forest wildlife, but there were shifts in small mammal diversity and reduction of deer browse (Taylor 1973). Around Columbia Falls, Montana, 37,000 ha of mixed coniferous forest were injured or killed by fluoride emissions from a nearby aluminum plant (Carlson and Dewey 1971). Although no wildlife assessment was made, forest-dwelling species were likely affected. In Tennessee, more than 2,800 ha of mixed deciduous-coniferous forest were killed by sulfur dioxide and heavy metal emissions from a smelter more than 70 years ago. The area still has not returned to its previous natural state (Witkamp et al. 1966). Other similar habitat losses were recorded at Trail, British Columbia; Wawa and Sudbury, Ontario; Redding, California; and Anaconda, Montana (Miller and McBride 1975). The loss of available habitat likely resulted in the loss and reduction of animal species. In spruce forests in Czechoslovakia damaged by industrial emissions, Flousek (1989) found the total density of breeding birds decreased more than 30 percent in a period of 4 years. Species composition also changed because of the more open forest.

Degradation of forest habitat from acid deposition can occur in a number of ways, including foliar injury, loss of sensitive plant species, and loss of productivity. Generally, the diversity of animal life is associated with the stratification and growth forms of the plants (MacArthur and MacArthur 1961). Any structural simplification of forest ecosystems that reduces the number of niches has an implied effect on the wildlife. Damage to vegetation may interfere with the specific ecological requirement of a species, e.g., the northern parula warbler (*Parula americana*) requires a certain type of lichen for nesting material. Lichens (*Usnea* sp.) are extremely sensitive to air pollution, and the reduction of this lichen group in eastern North America appears related to acid precipitation (Arbib 1980). The parula warbler distribution has been reduced with the reduction of the lichen, and it is proposed that acid precipitation has been a contributing factor.

Studies in Canada suggest that the avifauna change their foraging areas from the canopy to the more heavily foliated shrub stratum partly in response to atmospheric pollution. A number of insectivorous birds living in the overstory were less abundant in dieback-affected maple stands (DesGranges, Mauffette, and Gagnon 1987). Other terrestrial wildlife most likely experience acidification through changes in soils and vegetation. For example, the distribution and abundance of salamanders may be influenced by soil pH (Wyman and Hawksley-Lescault 1987).

Acid precipitation has been considered a contributing factor, along with ozone and other factors, in the forest dieback in Europe and North America affecting spruce, maple, beech, and oak (e.g., Ulrich 1982; Vogelmann 1982). This habitat change has been observed on mountaintops in Vermont (Siccama, Bliss, and Vogelmann 1982), New Hampshire, and New York (Johnson et al. 1984). These tree species have an important value to wildlife (Martin, Zim, and Nelson 1951). A reduction in their abundance could influence the presence and abundance of a number of animal species.

In a Pennsylvania oak forest, emissions from zinc smelters adversely affected frogs, toads, and salamanders as well as some songbirds, small mammals, and white-tailed deer (Beyer et al. 1985). However, it is uncertain in some species whether the effects were caused by ecological changes at the site, physiological effects due to the high metal concentrations, or a combination of effects.

The effects of air emissions on wildlife habitat cannot be generalized because of the specificity of the responses of plants to air pollution (Hindawi 1970; Smith 1974; Davis and Wilhour 1976). However, some common patterns seem to exist. Injurious air emissions will reduce vegetative species diversity and species numbers. Higher order ecosystems such as forests, especially coniferous forest, appear more sensitive to air emissions than lower order systems, such as shrub ecosystems, which may be more sensitive than herbaceous ecosystems (Woodwell 1970; Smith 1990). Thus, the biodiversity of habitat-dependent animal species will show a similar response.

If these patterns are correct, then there are several implications for wildlife. Injurious air emissions will change wildlife habitats and, ultimately, animal species diversity. Air pollution has changed animal species diversity in grassland, deciduous forest, and mixed coniferous forest habitats (Table 10-6). Effects on species diversity have included reduction not only in numbers but also in kinds of animals from the production of SO_2, fluorides, and oxidants. Insectivorous birds inhabiting air-polluted environments frequently decline in species numbers and density, but it is not known whether these birds are responding directly to air emissions or indirectly through a reduction in prey species (Newman, Novakova, and McClave 1985). A vertebrate-habitat-dependency matrix was developed by Miller and others to assess the potential change in wildlife species composition from the effects of oxidant emissions on mixed coniferous forests in southern California (Light 1973). Wildlife species dependent upon forest habitat are likely to be more affected than wildlife species in simpler habitat types. In some cases, habitat requirements of forest animals are considered in attempts to manage forests damaged by air pollution (Naef-Daenzer and Blattner 1989).

The direct effects of air pollutants on the populations of terrestrial animals have been observed. These effects did not appear to be related to changes in habitat. Forest insect species diversity can be affected by air pollution. Air pollutants can act synergistically to stimulate natural pest infestation (e.g., Miller and McBride 1975). Forest pest outbreaks have occurred in areas of high air pollution associated with vegetation damage. The damaging insects are often secondary pest species normally found in low numbers. The removal of natural biological controls (e.g., birds and other insects for these secondary pests) as a result of air pollution is often the cause for outbreaks of forest pest infestations.

Similar population effects have been reported in vertebrate animals. For example, rodent populations were lower than expected in an asbestos mining area of South Africa (Webster 1963). In mountainous areas of southern California with the high oxidant concentrations, population densities of small mammals were less than in cleaner areas (Kolb and White 1975). Similarly, a reduction in passerine bird populations was reduced in the vicinity of an aluminum-producing plant in Czechoslovakia (Feriancova-Masarova and Kalivodova 1965; Newman 1977). Miyamoto (1975) reported the decline of lark populations in Japanese cities with heavy air pollution. In London over the last 40 years, there has been a decline and subsequent return of bird populations, especially hirundines, to the inner city.

TABLE 10-6. Changes in Animal Species Diversity from Air Pollution Stress

Change in Species Diversity	Pollutants	Ecosystem (location)
Insects	SO_2	Grassland (Montana)
Birds	Fluorides	Deciduous forest (Czechoslovakia)
Mammals, insects	Oxidant, SO_2	Mixed coniferous forest (California, Canada)
Birds	SO_2, particulates	Urban (Japan, England)
Fish, birds, frogs, invertebrates	Acid deposition	Lakes (Sweden, Canada, USA)
Insects	SO_2, particulates	Urban woodlands (England, France)

Source: adapted from Newman and Schreiber 1984.

This has been attributed to the reduction of the once-high levels of smoke and other pollutants in central London (Cramp and Gooders 1967; Gooders 1968).

Changes in wildlife behavior attributed to air pollution also have been observed. Animal groups such as insectivorous birds and mice appear to avoid areas of higher emission levels, compared to areas with lower emission levels (Chilgren 1979). In aquatic systems, some species may be affected during certain life stages. For instance, Steele, Strickler-Shaw, and Taylor (1989) found significantly greater variability in activity of bullfrog tadpoles (*Rana catesbeiana*) exposed to lead than in control animals. However, interpreting the ecological significance of changes in activity and behavior of animals resulting from sublethal exposure to pollutants remains mostly speculative.

One of the most significant effects of the alteration of animal habitats by air pollution has been the effect on the genetic diversity of animals. Change in selection pressure and subsequent change in the population genetics of species is one of the earliest reported responses of animals to air pollution. Industrial melanism has been reported in a number of species of insects in England and on the European continent. In the peppered moth and other surface-resting species, melanism is primarily caused by selection of adaptive dark coloration in response to predation. This has occurred in regions where the coloration of its resting habitat (e.g., tree trunks) has changed due to air pollution; SO_2 and other pollutants have destroyed the green epiphytic cover (lichens) on the trunks of trees, resulting in a darkening of background coloration. Visual fitness for the two forms (light form in an unpolluted environment and dark form in a polluted environment) is estimated to range from 10 to 60 percent (Bishop, Cook, and Muggleton 1978). In some species, for example, the two-spot ladybird beetle (*Adalia bipunctata*), the exact mechanism may be related to the amount of light and the thermal advantage of the melanic form has over the typical lighter colored form (Brakefield and Lees 1987). Air pollution has indirectly caused genetic and phenotypic changes in a number of invertebrate species throughout northern Europe (Table 10-7). There is evidence that the process of industrial melanism reverses when the air becomes cleaner (i.e., air quality control). British air pollution control laws of the 1950's and 1960's have reduced smoke and SO_2 pollution. As a result, the range of the melanic form of the peppered moth has been reduced (e.g., Cook, Mani, and Varley 1986; Liebert and Brakefield 1987).

Genetic effects of air pollution also occur in vertebrates. Murton et al. (1973) reported phenotypic changes in pigeons in highly polluted areas. Richkind (1979) indicated that chronic exposure of small rodents (*Peromyscus californicus*) to ozone can cause genetic changes. These studies indicate that some vertebrate species are capable of rapid evolutionary response to certain air pollutants.

Effects to Aquatic Animals

Physiological Effects

In aquatic systems, the chemical environment can change rapidly in response to episodic acidic precipitation (e.g., snowmelt, storm event) (Baker et al. 1990). The decline in pH can cause physiological damage to biological organisms, particularly those at a sensitive life stage. Wiener (1987) concluded that "salmonid gametogenesis and early ontogeny were likely to be affected by environmental acidification which results in pH changes to levels below 5.5."

Eilers, Lien, and Berg (1984) showed that a large number of aquatic fauna (insects, insect larvae, leeches, gastropods, and mollusks) exhibit negative effects when pH has

TABLE 10-7. Examples of Industrial Melanism

Location	Species	Comments	Reference
		<u>BIRDS</u>	
England	*Columbia livia* (rock dove)		Murton, Westwood, and Thearle 1973
		<u>INSECTS</u>	
Czechoslovakia	*Philaenus spumarius* (spittel bug)	Melanic form associated with regions of high air pollution	Novak and Spitzer 1986
	Adalia bipunctata (two-spot ladybird beetle)	Melanic form found in Bohemia with high air pollution	Honek 1975
	Biston betularia (peppered moth)	Melanic form associated with high air pollution	Honek 1975
Denmark	*Biston betularia*		Douwes et al. 1976
England	*Biston betularia*	Associated with level of SO_2	Kettlewell 1973
	Adalia bipunctata	Melanics associated with level of smoke, has declined with smoke control	Brakefield and Lees 1987
	Mesopsocus unipunctatus (bark louse)	Melanic form found in industrial area	Popescu, Broadhead, and Shorrocks 1978
	Mesopsocus unipunctatus	Two gene control of melanism	Popescu 1979; Popescu, Broadhead, and Shorrocks 1978
	Philaenus spumarius (spittel bug)	Local melanism 95% within few kilometers of industry	Lees and Dent 1983
	Philgalia pilosauria monacharia and *P. p. intermedia*	Associated with *B. betularia*	Lees 1974
	Diurnea fagella dormoyella	Single gene control of melanism	Steward 1977
	Allophyes oxacanthae capucina	Melanic morph existed before industrialization	Steward 1976
	Salticus scenicus (jumping spider)		Machie 1960
	Oastearius melanopyginus (spider)		Machie 1960
	Drapetisia socialis		Machie 1960

TABLE 10-7. Examples of Industrial Melanism (continued)

Location	Species	Comments	Reference
		INSECTS (continued)	
Finland	*Biston betularia*		Douwes et al. 1976
	Xestia gelida	Subarctic species affected by transboundary air pollution	Mikkola 1989
	Adalia bipunctata	10% to 12% melanics in Helsinki	Mikkola and Albrecht 1988
	Oligia latruncula (noctuid moth)		Mikkola 1975
	Biston betularia		Mikkola 1984
Germany	*Adalia bipunctata*	Melanics comprise up to 20% in Dresden	Klausnitzer and Schummer 1983
	Biston betularia		Schummer 1976
	Biston betularia		Cleve 1970
Netherlands	*Biston betularia*		Kettlewell 1973
Norway	*Biston betularia*		Mikkola 1984
	Oligia latruncula	Melanics comprise 30% of polluted sites	Andersen and Ellefsen 1982
Sweden	*Biston betularia*		Douwes et al. 1976
U.S.S.R.	*Adalia bipunctata*	85% melanics in Leningrad	Mikkola and Albrecht 1988
U.S.S.R. (Estonia)	*Adalia bipunctata*	27% melanics in Tallinn	Mikkola and Albrecht 1988

declined to near 6. Because assimilation of calcium is affected by lower pH, crayfish develop thin or incomplete exoskeleton when pH declines. In fishes, the high levels of aluminum mobilized by the acidic conditions can damage the gills, causing osmotic stress and respiratory failure.

The magnitude of such physiological effects is exemplified by the interactive effects from acidifying pollutants on ponds and temporary melt-water pools that commonly form in forests in early spring. In the United States, approximately 50 percent of the species of frogs and toads and 30 percent of the species of salamanders use ephemeral forest ponds for reproduction (Pough and Wilson 1977). These small pools and ponds can be acidic because they receive snow melt and spring rains that have little contact with the soil buffering system. Studies have shown that a number of amphibian species in the northeastern United States are adversely affected by pH depression (Tome and Pough 1982; Freda 1986; Freda and Dunson 1986; Albers and Prouty 1987). Low pH reduces the reproductive capacity of amphibians by embryonic mortalities and deformities, decreased viable egg masses, delayed development, abnormalitites, decreased sperm motility, and ion regulatory failure (Freda 1986). The long-term result could be changes in populations, distribution, and species diversity.

Acidification increases the potential bioavailability of other toxic metals to fishes and other organisms (Spry and Wiener 1991). Levels of aluminum toxic to amphibians have been associated with acidic conditions. Freda (1986) stated that aluminum toxicity is an extremely complex phenomenon which is influenced by the hardness, pH, dissolved organic acids of the water, and the developmental stage and species of amphibian.

In aquatic birds, acid rain may reduce eggshell quality (Glooschenko et al. 1986; Omerod et al. 1988; Drent and Woldendorp 1989). It is suspected that calcium is less available in acidic habitats. In Scandinavia, eggshell impairment has been reported in songbirds that feed on aquatic invertebrates contaminated by aluminum mobilized under acid conditions during spring melt (Nyholm and Myhrberg 1977; Nyholm 1981); however, other studies have not duplicated this finding. In Ontario, a 3-year-old study of the eastern kingbird (*Tyrannus tyrannus*), an insectivore, demonstrated that a relationship between lake acidity and kingbird reproduction was measurable but was considered minor compared with the influence of genetic and other variables (Glooschenko et al. 1986). Ring doves (*Streptopelia risoria*) fed a diet supplemented with 0.1 percent aluminum were not affected in egg production, fertility, or hatchability (Carriere et al. 1986).

Elevated levels of metals such as mercury, cadmium, and lead have been found in lake waters and in biota from acidified areas of Ontario (Scheider, Jeffries, and Dillon 1979; Suns, Curry, and Russel 1980; Johnson 1987) and in the Adirondack region of New York (Schofield 1978; Bloomfield et al. 1980). In regions remote from direct sources of contamination, levels of metals in tissues of wildlife are higher in acid-stressed than in unstressed habitats.

There is a relationship between high mercury levels in fishes and lake acidity. Fishes accumulate mercury directly by absorption through the gills or by ingesting it with their food. Because mercury is eliminated very slowly, the highest concentrations are found in the larger, older fishes and may pose the greatest threat to top predators, including humans. Local environmental processes probably regulate the bioavailability of the mercury. Eggs of the common goldeneye (*Bucephala clangula*) from an acidified region of Sweden had higher mercury concentrations than those from nearby buffered lakes (Eriksson et al. 1980). St. Louis, Breebaart, and Barlow (1990) found that concentrations of both copper and zinc were significantly higher in the livers of tree swallow nestlings near acidic lakes than in those near reference lakes. The effect of this physiological stress on reproductive success is unknown. Mammals that reside in acidified lake regions and utilize the surface waters and aquatic resources might be expected to show higher than normal body burdens. Raccoons (*Procyon lotor*) from an acidified area of Ontario had liver mercury levels five times greater than raccoons from a nonacidified area (Wren et al. 1980). Mercury levels in tissues of mink and otter *(Lutra canadensis)* from Ontario followed a pattern similar to that of fishes and crayfishes, indicating bioaccumulation through the aquatic food chain (Wren, Stokes, and Fischer 1986). On a watershed basis, Organ (1989) found a strong relationship between mercury residues in fishes and otters.

Although the mechanisms are poorly understood, Wiener (1987) concluded that surface water acidification may enhance the accumulation of methylmercury, cadmium, and lead by fishes in affected areas. Studies with carnivorous (yellow perch) and omnivorous (white sucker) fishes have shown fishes from acidic lakes have higher concentrations of lead and cadmium in some of their tissues and that concentrations of aluminum were elevated in gill tissues (Stripp et al. 1990). The consequences of metal contamination of fishes in low-pH lakes and the implications for piscivorous wildlife suggest that acidification of poorly buffered lakes may increase the dietary uptake of methylmercury, cadmium, and lead in wildlife consuming fishes from low-alkalinity lakes. This is especially true for methylmercury, which has a greater biomagnification potential than cadmium and lead (Jenkins 1980; Wren, MacCrimmon, and Loescher 1983; Wiener 1987). Minks, otters, and other fish-eating mammals in regions affected by acid rain, such as New England and Eastern Canada, are at risk of contamination from mercury and possibly other trace metals mobilized by acidic precipitation. Both minks and otters accumulated 10 times more mercury than predatory fishes from the same drainage areas in Manitoba (Kucera 1983). High levels of mercury and toxic effects have been reported in mammals living downstream from industrial sources (Wren 1986; Blus, Henny, and Mulhern 1987). Minks are considered sensitive to mercury contamination (Linscombe, Kinler, and Aulerich 1983) and indicators of mercury contamination (Kucera 1983).

In piscivores, mercury concentrations in prey may reach levels known to cause reproductive impairment in birds and mammals. Insectivores and omnivores may, under certain conditions, experience increased exposure to toxic metals in some acidified environments. These species may experience a decrease in availability of dietary calcium. Herbivores may risk increased exposure to aluminum and lead, and perhaps cadmium, in acidified environments because certain macrophytes may accumulate high concentrations of these metals under acidic conditions (Scheuhammer 1991).

Finally, aquatic animals may show genetic effects from acidification. Andren, Marden, and Nilson (1989) found selection pressure toward higher embryonic acid tolerance in a population of moor frogs (*Rana arvalis*) exposed to critically low pH levels for about 15 generations. An adaptation to acid conditions may be possible during 30 to 40 years.

Ecological Effects

Acid-sensitive aquatic species occur in all major groups of organisms. Changes in species richness (the number of species) vary most in the pH range of 3 to 6. With increasing acidity, acid-sensitive species are lost and species richness declines. Species of minnows, zooplankton, mollusks, and mayflies are adversely affected at episodic pH levels between 5.5 and 6.0. Fish species diversity and total biomass were much lower in acid lakes than in nonacid lakes surveyed in Quebec (Frenette, Richard, and Moreau 1986). Important sport fishes including lake trout, rainbow trout, walleye, and small mouth bass are affected by pH from 5.5 to 6.0. Brook trout decline below pH 4.8 to 5.0. Few fish species are able to survive below pH 4.5 (Baker 1991).

Loss of aquatic species from acidified lakes, ponds, and wetlands causes loss or change in food resources, resulting in shifts in overall predator-prey relations and community structure (Yan et al. 1991). Some aquatic groups such as water boatmen (Corixids) and some dragonfly nymphs may increase because of their acid tolerance.

In Canada, Schindler et al. (1985) gradually acidified a lake and found that many aquatic organisms were affected as the pH and alkalinity gradient changed. Key organisms in the food chain in a lake trout system were affected first, and prey items such as freshwater shrimp (*Mysis relicta*) and fathead minnows (*Pimephales promelas*) sharply decreased between pH 5.9 and 5.6.

The change in species interactions following disturbance by air pollution can be complex. For example, Bendell and McNicol (1987) found that fish predation was the most immediate factor structuring aquatic insect assemblages in small lakes, and change of predator-prey relationships is coincident with lake acidification.

Acidity may reduce insect diversity, affecting some avian insectivores, but this may be partially balanced by less predation of fish populations that have also been reduced or eliminated. For example, until the level of acidity reduces or eliminates the fish, they compete with duck species for the limited invertebrate prey (DesGranges and Rodrigue 1986).

Birds exhibit a variety of feeding habits (divers, surface feeders, aerial flycatchers, piscivores, insectivores) and can be adversely affected in several ways (see reviews by Mitchell 1989, Goriup 1989). For example, recent research in Wales has demonstrated that the distribution of breeding dippers (*Cinclus cinclus*) relates to the abundance of stream macroinvertebrates, which, in turn, are reduced in acidic streams (Omerod et al. 1986). Delayed clutch initiation and significantly smaller clutch and brood sizes were found in dippers breeding along acidic streams (Omerod and Tyler 1987). Loons, herons, and mergansers prey on fishes and aquatic invertebrates, and are another group of birds that bridge aquatic and terrestrial habitats and may be affected by acid precipitation.

Acidification of forest ponds and lakes that affects the availability of food organisms for wildlife may also alter habitat used for cover and nesting sites (for waterfowl, see review by Hansen 1987). Some forest lakes and wetland areas that are important breeding areas for waterbirds are vulnerable to acidification (McNicol, Bendell, and McAuley 1987; McNicol, Bendell, and Ross 1987; McNicol, Blancher, and Bendell 1987). Lake acidity is important in determining waterbird habitat quality and nesting site selection by different species (DesGranges and Darveau 1985). Longcore, Ross, and Fischer (1987) recently used risk criteria to evaluate acidification and its severity on migratory birds. They estimated that approximately 17 percent of the breeding range of seven species of waterbirds in eastern North America occurred in areas sensitive to acidification and that 8 to 17 percent of the breeding pairs occupied that range.

Blancher and McAuley (1987) indicated that reproductive success of birds with diverse feeding habits was adversely affected by acidification of wetlands. Reduced food abundance appeared to be the primary factor related to poorer reproduction in the low-pH wetlands. Observational data from acid lakes in Ontario indicate that populations of the common loon may be reduced because of poor food supply (Alvo 1987). In southern Quebec, both the black duck and common loon were found to avoid acid lakes (DesGranges and Darveau 1985). In Ontario, the number of broods of the piscivorous common loon and common merganser was lower in the acid-stressed area than in the unaffected area (McNicol, Bendell, and McAuley 1987).

Limited research on artificial wetlands (lacking fishes) suggests that black duck ducklings have a higher mortality in acidified than in circumneutral wetlands (Haramis and Chu 1987). Reduced phytoplankton, algal growth, and invertebrate biomass on the acidified wetlands affected duckling growth, physiological condition, and survival (Rattner et al. 1987). Productivity, acidity, and competition with fishes may affect duckling growth and survival (DesGranges and Hunter 1987), but present data do not permit clear distinction of the importance of each factor. Investigations continue in the United States and Canada on the effect of acidification on waterfowl productivity due to changes in water chemistry and the availability of food items.

In the Hubbard Brook Experimental Forest of New Hampshire, salamanders disappeared from the study area when a section of the forest stream was artificially acidified to a mean pH of 4.0 (Hall and Likens 1980). Examples of other affected amphibian populations include the natterjack toad (*Bufo calamita*), which was not found in ponds below pH 5.0 (Beebee and Griffin 1977), and the smooth newt (*Triturus vulgaris*), which was rarely found in ponds below pH 6.0 (Cooke and Frazer 1976). In Sweden, Hagstrom (1977) observed that the common toad (*Bufo bufo*) and common frog (*Rana temporaria*) disappeared when the pH levels fell below 4.5 to 4.0. Acidification has been suggested as one factor possibly contributing to the decline of certain amphibian species in the Colorado Rocky Mountains (Corn and Bury 1987; Harte and Hofman 1989), although the current impact of acidification in this area appears minimal (Gibson et al. 1983).

In forest ecosystems, changes in amphibian populations from the physiological effects of acidity could have a number of ramifications. Amphibians are important predators and herbivores (i.e., tadpoles) in these communities and alterations in their abundance or diversity could affect both higher and lower trophic levels. Amphibians, specifically salamanders, represent a significant component of the animal biomass in the forest (Burton and Likens 1975). They are primary energy movers in the ecosystem and provide a high-quality food resource for a variety of vertebrate predators, including snakes, birds, and mammals. A reduction in amphibian populations would presumably affect such tertiary consumers and the subsequent cycling of nutrients.

Few studies relate the ecological effects of acidity on mammals. Feral mink (*Mustela vison*) populations in Norway may have declined because acid precipitation has reduced fish populations, a primary food resource (Bevanger and Albu 1986). In Scotland, significantly fewer signs of otters (*Lutra lutra*) occurred where river pH was low enough to harm fish populations (Mason and MacDonald 1989).

THREATS TO ANIMAL BIODIVERSITY

Introduction

Threats to animal biodiversity from air pollution can be viewed from species and geographic points of view. At the species level, there are some animal species at risk because of their known or likely sensitivity to air pollution. At a geographic level, there are areas in the United States and other countries considered at risk because they are either currently affected by air pollution and have shown effects to species and habitats or, because of projected future development, will become sources or receptors of air pollution and contain species and habitats that are sensitive to air pollution.

Species at Risk

Species at risk are those that, because of their habits, habitats, and known sensitivity, are most susceptible to harm from air pollution. The term "risk" as used here implies potential exposure and vulnerability to air pollutants. There are certain physiologically sensitive species that are more vulnerable to air pollutants (e.g., gill-breathing animals such as fishes, and skin-breathing animals such as amphibians) than other species (e.g., mammals). Also, there are ecologically sensitive species whose habitats or food resources are exposed to air pollution and who are sensitive to these changes. Within this group, there are species that have relatively restricted food requirements so that any change in the quantity or quality of their food resources from air pollution (or other factors) would affect their survival as a population or as a species. There are also species that have narrow habitat requirements or are habitat sensitive. A loss or change in these habitats due to air pollution (factors) would affect the existence or distribution of these species. Also, there are animal species whose habitats have a greater exposure potential to air pollution than other species (e.g., aerial-feeding birds compared with soil burrowing mammals, freshwater fishes from first-order streams compared with estuarine and marine fishes).

Physiologically Sensitive Species

Numerous species of freshwater animals are considered physiologically sensitive to acidifying air pollutants. In general, higher order organismal groups (up to fish) are less diverse and less able to adapt to acidification. This sensitivity or risk to acidity for community-level effects (i.e., decrease in total abundance or productivity) has been characterized as follows: fish > invertebrates (benthic invertebrates and zooplankton) > algae > microbes (Baker 1991).

The sensitivity or the risk to fishes from acidification varies. Some species are at risk at pH levels as high as 6.0 to 6.5, while other species are not affected at pH levels of 4.0 or below. Cyprinid and dace species are at risk as pH decreases from 6.0 to 5.5. Many important sport fish species, e.g., lake trout, are at risk in the pH range of 5.0 to 5.5 (Baker et al. 1990).

Table 10-8 presents the relative risk of different aquatic taxa based on their pH tolerance. Amphibians have been proposed as a group of animals that are physiologically sensitive to acidifying pollutants (Beebee and Griffin 1977; Hagstrom 1977; Corn and Bury 1987; Wymen and Hawksley-Lescault 1987; Freda 1991). Although few systematic studies have been conducted, an international workshop concluded that some of the world's amphibians are declining at an alarming rate. A species decline was reported in Costa Rica, Australia, Europe, and the United States (Phillips 1990). In Colorado, Harte and Hoffman (1989) suggested that long-range transport of acid deposition from Arizona smelters and California smog had increased the acid content of ponds used by salamanders for breeding, and the higher acidity caused a reduction in normally developed eggs. The highest risk for sensitive amphibians occurs in species that use ephemeral ponds or pools of water (Table 10-9).

Ecologically Sensitive Species

Forest habitats are exposed to air pollution in many regions in the world. When these habitats and foods are adversely affected, forest-dependent animal species have been affected. For example, studies in the San Bernardino forest of California suggest that the abundance and distribution of small mammals were changed by the effects of air pollution on the quality of key vegetation and soil habitat requirements (Miller and Eldemman 1977).

TABLE 10-8. Relative Physiological Sensitivity of Different Aquatic Taxa Based on Median Minimum pH Tolerances

Taxonomic Group	Median pH Tolerance
Insects	
Odonata (dragonflies)	6.4
Trichoptera (caddisflies)	6.3
Ephemeroptera (mayflies)	6.0
Hemiptera (bugs)	6.0
Diptera (true flies)	5.6
Coleoptera (beetles)	5.5
Plecoptera (stoneflies)	5.2
Miscellaneous groups	
Pelecypoda (bivalves)	6.6
Gastropoda (snails)	6.6
Hirudinea (leeches)	6.5
Porifera (sponges)	5.5
Crustacea (crustaceans)	5.2
Vertebrates	
Teleostei (fish)	4.9
Anura (frogs)	4.1

Source: adapted from Eilers, Lien, and Berg 1984.

TABLE 10-9. Relative Risk of Amphibian Species in Northeastern North America Whose Range Overlaps Areas Receiving Acidic Deposition

Relative Risk of Air Pollution Effects	Common Name	Habitat
High	Yellow-spotted salamander (*Ambystoma maculatum*)	Meltwater pools
	Blue-spotted salamander (*Ambystoma laterale*)	
	Tremblay's salamander (*Ambystoma tremblayi*)	
	American toad (*Bufo americanus*)	
	Chorus frog (*Pseudacris triseriata*)	
	Wood frog (*Rana sylvatica*)	
	Northern leopard frog (*Rana pipiens*)	
	Northern spring peeper (*Hyla crucifer*)	
	Gray tree frog (*Hyla versicolor*)	
Moderate	Mudpuppy (*Necturus maculosus*)	Permanent ponds
	Red-spotted newt (*Notophthalmus viridescens*)	
	American toad (*Bufo americana*)	
	Gray tree frog (*Hyla versicolor*)	
	Chorus frog (*Pseudacris triseriata*)	
	Bullfrog (*Rana catesbeiana*)	
	Green frog (*Rana clamitans*)	
	Northern leopard frog (*Rana pipiens*)	
	Mink frog (*Rana septentrionalis*)	
	Northern two-lined salamander (*Eurycea bislineata*)	Streams
	Mudpuppy (*Necturus maculosus*)	
	Bullfrog (*Rana catesbeiana*)	Lakes
Low	Four-toed salamander (*Hemidactylium scutatum*)	Bogs
	Red-backed salamander (*Plethodon cinereus*)	Logs and stumps

Source: modified from Clark and Fischer 1981.

The loss of fishes and invertebrates in aquatic habitats impacted by air pollutants may result in risks to birds and mammals that are dependent upon these animals as food resources (Scanlon 1990). Waterfowl that are dependent upon fishes and invertebrates for nutrition are at risk from acidification that causes losses in their food resources. Survival rates and growth rates have been affected. Also at risk are passerine species in Europe that consume aquatic invertebrates and have exhibited decreased reproduction associated with elevated aluminum concentrations in acidic waters (Baker et al. 1990). Mammals such as shrews, mink, and otters, that are dependent upon aquatic food resources and whose range or distribution includes areas known to be affected or susceptible to acid deposition, are potentially at risk.

Water shrews (*Sorex palustris*) occur in regions of North America (Blue Ridge, New England, eastern Canada, and Rockies) that have experienced aquatic effects from acid deposition or occur in regions considered to be potentially sensitive to acid deposition (Banfield 1974; Hall and Kelson 1975). Because shrews have very high metabolic rates, consuming more than their body weight in food per day, they may be particularly vulnerable to loss of prey items. The water shrew is more restricted in its habitat requirements than most other species of shrews. It occurs along borders of ponds and streams in meadows, marshes, and woods and is not a common species (Hooper 1942; Johnson 1951; Wrigley 1969). The primary food of the water shrew is insects, with a high percentage of aquatic larvae that are acid-sensitive (stoneflies, caddisflies, and mayflies) and other aquatic organisms, such as amphibians and small fishes.

The river otter is an aquatic carnivore that may be more at risk from the indirect effects of acid deposition than the mink. It is more dependent on fishes and other aquatic organisms for food than the mink and is more restricted in its habitat. Food habit studies show that fishes can make up 50 to 100 percent of its diet (Greer 1955; Sheldon and Toll 1964; Knudsen and Hale 1968). It has a wider geographical distribution than the mink but is more restricted to stream and lake habitats. It is also found in regions of North America affected by acid deposition, including New England, New York, the Smoky Mountains, Nova Scotia, eastern Canada, and the Rocky Mountains.

Similar risks have been identified for birds (Longcore et al. 1987; McNicols, Bendell, and McAuley 1987). Based on feeding strategies, DesGranges (1987) projected the effects of forest dieback on associated avifauna. Studies suggest the reduction of preferred prey species for certain insectivorous birds could affect their distribution (DesGranges, Maufette, and Gagnon 1987). Table 10-10 lists those North American avian species considered sensitive to reduction in food resources from acid precipitation.

Lichens are one plant group that is sensitive to air pollution such as SO_2 and possibly acid rain and exhibit geographical declines (Fritz-Sheridan 1985; Gilbert 1986; Sigal and Johnston 1986a, b). Richardson and Young (1977) reviewed the relationship of lichens and vertebrates. A few species of mammals are dependent on lichens for food; for example, caribou (*Rangifer tarandus*) are dependent upon lichens for winter food (Thompson and McCourt 1981). The distribution of caribou in eastern Canada (Hall and Kelson 1975) overlaps areas affected by acid rain. Scott and Hutchinson (1987) have demonstrated short-term declines in photosynthesis to caribou-forage lichens, *Cladina stellaris* and *C. rangiferina*, by simulating acid deposition episodes. Although widespread declines in lichens because of acid deposition have not been reported in these regions, if acid precipitation continues there is a potential risk to caribou from the loss of its primary winter food (Singer and Fischer 1984).

TABLE 10-10. Ecologically Sensitive Avian Species Most Likely To Be Influenced by a Reduction in Food Resources Due To Acid Precipitation

Species	Feeding Habitat	Potential Effect from Air Pollution
Common loon (*Gavia immer*)	Lakes	Reduced biomass of fish, aquatic invertebrates, amphibians
Osprey (*Pandion haliaetus*)		
Great blue heron (*Ardea herodias*)	Littoral zone	
American bittern (*Botaurus lentiginosus*)		
Belted kingfisher (*Ceryle alcyon*)		
Hooded merganser (*Lophodytes cucullatus*)		
Common merganser (*Mergus merganser*)		
Common goldeneye (*Bucephala clangula*)	Littoral zone	Reduced biomass of aquatic invertebrates
Red-breasted merganser (*Mergus serrator*)		
Ring-necked duck (*Aythya collaris*)		
Black duck (*Anas rubripes*)		
Virginia rail (*Rallus limicola*)		
Spotted sandpiper (*Actitis macularia*)	Riparian	
Eastern kingbird (*Tyrannus tyrannus*)	Wetlands	Reduced biomass of aquatic invertebrates with adult stage terrestrial invertebrates
Eastern phoebe (*Sayornis phoebe*)		
Tree swallow (*Tachycineta bicolor*)		
Barn swallow (*Hirundo rustica*)		
Bank swallow (*Riparia riparia*)		
Yellow-rumped warbler (*Dendroica coronata*)		
Blackpoll warbler (*Dendroica striata*)		
Palm warbler (*Dendroica palmarum*)		
Common yellowthroat (*Geothlypis trichas*)		

Source: modified from Clark and Fischer 1981.

Geographical Threats

Local and Regional Threats

Animals are threatened at local and regional levels. Animals in the vicinity of air pollution sources may be at risk depending upon the type and magnitude of air emissions. In the past (see Tables 10-1 and 10-2), the risk was high to animals in the vicinity of air pollution sources. Adoption of stringent air pollution controls in developed countries has generally reduced this type of risk. However, bioaccumulation and biomagnification of metals in animals still occurs in the vicinity of major industrial sources (e.g., Beyer et al. 1985). Also, accidental releases still occur and harm animals and their habitat. Unique meteorologic conditions (e.g., acidic fogs coinciding with migrations of birds) can create an exposure and hazard that may place certain animal populations at risk. Certain industrial activities, such as oil and gas production, continue to harm small numbers of animals near these facilities. In developing countries and countries in Eastern Europe without proper air pollution controls, the threat to animals from local air pollution sources is high. Regional threats to animals from air pollution is the most common and widespread problem in developed countries. In North America, acidification effects to aquatic organisms have been reported for lakes and streams in the Adirondacks of New York and the LaCloche Mountain region of Ontario (Sullivan 1991). Physical chemical changes have been reported for headwater streams in Pennsylvania, North Carolina, the Muskoka-Halburton region of Ontario, and coastal areas of Nova Scotia (Altshuller and Linthurst 1984). Sensitive aquatic systems have been identified in much of eastern Canada and New England, parts of the Allegheny, Smoky, Rocky, and Sierra Nevada Mountains, and the northwest and north-central United States (Sullivan 1991). Since the sensitivity of surface waters to acidification is partly a function of the local soils and watershed characteristics, the response of species and communities is often most evident at the local scale.

The estimated percentage of National Surface Water Survey (NSWS) waters with acid-base chemistry unsuitable for survival of acid-sensitive fish species range from less than 5 percent in areas in the Upper Midwest to nearly 60 percent for upper stream reaches in the Mid-Atlantic Coastal Plain. Twenty-three percent of the Adirondack Lakes and 18 percent of the Mid-Appalachian streams have waters unsuitable for brook trout survival (Baker 1991).

Regional threats become problems as lesser developed countries industrialize (e.g., China). Acid deposition is considered the greatest regional air pollution threat, affecting aquatic animals in both temperate and some tropical regions of the world (Rodhe and Herrera 1988). Acid deposition and chronic fumigation by sulfur dioxide, other acidifying pollutants, and particulates have caused, and will continue to cause, a great regional threat to animal populations in Czechoslovakia, eastern Germany, and Poland until adequate pollution control is implemented. Regional losses of species or species attributes have caused potential management problems. For example, in Nova Scotia, the concern over loss of Atlantic salmon has promoted liming to create deacidified refuges to protect and preserve genetic diversity (Watt 1987).

Interspecific and intraspecific differences in the response of species to acidic deposition may result in changes in local populations and regional distributions. For example, studies have shown wide differences in the embryonic tolerances of amphibians to the toxicity of acidic waters (e.g., Dale, Freedman, and Kerekes 1985; Pierce 1985; Freda 1986; Pierce and Harvey 1987). Corn and Bury (1987) found that western populations of wood frogs were more sensitive to acidity than their eastern counterparts. Over time, these types of species-specific traits may provide enough competitive advantage that some species populations and distributions will change. Distribution of amphibian species also may be influenced by microhabitat placement of egg masses and timing of oviposition in ponds near the edge of a species-tolerance range and by soil acidity (Wyman and Hawksley-Lescault 1987), both of which may be influenced by atmospheric deposition.

Transboundary Threats

Significant transboundary ecological effects from air pollution occur in Europe and North America. Scandinavian countries are affected by air pollution from England and especially from Eastern Europe. Air pollution from industrial areas of Russia is transported to the Arctic. Eastern European countries are both exporting and importing air pollution, which is having significant ecological effects on terrestrial and aquatic habitats in the region. The United States and Canada are both contributing to cross-border acid deposition. Air pollution from industries and pesticides applied in Mexico are transported to the United States.

Although a complete assessment of the transboundary habitat effects of air pollution from Czechoslovakia is not available at this time, there are major ecological problems. In 1982, 22.4 percent of the spruce forest on the Bavarian side of the Czechoslovak-German border was damaged by air pollution from Czechoslovakia. The Krusne Hory forests along the border of the former East Germany and Czechoslovakia are severely damaged by air pollution from both countries. As of 1984, 50 percent of the forest of the Krkonose National Park of Czechoslovakia, which lies along the border with Poland, was estimated to be damaged by air pollutants, including SO_2. About 47 percent of the sulfur deposition in this region came from eastern Germany and 26 percent from Poland. Western Slovak Republic forests are affected by air emissions from sources in Poland's Upper Silesia and the Czech Republic (Czechoslovak Academy of Science 1989). The bird diversity in both these forests has been significantly degraded (Stastny and Bejcek 1983; Stastny, Bejcek, and Barta 1987).

Long-range transport of air pollutants makes their effects and control an international problem. Early concerns about declining fishery resources in Scandinavian countries drew public attention to the issues (Drablos and Tollan 1980). Decline in top carnivores has been related to pollutants that reach remote areas by airborne transport (Wren, Stokes, and Fischer 1980).

Various initiatives and treaties to reduce air pollution have been proposed. The United States, Canada, and the European Economic Communities (EEC) have signed agreements to reduce SO_2. The proposed reductions, however, are based primarily on political criteria rather than ecological criteria. Also, most Eastern European countries that are signatories to the EEC treaty (e.g., Poland and Czechoslovakia) will be unable to meet these reductions because of economic and political conditions. Therefore, the ecological effectiveness of these reductions, even if they occur, are not likely to reduce the threat to biodiversity of Central Europe and Scandanavia in the near future. The continuing industrial development of lesser developed nations occurs with minimal air pollution control systems. Scrubbers and flue gas desulfurization equipment required in the Western developed countries are often too expensive and complicated for the developing countries to maintain. Thus, acidification and other results of long-range transport of byproducts from fossil fuel combustion will continue to occur at national and international scales.

The International Council of Scientific Unions, specifically the Scientific Committee on Problems of the Environment (SCOPE), recently evaluated the problem of acidification in tropical countries (Rodhe and Herrara 1988). They concluded that large areas of southern

China, southwestern India, southeastern Brazil, northern Venezuela, and parts of equatorial Africa have soils that are sensitive to acidification. Based on the acidification effects observed in North America and Europe (Rodhe and Hemera 1988), the loss of sensitive stream fauna and subsequent effects to other water-dependent species can be expected from acidification in these regions.

The transboundary effect of radiation is of great concern, not only to human health but also to ecological health. This is best exemplified by an explosion at the nuclear reactor facility at Chernobyl in the Russian Ukraine on April 26, 1986, resulting in the release into the atmosphere of 3.5 percent of the reactor's radioactive material. This corresponds to 2×10^{18} Bq. Elevated radiation levels were recorded in Sweden the next day. Terrestrial and aquatic ecosystems over 1,000 km away were contaminated by radioactive fallout within a matter of days. Bioaccumulation of radioactive elements was observed in fishes, shellfishes, and aquatic birds. Although radioactive contamination occurred, Petersen, Landner, and Blanck (1986) concluded that the observed levels in Sweden were 20 times below the level reported to cause adverse effects in the short term.

The transboundary effects of the Chernobyl accident also were observed in Finland. Air currents carried the nuclear cloud from Chernobyl in the Ukraine to Finland in a matter of days. The air currents also were favorable for the migration of lepidoptera, and sampling by Mikkola and Albrecht (1986) revealed both migrating species (*Plutella xylostella*) and local species (*Achlys flavicornis*) of moths had detectable radiation. It was concluded that the local species became contaminated by radiation on the resting surface of trunks and branches. The migrating species was thought to have been contaminated closer to Chernobyl. Still, the amount of radiation observed was very low and not considered harmful. Also, the authors concluded that the mutation rate may be higher than normal but, because natural selection will eliminate the most harmful mutation, this effect would not be observable. In another case of transboundary contamination, Kettlewell and Heard (1961) captured a lepidopterian species, *Nomophila noctuella*, that was contaminated by radiation and was considered to have been exposed to a French atomic bomb experiment in the Sahara.

The transboundary effect from pesticides, another class of airborne pollutants, is a continuing threat to animal species. Atmospheric transport and organic derivations of pesticides have been reported in biota from remote areas of eastern North America and Sweden. DDT isomers, along with other organic compounds, continue to be found in the tissues of fishes, e.g., trout (*Salvelinus* sp.), sucker (*Catostomus* sp.), pike (*Esox lucius*), and muskelunge (*Esox masquinongy*) (Lockerbie and Clair 1988). Similar contamination has been reported in lake trout (*Salvelinus namaycush*) and whitefish (*Coregeonus culpeaformis*) from a remote lake on Isle Royale, Lake Superior (Swackhammer and Hites 1988). Bioaccumulation has also been found in fishes from remote lakes in Maine, New Hampshire, and Vermont (Haines 1983).

Other Threats

Other air pollution-related effects have animal biodiversity implications, including stratopheric ozone depletion and global warming. The significance of these effects is being debated and, at this time, a scientific consensus does not exist on specific mechanisms and magnitude of effects. Air emissions from industrial and urban sources contribute to the increase in greenhouse gases. The consensus is that increased greenhouse gas concentration will change global climate; the rate and magnitude of the change are not certain. The ecological effects and particularly the effects on animal biodiversity could be significant. The following is a summary of some of the predicted effects on animal biodiversity from global warming. This information was developed from recent reviews (Smith and Tirpak 1989; Marshall-Forbes 1991).

- The distribution of natural ecosystems, such as forests, will change both latitudinally and altitudinally, and the animals inhabiting these natural systems will have their ranges and distributions changed. For example, the southern boundaries of sugar maple and hemlock forests in the eastern United States are expected to shift northward by 600 to 700 km. The northern boundary will not expand as far or as rapidly, and the size and geographical distribution of these forests will decline.
- Global warming could result in a 26 to 66 percent loss of coastal wetlands from a 1-meter rise in sea level. Habitat loss will impact fisheries, water birds, and associated wetland biota.
- The extinction of animal species will increase. Effects to freshwater fishes will vary. Some species will decline (e.g., cold-tolerant species) and others increase (e.g., warm water species). Marine species will be similarly affected. With the loss of coastal wetlands, marine and estuarine species that depend upon these habitats for food and nursery areas will decline.
- Migratory birds will be affected both positively and negatively. Some arctic-nesting herbivores will increase with increased temperatures and arctic productivity, but continental shore birds and nesters will decline from the loss of habitat. If the inland prairie potholes change because of midcontinental dryness, abundance of waterfowl will decline.

Overall, the indirect effects of climatic change on habitats, food availability, and predator/prey relationships may be more profound than the direct physiological effects of climatic change. Several species are potentially susceptible to these changes.

1. Blue grouse (*Dendragapus obscurus*) populations in the southwestern United States may increase from predicted increased productivity in needleleaf forest.
2. Coldwater rainbow trout (*Oncorhynchus mykiss*), already at the edge of its range, may be reduced or extirpated by increases in water temperatures.
3. Brine shrimp would be affected by changes in salinity.
4. Animals with poor dispersal powers, like snails and flightless beetles, may not be able to migrate fast enough to follow rapid climatic changes.
5. The productivity of canvasback ducks (*Aythya valisineria*) may increase with the increase in arctic breeding areas.

A number of mammals in New Mexico that are considered sensitive to altitudinal habitat changes could be affected, including the marmot (*Marmota* sp.), whitetailed jack rabbit (*Lepus townsendii*), red squirrel (*Tamiasciurus hudsonicus*), short-tail weasel (*Mustela* sp.), pine martin (*Martes americana*), tassel-eared weasel, mountain goat (*Oreamnos americanus*), elk (*Cervus elaphus*), and bighorn sheep (Marshall-Forbes 1991).

Botkin, Woodby, and Nisbet (1991) indicated that the Kirtland's warbler (*Dendroica kirtlandii*) may be very susceptible to climatic warming and act as an early indicator of such changes in North America. The Kirtland's warbler nests only in the lower peninsula of Michigan in young (6 to 21 years old) jack pine stands (*Pinus banksiana*). Climatic projections indicate that jack pine might decline rapidly with global warming, resulting in irreversible loss of breeding habitat for the Kirtland's warbler.

MONITORS OF AIR POLLUTION EFFECTS ON BIOLOGICAL DIVERSITY

Monitoring of Animal Biodiversity

Monitoring the effects of air pollution on animals is necessary to determine the effects to a species population from a particular air quality situation, and to determine the effects to animal biodiversity. Basic questions that need to be answered to fully understand the significance of the effects to species populations and, ultimately, animal biodiversity include:

1. What air pollutant(s) is (are) causing the effects?
2. What is the ambient concentration of the pollutant(s)?
3. What is the source of the air pollution?
4. What are the exposure conditions?
5. What species are affected by air pollution?
6. What is the nature of the effects?
7. Are other stressors contributing to the effects?
8. Do these effects threaten the existence of the species locally, regionally, or globally?
9. What corrective actions can be taken to minimize or eliminate the effects?
10. What is the most appropriate monitor or ecological indicator of the future status of the species?

In the past, most reports on the effects of air pollutants on animals have primarily provided information on the type of air pollutant (Question 1) and primary species affected (Question 5). Answers to the other questions generally have not been provided.

The U.S. Environmental Protection Agency (EPA), as part of its Environmental Monitoring and Assessment Program (EMAP), is developing recommendations on ecological monitors for various environmental perturbations, including air pollution (US EPA 1990). For monitoring the status of an organism, species abundance and distribution can be estimated directly (e.g., population density) or indirectly (e.g., habitat quality, chemical or physiological measures such as body burdens). For those species that are difficult to monitor, the status and trends of other species that are equally or more sensitive may be appropriate to monitor, e.g., sentinel organisms and bioassay organisms (Newman 1980). Small mammals may be particularly suitable for monitoring specific groups of pollutants (Talmage and Walton 1991).

Furthermore, the EPA has concluded that a number of traditional measures of species diversity, including species richness, information theory diversity indices, and traditional indicator species have limited value for monitoring some types of ecological conditions (Hunsaker and Carpenter 1990). These methods may also be of limited use as measures of animal biodiversity if used alone. For example, the Shannon-Weaver index for species diversity combines species richness with a measure of evenness of relative abundance, but species identity is not reflected in the measure (Noss and Harris 1986). Relative abundance or presence or absence of select species may be more appropriate (Hunsaker et al. 1990). As with individual species, indirect measures of abundance and distribution through measurement of habitat quality may be important in evaluating the effects of air pollution on animal biodiversity in certain areas.

Animals as Monitors of Ecosystem Diversity

Animals may be monitors of ecosystem diversity as affected by air pollution (Newman and Schreiber 1984). The response of species and their populations likely serve as indicators of air pollution stress to ecosystems. For example, the house martin (*Delichon urbica*), an insectivorous migratory bird, is a sensitive biological indicator of acidification and particulate air pollution in urban areas of Europe and Britain. Population censuses of the house martin in Czechoslovakia have shown the species to be rare or absent in areas with heavy fluoride, sulfur dioxide, fly ash, cement dust, or nitrogen oxide pollution (Newman 1980). Its numbers were 60 percent lower in moderately polluted areas than in control areas (Newman 1977; Newman and Novakova 1979). As a migratory species, it has avoided potentially suitable nest sites in areas of heavy pollution in favor of cleaner areas. Studies on the sensitivity of the house martin to air emissions suggest that animals found in less "optimal" habitats may be more sensitive to air emissions than portions of the same population found in more "suitable" parts of their habitats (Newman and Novakova 1979).

Animals may also be monitors of the status of ecosystem structure and function (Newman and Schreiber 1984). For example, birds, mammals, fishes, and insects can indicate the alteration of community energetics from acidification and particulate pollutants including SO_2 and lead. Alteration of community energetics has been observed in grasslands, coniferous forest, mixed deciduous-coniferous forest, and urban, agricultural, and lake ecosystems in North America and Europe. The response of insects and arthropods to SO_2 has reflected altered energy ratios in grasslands, agroecosystems, and lakes. Effects of air emissions observed in fishes and aquatic invertebrate food chains and webs have caused food-chain simplification and the favoring of herbivores over carnivores. For example, Hillman and Benton (1972) observed declines in secondary consumers (parasitic wasps) and increases in primary consumers (aphids) with increasing levels of SO_2. They indicate that indirect effects at the primary producer levels cause a decline in pollinators and pollen collection with increasing SO_2.

Because acidity influences species composition and distribution in aquatic invertebrates, individual species and species assemblages have been used as biological indicators of the aquatic ecosystem condition (e.g., Engblom and Lingdell 1984). Changes in aquatic ecosystem productivity have been indicated by the response of fish populations to acid precipitation in North America and Europe. The decrease of acid-intolerant or increase of acid-tolerant fish species in an acidified lake or stream can affect the productivity of the system. Because fishes influence invertebrate size distribution, diversity, and number of species, they have an important role in directing the population dynamics of the supporting food chain (Hall and Likens 1980). When fishes disappear from acidified aquatic ecosystems, they are replaced by invertebrate predators with different feeding strategies (Fritz 1980). Thus, a selective impact on the aquatic prey species occurs with potentially changing productivity in the system (Henrikson, Oscarson, and Stenson 1980; Overrein, Seip, and Tollan 1980). The consequences of this change in productivity can be both biological and socioeconomic, as exemplified by the reduction in productivity of Atlantic salmon in acidified waters (Sochasky 1981).

Often pollution effects will be most noticeable in carnivores because of bioaccumulation/biocontraction through the food chain. This is particularly evident with airborne pollutants such as pesticides (Stickel 1975). The alteration of ecosystems by acidification and particulate pollutants is most noticeable through changes in age-class structure, niche specialization, behavioral interactions, life cycles, reproductivity status, and morbidity/mortality rates of birds, mammals, fishes, insects, and amphibians (Newman and Schreiber 1984).

Animals can serve as monitors of the fate and transport of metals through ecosystems. Arsenic, cadmium, chromium, lead, selenium, tin, vanadium, and mercury (including methylmercury) have shown a potential for biotransformation, and associated changes in metal loading have been observed in fishes, insects, birds, and mammals. For instance, nickel levels in the pelage of red squirrels (Lepage and Parker 1988) and the primary feathers of ducks (Ranta, Tomassini, and Nieboer 1978) is a potential diagnostic in relating distance from pollution source and duration of environmental exposure. Studies in West Germany have found that concentrations of lead and cadmium in the moulted primary feathers of breeding female goshawks (*Accipiter gentilis*) are correlated with wet deposition rates of the metals and, therefore, may serve as biomonitors of landscape pollution (Dietrich and Eilenberg 1986). Hahn, Hahn, and Ellenberg (1989) found feathers integrate heavy metals from the atmosphere in proportion to time and space of a bird's home range. Studies of nestlings of insectivorous birds, such as the pied flycatcher (*Ficedula hypoleuca*), have shown heavy metal concentrations that correlate well with the distance of their nests from a pollutant source (Nyholm 1987). Changes in species composition of benthic invertebrates have been used to develop acidification indices (Fjellheim and Raddum 1990).

LAWS AND PROGRAMS TO CONTROL THE EFFECTS OF AIR POLLUTION ON ANIMAL BIODIVERSITY

The continuing loss of biodiversity is a global crisis (Office of Technology Assessment 1987). Protection of animal biodiversity from air pollution likely will be from existing national and international laws and policies.

International Laws and Programs

Present international laws and policies governing animal biodiversity are generally narrow, focusing on single species or groups of species whose existence is threatened, often by overexploitation. In addition, most countries have endangered species protection policies, but not all countries provide adequate enforcement of such policies. We are not aware of any countries with specific policies or laws that protect animals from air pollution effects, although the Clean Air Act of the United States does consider ecological effects related to biodiversity.

International policies on air pollution (e.g., Convention on Long-Range Transboundary Air Pollution on the Reduction of Sulfur Emissions, Montreal Protocol on Substances That Deplete the Ozone Layer) do not specifically address animal biodiversity. However, a number of international laws, conventions, and declarations protect animal species and could be used as tools for such protection, including:

1. Convention on International Trade in Endangered Species of Wild Fauna and Flora (CITES), 1973;
2. Convention on Wetlands of International Importance Especially Waterfowl Habitat, known as the Ramsar Convention, 1971;

3. Convention Concerning the Protection of the World Cultural and Natural Heritage, 1972;
4. Convention on the Conservation of Migratory Species of Wild Animals, 1979;
5. The Declaration on the Human Environment, especially Principle 2, which addresses fauna as well as other biological resources, 1972;
6. World Conservation Strategy (WCS), prepared by International Union for Conservation of Nature (IUCN), initiated in 1980; and
7. World Charter for Nature adopted by the United Nations General Assembly in the 1982 (United States dissented), which has several principles related to biodiversity.

Several treaties protect specific birds (e.g., Migratory Bird Act), if the habitat preservation aspects are emphasized. Although these conventions, policy statements, and treaties do not directly address air pollution, they have as their thrust the protection of animals from human impacts (including pollution), and call for the preservation and management of animal biodiversity. For air pollution problems that have transboundary effects on animal biodiversity, the signatories of the agreements have an obligation to prevent such effects from occurring, either by taking corrective action or notifying the world community of such conditions.

Another application of these agreements for protection of animal diversity involves the future development of emission sources that are funded by bilateral [e.g., U.S. Agency for International Development (USAID)] or multilateral (e.g., World Bank) lending agencies. These lending agencies are bound by environmental policies (e.g., USAID and the National Environmental Policy Act and Executive Order 12114) or environmental directives (e.g., the World Bank and Operational Directive 4:00) to assure such projects are compatible with international environmental policies. These lending institutions have adopted specific policies for the protection of biodiversity.

Because there is generally no international convention for protection of animal biodiversity, adherence to these other conventions, agreements, and policy statements is not consistent. In recent years, there has been a growing international environmental awareness, and conservation organizations are becoming more involved in monitoring international conditions and development programs. Preserving and protecting biological diversity on U.S. Federal lands has become a policy issue (Keystone Report 1991).

National Laws and Programs

In the United States, certain types of animals as well as other ecological resources are protected from air pollution effects by the Clean Air Act (CAA) of 1970 and subsequent amendments (Avery and Schreiber 1979). Revisions to the CAA were made in 1990. The CAA Amendments now include revisions that, in part, have a basis in protecting animal and plant biodiversity. These include programs to control acid deposition (SO_2 and NO_x emissions) and ozone depletion (phasing out of CFC and halons), to set regulations to control air pollution from Outer Continental Shelf oil and gas production activities, and to establish a program to monitor transboundary air pollution along the border of Mexico and the United States.

The CAA provides for the Federal land manager (e.g., U.S. Fish and Wildlife Service, National Park Service, U.S. Forest Service) responsible for certain designated parks, wilderness areas, and other preserves known as Class I areas to take an affirmative responsibility in protecting air quality-related values (AQRV) for such areas. The U.S. Department of the Interior in 1978 administratively defined AQRV to be:

All those values possessed by an area except those that are not affected by changes in air quality and include all those assets of an area whose vitality, significance, or integrity is dependent in some way upon the air environment. These values include visibility and those scenic, cultural, biological, and recreational resources of an area that are affected by air quality.

Important attributes of an area are those values or assets that make an area significant as a national monument, preserve, or primitive area. They are the assets that are to be preserved if the area is to achieve the purposes for which it was set aside [Federal Register 43(69):15016. 1978].

Except for visibility, AQRVs have not been specifically defined. However, land managers have identified odor, soil, flora, fauna, cultural resources, geological features, water, and climate as AQRVs. Specific AQRVs need to be defined for each Class I area (US EPA 1990). Inherent in the concept of AQRV are animals as indicators and integrators of ecosystem response to air quality change. Although CAA requires consideration of ecosystem impacts by air pollution sources, often these impacts are only cursorily discussed in most permitting processes. Likewise, the concept of biodiversity may be implied in the consideration of AQRVs, but it is not specifically addressed in current air quality management.

Another major United States Federal legislative action directed at management and regulation of the air resource is the Acid Precipitation Act of 1980 (42 USC 8901 et seq.). This legislation provided for the development of a 10-year research program on acid deposition, specifically including aquatic and terrestrial effects, monitoring, and assessment. In particular, it emphasized research on fish resources, which are among the most important aquatic ecosystem components to indicate biological damage from the acidification process (Interagency Task Force 1982). Much research authorized by the 1980 amendments to the Act has been completed (NAPAP 1990). The 1990 Amendments also limit the emissions of acid precursors and implemented the development of a program to assess the status and effectiveness of the controls as well as authorized additional acid deposition research and ecological assessment to determine the effects of the emission controls. The 1990 Amendments to the CAA (Section 405) also require the EPA to create a National Acid Lake Registry that will list all lakes known to be acidified. When a registry is developed, it may be an indirect measure of the effects of acidifying pollutants on aquatic animal biodiversity.

These acts are supported by other relevant legislation, such as the Endangered Species Act, which provides mechanisms for analysis and review by Federal Agencies of ecological issues relating to actual or potential impacts by air pollution. The National Park Service, U.S. Fish and Wildlife Service, and the U.S. Forest Service over the past 10 years have become more active in evaluating the ecological effects of air pollution.

Air Quality Standards and Animal Effects

In the United States, major air pollutants are controlled by the CAA and its 1990 amendments. Air emissions are regulated to control health effects either directly, by specific ambient air quality standards (NAAQS) or indirectly, by New Source Performance Standards (NSPS), which place emission limits for air pollution sources. Air pollutants are classified into criteria pollutants for which air quality ambient standards have been established and noncriteria pollutants for which ambient standards have not been promulgated. Currently only six criteria pollutants (carbon dioxide, nitrogen dioxide, sulfur dioxide, particulate matter, ozone, and lead) are regulated not only on the basis of emission rates but also as ambient concentrations (ambient air quality standards).

The 1990 Amendments to the CAA (Title I, Section 112) list 189 toxic air pollutants whose emissions must be reduced within 10 years. Toxic air pollutants or air toxics are defined by Section 313 of the Superfund Amendments and Reauthorization Act (SARA) of 1986. They include 308 chemicals and 20 chemical categories. The chemicals include a large number of organics (e.g., benzene, ethylene, vinyl chloride), as well as pesticides (e.g., lindane and aldrin, and 2,4-D), inorganics (e.g., nitric acid, sulfuric acid), and metals (e.g., arsenic, mercury, cadmium). The 20 chemical categories primarily are metallic compounds, along with polychlorinated biphenyls (PCBs), chlorophenols, and glycol ethers. EPA has published a list of industrial processes and sources that emit air toxics, including chemical plants, chrome-plating facilities, coal-fired power plants, dry cleaners, and nonpoint sources, such as landfills, mines, and residential wood combustion.

For regulated air pollutants, the Federal and State governments set two types of standards: primary standards for the protection of human health and secondary standards for the protection of the public welfare. The public welfare includes protection of animals (Avery and Schreiber 1979). Secondary standards to protect the public welfare have been set in the United States for only six air pollutants (Table 10-11). Also, secondary standards have only emphasized direct effects from short- and long-term exposures. A review of the animal effects studies reveals that biological effects can occur at and even below the promulgated standards (Newman and Schreiber 1988) and these standards do not ensure protection for animals and their habitats (Avery and Schreiber 1979).

In industrialized countries, most effects from these regulated pollutants are from chronic exposure. In developing countries, where there are inadequate or no standards or pollution controls, both chronic and acute exposure can occur. For the remaining and majority of air pollutants that are unregulated, both acute and chronic exposure are possible. Economics often do not permit the necessary air quality protection.

For fluoride, specific secondary standards have been set for food-web effects to animals. These standards have been adopted in some U.S. States and in a number of other countries (e.g., Canada and Poland). Standards for fluoride levels in forage have been set to protect domestic animals; however, they are designed to protect the economic value of the animal rather than to prevent fluorosis, a debilitating disease in animals, including wildlife, that can occur even if the standards are not exceeded (Newman 1984).

In the United States, monitoring data indicate that a number of criteria for pollutants are being exceeded. For regulatory purposes, these are designated by EPA as nonattainment areas. Large areas in the United States are exceeding the standards (US EPA 1991). For example, much of urban California has been unable to meet ozone standards. Studies in these areas as early as 1970's have shown negative effects to animals and habitats (Light 1973; Miller and Eldemman 1977; Richkind 1979).

High levels of lead have been found in rock doves in urban areas that do meet the secondary ambient air quality standard for lead (Tansey and Roth 1970). Secondary lead exposure occurs in urban peregrine falcon that prey on urban rock doves (DeMent et al. 1986). Of the trace metals, only lead has an ambient air quality standard, and that is a quarterly average concentration (Table 10-11). The U.S. Fish and Wildlife Service (e.g., Eisler 1985a, b, c) has published a series of reviews on various contaminants that included recommendations for ambient levels of cadmium, chromium, and selenium in air and

Table 10-11. National Ambient Air Quality Standards (NAAQS)

	Primary (Health-Related) Criteria		Secondary (Welfare-Related) Criteria	
	Standard Level		Standard Level	
Pollutant	Averaging Time	Concentration[a]	Averaging Time	Concentration
PM10	Annual Arithmetic Mean[b]	50 $\mu g/m^3$		Same as Primary
	24-hour[b]	150 $\mu g/m^3$		Same as Primary
SO_2	Annual Arithmetic Mean	(0.03 ppm) 80 $\mu g/m^3$	3-hour[c]	1,300 $\mu g/m^3$ (0.50 ppm)
	24-hour[c]	(0.14 ppm) 365$\mu g/m^3$		
CO	8-hour[c]	(9 ppm) 10 mg/m^3		No Secondary Standard
	1-hour[c]	(35 ppm) 40 mg/m^3		No Secondary Standard
NO_2	Annual Arithmetic Mean	(0.053 ppm) 100 $\mu g/m^3$		Same as Primary
O_3	Maximum Daily 1-hour Average[d]	(0.12 ppm) 235 $\mu g/m^3$		Same as Primary
Pb	Maximum Quarterly Average	1.5 $\mu g/m^3$		Same as Primary

[a] Parenthetical value is an approximately equivalent concentration.

[b] TSP was the indicator pollutant for the original particulate matter (PM) standards. This standard has been replaced with the new PM10 standard and is no longer in effect. New PM standards were promulgated in 1987 using PM10 (particles less than 10u in diameter) as the new indicator pollutant. The annual standard is attained when the expected annual arithmetic mean concentration is less than or equal to 50 $\mu g/m^3$; the 24-hour standard is attained when the expected number of days per calendar year above 150 $\mu g/m^3$ is equal to or less than 1, as determined in accordance with Appendix K of the PM NAAQS.

c Not to be exceeded more than once per year.

d The standard is attained when the expected number of days per calendar year with maximum hourly average concentrations above 0.12 ppm is equal to or less than 1, as determined in accordance with Appendix H of the ozone NAAQS.

Source: US EPA 1991.

tissues. These recommendations can be used for evaluating the risks to fishes and wildlife from air pollution.

In less developed countries and in countries of Eastern Europe, air pollution legislation and controls either are not enforced or do not exist. In Czechoslovakia, for example, air quality standards exist that in some cases (e.g., SO_2) are more stringent than United States standards. However, there has been no enforcement. In these countries, severe ecological problems from air pollution are occurring but effects on biodiversity can only be postulated at this time.

CONCLUSIONS

Air pollution has been an increasing threat to animal biodiversity since the advent of the industrial revolution 150 years ago. Air pollution effects were first recorded as a genetic phenomena named "industrial melanism." Over this period of time, the nongenetic effects to animals have been recognized to range from local population extinctions and changes in animal species assemblages to regional extinctions and changes in animal biodiversity. Initially, the threats to animal biodiversity involved primarily terrestrial species; now major threats involve aquatic species. Both vertebrate and invertebrate species have been affected.

Animal biodiversity is affected by acidifying air pollutants, including: sulfur dioxide and derivative acids, particulate matter (including radioactive particulates), photochemical oxidants, and organic compounds (e.g., airborne pesticides). The effects to animals from these pollutants can be direct (e.g., bird mortality from inhalation of hydrogen sulfide) or indirect (e.g., change in pH of streams and adverse physiological effects to fishes). The effects can be physiological effects (e.g., lower reproduction) or ecological effects (e.g., loss of food resources). As a result of the exposure and hazard to animal species from air pollutants, there are species at risk to air pollution because they are more physiologically sensitive to air pollution than other species (e.g., embryonic mortality in amphibians), as well as other species at risk because they are ecologically sensitive to air pollution effects (e.g., change in local distribution in the river otter).

No specific international laws or policies address the effects of air pollution on animal biodiversity, although more general laws or policies exist that can be used to address this problem. In the United States, the Clean Air Act of 1970 and its amendments have provisions for addressing air pollution effects on animals that occur in specially designated national parks, wildlife refuges, and wilderness areas called Class 1 areas. This legislation can be used to address the broader concept of animal biodiversity.

Presently, the major threats to animal biodiversity are regional threats from nitrogen inputs, acidifying pollutants and, possibly, airborne pesticides. Without controls, the animal biodiversity of sensitive aquatic ecosystems will be lost, and extinction of species with limited ranges is likely. These problems already exist in developed countries and are now being observed in developing countries as they industrialize. Transboundary threats are associated with these conditions. Local threats to animal biodiversity are most pronounced in Eastern Europe (where inadequate pollution control exists) and in other developing countries (which also do not have adequate air pollution control).

REFERENCES

1. Albers, P.H., and R.M. Prouty. 1987. Survival of spotted salamander eggs in temporary woodland ponds of coastal Maryland. *Environ. Pollut.* 46:45-61.
2. Allaway, W.H. 1970. Sulphur-selenium relationships in soils and plants. *Sulphur Inst. J.* 6 (3): 3-5.
3. Altshuller, A.P., and R.A. Linthurst, eds. 1984. *The acidic deposition phenomena and its effects: Critical assessment review papers.* Vol. 2, *Effects sciences.* NCSU Acid Precipitation Program, Raleigh, NC. US EPA Office of Research and Development: EPA Report No. 600/8-83-0168BF.
4. Alvo, B. 1987. The acid test. *Living Bird Q.* 6:25-30.
5. Andersen, T., and G.E. Ellefsen. 1982. *Industrial melanism in* Oligia-latruncula *Lepidoptera Noctuidae in the greenland area southeastern Norway.* Copenhagen: Zoologisk Museum.

6. Andren, C., M. Marden, and G. Nilson. 1989. Tolerance to low pH in a population of moor frogs, *Rana arvalis*, from an acid and a neutral environment: A possible case of rapid evolutionary response to acidification. *Oikos* 56:215-23.
7. Arbib, R. 1980. The blue list for 1980. *Am. Birds* 33:830-35.
8. Avery, M., and R.K. Schreiber. 1979. *The Clean Air Act: Its relation to fish and wildlife resources*. US Fish and Wildlife Service, Biological Services Program, National Power Plant Team: FWS/OBS-76/20.8.
9. Babinska-Werka, J., and K. Czarnowska. 1988. Heavy metals in roe-deer liver and alimentary tract and their content in soil and plants in central Poland. *Acta Theriol.* 15 (33): 219-30.
10. Baker, J.P. 1991. Biological effects of changes in surface water acid-base chemistry. In *Acidic deposition: State of science and technology—Summary report of the U.S. National Acid Precipitation Assessment Program*, ed. P.M. Irving. Washington, DC: National Acid Precipitation Assessment Program.
11. Baker, J.P., D.P. Bernard, S.W. Christensen, M.J. Sale, J. Freda, K. Heitcher, D. Marmorek, L. Rowe, P. Scanlon, G. Suter, W. Warren-Hicks, and P. Welbourn. 1990. *Biological effects of changes in surface water acid-base chemistry*. Washington, DC: National Acid Precipitation Assessment Program (NAPAP) Report 13.
12. Baker, J.P., and C.L. Schofield. 1985. Acidification impacts on fish populations—A review. In *Acid deposition: Environmental, economic, and policy issues*, ed. D.D. Adams and W.P. Page. New York: Plenum Publishing.
13. Balazova, G., and E. Hluchan. 1969. Der einfluss von fluorexhalten auf die tiere in der umgebung einer aluminiumfabrik (The influence of fluoride emissions on the animals in the vicinity of an aluminum factory). In *Proceedings of the First European Congress on the Influence of Air Pollution*. Wageningen, Netherlands: Centre for Agricultural Publishing.
14. Banfield, A.W.F. 1974. *The mammals of Canada*. National Museum of Natural Sciences, National Museum of Canada. Toronto: Univ. of Toronto Press.
15. Bassus, W. 1968. Uber wirkungen von industrieexhalton auf den nematodenbesatz im boden von kiefernwaldern (On the effects of industrial emissions on the population of nematoda in the soil of pine forests). *Pedobiologia* 8:289-95.
16. Beebe, W. 1906. *The bird, its form and function*. New York: H. Holt and Co.
17. Beebee, T.J.C., R.J. Flower, A.C. Stevenson, S.T. Patrick, P.G. Appleby, C. Fletcher, C. Marsh, J. Natkanski, B. Rippey, and R.W. Battarbee. 1990. Decline of the natterjack toad *Bufo calamita* in Britain: Paleocological, documentary and experimental evidence for breeding site acidification. *Biol. Conserv.* 53:1-20.
18. Beebee, T.J.C., and J.R. Griffin. 1977. A preliminary investigation into natterjack toad (*Bufo calamita*) breeding site characteristics in Britain. *J. Zool. (Lond.)* 181:341-50.
19. Bendell, B.E., and D.K. McNicol. 1987. Fish predation and the composition of aquatic insect assemblages. *Hydrobiologia* 150:193-202.
20. Bengtsson, G., S. Nordstrom, and S. Rundgren. 1983. Population density and tissue metal concentration of Lumbricids in forest soils near a brass mill. *Environ. Pollut. Ser. A Ecol. Biol.* 30:87-108.
21. Bernes, C., and E. Thornelof. 1990. Present and future acidification in Swedish lakes: Model calculations based on an extensive survey. In *Impact models to assess regional acidification*, ed. J. Kamari. Portrecht, Netherlands: Kluwer Academic Publishers.
22. Bevanger, K., and O. Albu. 1986. Decrease in a Norwegian feral mink, *Mustela vison*, population—A response to acid precipitation. *Biol. Conserv.* 38:75-78.
23. Beyer, W.N., and A. Anderson. 1985. Toxicity to wood lice of zinc and lead oxides added to soil litter. *Ambio* 14:173-74.

24. Beyer, W.N., L.H. Pattee, L. Sileo, D.J. Hoffman, and B.M. Mulhern. 1985. Metal contamination in wildlife living near two zinc smelters. *Environ. Pollut. Ser. A Ecol. Biol.* 38A:63-86.
25. Bicknell, W.B. 1984. *A cooperative hydrogen-sulfide monitoring study: The Lone Butte oil field, McKenzie County, ND.* US Fish and Wildlife Service, Habitat Resources Field Office, Bismarck, ND.
26. Bishop, J.A., L.M. Cook, and J. Muggleton. 1978. The response of two species of moths to industrialization in northwest England. II. Relative fitness of morphs and population size. *Philos. Trans. R. Soc. Lond.* 13:517-42.
27. Bjorge, R.R. 1987. Bird kill at an oil industry flare stack in northwest Alberta. *Can. Field-Nat.* 101 (3): 346-50.
28. Blancher, P.J., and D.G. McAuley. 1987. Influence of wetland acidity on avian breeding success. *Trans. N. Am. Wildl. Nat. Resour. Conf.* 52:628-35.
29. Bloomfield, J.A., S.O. Quinn, R.J. Scrudata, D. Long, A. Richards, and F. Ryan. 1980. Atmospheric and watershed inputs of mercury to Cranberry Lake, St. Lawrence County, New York. In *Polluted rain*, ed. T.Y. Toribara, M.W. Miller, and P.E. Morrow. New York: Plenum Press.
30. Blus, L.G., C.J. Henny, and B.M. Mulhern. 1987. Concentrations of metals in mink and other mammals from Washington and Idaho. *Environ. Pollut.* 44:307-18.
31. Botkin, D.B., D.A. Woodby and R.A. Nisbet. 1991. Kirtland's warbler habitats: A possible early indicator of climatic warming. *Biol. Conserv.* 56:63-78.
32. Brakefield, P.M., and D.R. Lees. 1987. Melanism in *Adalia* ladybirds and declining air pollution in Birmingham. *Heredity* 59:273-77.
33. Brush, T., and E.W. Stiles. 1986. Using food abundance to predict habitat use by birds. In *Wildlife 2000: Modeling habitat relationships of terrestrial vertebrates*, ed. J. Verner, M.L. Morrison, and C.J. Ralph. International Symposium, 7-11 October 1984, Stanford Sierra Camp, Fallen Leaf Lake, CA. Madison, WI: Univ. of Wisconsin Press.
34. Bull, K.R., R.D. Roberts, M.J. Inskip, and G.T. Goodman. 1977. Mercury concentrations in soil, grass, earthworms and small mammals near an industrial emission source. *Environ. Pollut.* 12:135-40.
35. Burton, T.M., and G.E. Likens. 1975. Salamander populations and biomass in the Hubbard Brook Experimental Forest, New Hampshire. *Copeia* 1975:541-46.
36. Carlson, C.E., and J.E. Dewey. 1971. *Monitoring fluoride pollution in Flathead National Forest and Glacier National Park.* USDA, Insect and Disease Branch, Division of State and Private Forestry, Forest Service, Missoula, MT.
37. Carriere, D., K. Fischer, D. Peakall, and P. Angehrn. 1986. Effects of dietary aluminum in combination with reduced calcium and phosphorous on the ring dove (*Streptopelia risoria*). *Water Air Soil Pollut.* 30:757-64.
38. Chilgren, J.D. 1979. Small mammal investigation at ZAPS: Demographic studies and responses to gradient levels of SO_2. In *The bioenvironmental impacts of a coal fired power plant*, 4th Interim Report, December 1978, Colstrip, MT, ed. E.M. Preston and T.L. Gullet. US EPA Environmental Research Laboratory, Corvallis: EPA Report No. 600/3-79-044.
39. Chlodny, J., I. Matuszczyk, B. Styfi-Bartkiewicz, and D. Syrek. 1987. Catchability of the epigeal fauna of pine stands as a bioindicator of industrial pollution of forests. *Ekol. Pol.* 35:271-90.
40. Clark, D.R. 1979. Lead concentrations: Bats vs. terrestrial small mammals collected near a major highway. *Environ. Sci. & Technol.* 13:338-41.

41. Clark, K.L., and K. Fischer. 1981. *Acid precipitation and wildlife*. Draft Report No. 43. Ottowa, ON: Environment Canada, Canadian Wildlife Service.

42. Cleve, K. 1970. Die erforschung der ursachen fur das auftreten melanistischer schmetterlingformen im laufe der letzen hundert jahre (Investigation of the reasons for melanic forms of butterflies in the course of the last hundred years). *Z. Angew. Entomol.* 65:371-87.

43. Cooke, L.M., G.S. Mani, and M.E. Varley. 1986. Postindustrial melanism in the peppered moth. *Science* 231:611-13.

44. Cooke, A.S., and J.F.D. Frazer. 1976. Characteristics of newt breeding sites. *J. Zool. (Lond.)* 178:223-36.

45. Corn, P.S., and R.B. Bury. 1987. The potential role of acidic precipitation in declining amphibian populations in the Colorado Front Range. In *Aquatic effects task group VI peer review summaries*. 17-23 May 1987, New Orleans, LA. Washington, DC: National Acid Precipitation Assessment Program.

46. Corn, P.S., and F.A. Vertucci. In press. Descriptive risk assessment of the effects of acid deposition on Rocky Mountain amphibians. *J. Herpetol.*

47. Cowling, E.B. 1982. Acid precipitation in historical perspective. *Environ. Sci. & Tech.* 16:110-23.

48. Cowling, E.B. 1989. Recent changes in chemical climate and related effects on forests in North America and Europe. *Ambio* 18:167-71.

49. Cramp, S., and J. Gooders. 1967. The return of the house martin. *Lond. Bird Rep.* 31:93-98.

50. Crete, M., R. Nault, P. Walsh, J.L. Benedetti, M.A. Lefebvre, J.P. Weber, and J. Gagnon. 1989. Variation in cadmium content of caribou tissues from northern Quebec. *Sci. Total Environ.* 80:103-12.

51. Cristaldi, M., E. D'Arcangelo, L.A. Ieradi, D. Mascanzoni, T. Mattei, and I. Van Axel Castelli. 1990. Cs determination and mutagenicity tests in wild *Mus musculus domesticus* before and after the Chernobyl accident. *Environ. Pollut.* 64:1-9.

52. Czechoslovak Academy of Science. 1989. *Stav a vyvoj zivotniho prostredi v Ceskoslovensku (State of the development of environment in Czechoslovakia)*. Cesky Svaz Ochrancu Prirody. Prague, Czechoslovakia: Creske Budejovice.

53. Dale, J.M., B. Freedman, and J. Kerekes. 1985. Experimental studies of the effects of acidity and associated water chemistry on amphibians. *Proc. Nova Scotian Inst. Sci.* 35:35-54.

54. Davies, E.B., and J.H. Watkinson. 1966. Uptake of native and applied selenium by pasture species. II. Effects of sulphate and of soil type on uptake by clover. *New Zealand J. Agric. Res.* 8:641-45.

55. Davis, D.D., and R.G. Wilhour. 1976. Susceptibility of woody plants to sulfur dioxide and photochemical oxidants. US EPA Environmental Research Laboratory, Corvallis: EPA Report No. 600/3-76-102.

56. DeMent, D.H., J.J. Chisolm, J.C. Barber, and J.D. Strandberg. 1986. Lead exposure in an "urban" peregrine falcon and its avian prey. *J. Wildl. Dis.* 22:238-44.

57. DesGranges, J.L. 1987. Forest birds as biological indicators of the progression of maple dieback in Quebec. In *The value of birds*, ed. A.W. Diamond and F.L. Filion. International Council for Bird Preservation (ICBP), Technical Publication 6. Cambridge, England.

58. DesGranges, J.L., and M. Darveau. 1985. Effect of lake acidity and morphometry on the distribution of aquatic birds in southern Quebec. *Holarct. Ecol.* 8:181-90.

59. DesGranges, J.L., and M.L. Hunter. 1987. Duckling response to lake acidification. *Trans. N. Am. Wildl. Nat. Resour. Conf.* 52:636-44.

60. DesGranges, J.L., Y. Mauffette, and G. Gagnon. 1987. Sugar maple forest decline and implications for forest insects and birds. *Trans. N. Am. Wildl. Nat. Resour. Conf.* 52:677-89.

61. DesGranges, J.L, and J. Rodrigue. 1986. Influence of acidity and competition with fish on the development of ducklings in Quebec. *Water Air Soil Pollut.* 30:743-50.
62. Dewey, J.E. 1973. Accumulation of fluoride by insects near an emission source in western Montana. *Environ. Entomol.* 2 (2): 179-82.
63. Dickson, W. 1985. Liming in Sweden. In Liming acidic waters: Environmental and policy concerns. Rochester, NY: Center for Environmental Information.
64. Dietrich, J., and H. Eilenberg. 1986. Habicht-Mauserfedern als hochintegrierende, standardisierte umweltproben. *Verh. Ges. Oekol.* (Hohenheim 1984) Band XIV 1986:413-26.
65. Douwes, P., K. Mikkola, B. Petersen, and A. Vestergren. 1976. Melanism in *Biston betularius* from north-west Europe (Lepidoptera: Geometridae). *Entomol. Scand.* 7:261-66.
66. Drablos, D., and A. Tollan, eds. 1980. Ecological impacts of acid precipitation. Proceedings of an International Conference, Sandefjord, Norway. SNSF Project, AS-NLH, Norway.
67. Drent, P.J., and J.W. Woldendorp. 1989. Acid rain and eggshells. *Nature* 339:431.
68. Dvorak, A.J., B.G. Lewis, P.C. Chee, E.H. Dettmann, R.F. Freeman III, R.M. Goldstein, R.R. Hinchman, J.D. Jastrow, F.C. Kornegay, D.L. Mabes, P.A. Merry, E.D. Pentecost, J.C. Prioleau, L.F. Soholt, W.S. Vinikour, and E.W. Walbridge. 1978. *Impacts of coal-fired power plants on fish, wildlife, and their habitats.* US Fish and Wildlife Service, Biological Services Program, National Power Plant Team: FWS/OBS-78/29.
69. Eilers, J.M., G.J. Lien, and R.G. Berg. 1984. *Aquatic organisms in acidic environments: A literature review.* Wisconsin Department Natural Resources Technical Bulletin 150.71.
70. Eisler, R. 1985a. *Selenium hazards to fish, wildlife, and invertebrates: A synoptic review.* Contaminant Hazard Reviews Report No. 5, Biological Report 85 (1.5). Washington, DC: US Fish and Wildlife Service.
71. Eisler, R. 1985b. *Cadmium hazards to fish, wildlife, and invertebrates: A synoptic review.* Contaminant Hazard Reviews Report No. 2, Biological Report 85 (1.2). Laurel, MD: US Fish and Wildlife Service.
72. Eisler, R. 1985c. Chromium Hazards to Fish, *Wildlife and Invertebrates: A Synoptic Review.* Contaminant Hazard Reviews Report No. 6, Biological Report 85 (1.6). Laurel, MD: US Fish and Wildlife Service.
73. Engblom, E., and P. Lingdell. 1984. The mapping of short-term acidification with the help of biological pH indicators. In *Report No. 61, Institute of Freshwater Research*, ed. L. Nyman. Drottningholm, Sweden: National Swedish Board of Fisheries.
74. Eriksson, M.O.G., L. Henrikson, B.I. Nilsson, G. Mymon, H.G. Oscarson, and A.E. Stenson. 1980. Predatory-prey relations important for biotic changes in acidified lakes. *Ambio* 9:248-49.
75. Fennelly, P.F. 1976. The origin and influence of airborne particulates. *Am. Sci.* 64:46-55.
76. Feriancova-Masarova, Z., and E. Kalivodova. 1965. Niekolko poznamok vplyve fluorovych exhalatov v okoli hlinikarne v Ziari nad Hronom na kvantitu hniezdiacich vtakov (The effects of exhalations from the aluminum plant in Ziar nad Hronom on the spectrum of bird species in the vicinity of the plant). *Biologia (Bratisl.)* 20:341-46.
77. Fjellhelm, A., and G.G. Raddum. 1990. Acid precipitation: Biological monitoring of streams and lakes. *Sci. Total Environ.* 96:57-66.
78. Flousek, J. 1989. Impact of industrial emissions on bird populations breeding in mountain spruce forests in Central Europe. *Ann. Zool. Fenn.* 26:255-63.
79. Flueck, W.T. 1990. Possible impact of emissions on trace mineral availability to free-ranging ruminants: Selenium as an example (in German). *Z. Jagdwiss.* 36:179-85.
80. Frank A., L. Petersson, and T. Morner. 1981. Lead and cadmium contents in organs of moose, roe deer, and hare (in Swedish). *Sven. Veterinartidn.* 33:151-56.

81. Freda, J. 1986. The influence of acidic pond water on amphibians: A review. *Water Air Soil Pollut.* 30:439-50.
82. Freda, J. 1991. The effects of aluminum and other metals on amphibians. *Environ. Pollut* 71:305-28.
83. Freda, J., and W.A. Dunson. 1986. Effects of low pH and other chemical variables on the local distribution of amphibians. *Copeia* 2:454-66.
84. Freitag, R., and L. Hastings. 1973. Ground beetle populations near a kraft mill. *Can. Entomol.* 105:299-310.
85. Frenette, J.J., Y. Richard, and G. Moreau. 1986. Fish responses to acidity in Quebec lakes: A review. *Water Air Soil Pollut.* 30:461-75.
86. Fritz, E.S. 1980. *Potential impacts of low pH on fish and fish populations*. US Fish and Wildlife Service, Biological Services Program, National Power Plant Team: FWS/OBS-80/40.2.
87. Fritz-Sheridan, R.P. 1985. Impact of simulated acid rains on nitrogenase activity in *Peltigera aphthosa* and *P. polydactyla*. *Lichenologist (Lond.)* 17:27-31.
88. Gibson, J.H., J.N. Galloway, C. Schofield, et al. 1983. *Rocky Mountain acidification study*. US Fish and Wildlife Service, Division of Biological Services, Eastern Energy and Land Use Team: FWS/OBS-80/40.17.
89. Gilbert, O.L. 1986. Field evidence for and acid rain effect on lichens. *Environ. Pollut. (Series A)* 40:227-31.
90. Gissel-Nielsen, G. 1973. Uptake and distribution of added selenite and selenate by barley and red clover as influenced by sulfur. *J. Sci. Food Agric.* 24:649-55.
91. Glooschenko, V., P. Blancher, J. Herskowitz, R. Fulthorpe, and S. Rang. 1986. Association of wetland acidity with reproductive parameters and insect prey of the eastern kingbird (*Tyrannus tyrannus*) near Sudbury, Ontario. *Water Air Soil Pollut.* 30:553-67.
92. Gooders, J. 1968. The swift in central London. *Birds (Lond.)* 32:93-98.
93. Gordon, C.C. 1969a. *Cominco American report II*. Department of Environmental Studies, Univ. of Montana, Missoula.
94. Gordon, C.C. 1969b. *East Helena report*. Department of Environmental Studies, Univ. of Montana, Missoula.
95. Goriup, P.D. 1989. Acidic air pollution and birds in Europe. *Oryx* 23:83-86.
96. Graveland, J. 1990. Effects of acid precipitation on reproduction in birds. *Experientia (Basel)* 46:962-70.
97. Greer, K.R. 1955. Yearly food habits of the river otter in the Thompson Lakes region, northwestern Montana, as indicated by scat analysis. *Am. Midl. Nat.* 54:299-313.
98. Grodzinska, K. 1977. Changes in the forest environment in southern Poland as a result of steel mill emissions. In *Vegetation science and environmental protection*, ed. A. Miyawaki and R. Tuxen. Proceedings of the International Symposium in Tokyo on Protection of the Environment and Excursion on Vegetation Science through Japan, 1974. Maruzen, Tokyo.
99. Grodzinska, K., W. Grodzinski, and S.I. Zeveloff. 1983. Contamination of roe deer forage in a polluted forest of southern Poland. *Environ. Pollut. (Series A)* 30:257-76.
100. Hagstrom, T. 1977. Grodornas forsvinnande i en forsuard sjo (The extinction of frogs in a lake acidified by atmospheric pollution). *Sver. Nat.* 11:367-69.
101. Hahn, E., K. Hahn, and H. Ellenberg. 1989. Schwermetallgehalte in federn von elstern (*Pica pica*): Folge exogener auflagerung aus der atmosphare? (Heavy metal in the feathers of the magpie (*Pica pica*): Consequence of exogenous deposition from the atmosphere). *Verh. Ges. Oekol.* (Essen 1988) Band XVIII.

102. Hahn, E., K. Hahn, and M. Stoeppler. 1989. Schwermetalle in federn von habichten (*Accipiter gentilis*) aus unterschiedlich belasteten gebieten (Heavy metals in the feathers of the hawk (*Accipiter gentilis*) from different contaminated regions). *J. Ornithol.* 130:303-9.
103. Haines, T.A. 1981. Acidic precipitation and its consequences for aquatic ecosystems: A review. *Trans. Am. Fish. Soc.* 110 (6): 79-93.
104. Haines, T.A. 1983. Organochlorine residues in brook trout from remote lakes in the northwestern United States. *Water Air Soil Pollut.* 20:47-54.
105. Haines, T.A., and J.P. Baker. 1986. Evidence of fish population responses to acidification in the eastern United States. *Water Air Soil Pollut.* 31:605-29.
106. Hais, K., and J. Masek. 1969. Vcinky nekterych exhalaci na hospodarska zvirata (Effects of some exhalations on agricultural animals). *Ochr. Ovzduzi* 3:122-25.
107. Hall, R.J., and K.R. Kelson. 1975. *The mammals of North America.* New York: Ronald Press Co.
108. Hall, R.J., and G.E. Likens. 1980. Ecological effects of whole-stream acidification. In *Atmospheric sulfur deposition: Environmental impacts and health effects*, ed. D.S. Shriner, C.R. Raymond, and S.E. Lindberg. Ann Arbor, MI: Ann Arbor Science.
109. Hansen, P.W. 1987. *Acid rain and waterfowl: The case for concern in North America.* Arlington, VA: Izaak Walton League of America.
110. Haramis, G.M., and D.S. Chu. 1987. Acid rain effects on waterfowl: Use of black duck broods to assess food resources of experimentally acidified wetlands. In *The value of birds*, ed. A.W. Diamond and F.L. Filion. International Council for Bird Preservation (ICBP) Technical Publication No. 6. Cambridge, England.
111. Harr, J.R. 1978. Biological effects of selenium. In *Toxicity of heavy metals in the environment*, ed. F.W. Ochme. New York: M. Dekker.
112. Harris, R.D. 1971. *Birds collected (die off) at Prince Rupert, British Columbia, September 1971.* Unpublished Final Report. Vancouver, BC: Canadian Wildlife Service.
113. Harte, J., and E. Hoffman. 1989. Possible effects of acidic deposition on a Rocky Mountain population of the tiger salamander (*Ambystoma tiginum*). *Conserv. Biol.* 3:149-58.
114. Henrikson, L., H.G. Oscarson, and J.A.E. Stenson. 1980. Does the change of predator system contribute to the biotic development in acidified lakes? In *Ecological impacts of acid precipitation*, ed. D. Drabloș and A. Tollan. Proceedings of International Conference, Sandefjord, Norway. SNSF Project, As-NLH, Norway.
115. Herbert, D.M., and I.M. Cowan. 1971. White muscle disease in the mountain goat. *J. Wildl. Manage.* 34 (4): 752-56.
116. Hillman, R.C., and A.W. Benton. 1972. Biological effects of air pollution on insects, emphasizing the reaction of the honeybee (*Apis mellifera* L.) to sulfur dioxide. *J. Elisha Mitchell Sci. Soc.* 88:195.
117. Hindawi, I. 1970. *Air pollution injury to vegetation.* US Public Health Service, Environmental Health Service, National Air Pollution Control Administration, Raleigh, NC.
118. Hirao, Y., and C.C. Patterson. 1974. Lead aerosol pollution in the High Sierra overrides natural mechanisms which exclude lead from a food chain. *Science* 184:989-92.
119. Honek, A. 1975. Color polymorphism in *Adalia bipunctata* Coleoptera Coccinellidae in Bohemia Czechoslovakia. *Entomol. Ger.* 1 (3-4): 293-99.
120. Hooper, E.T. 1942. The water shrew (*Sorex palustris*) of the southern Allegheny Mountains. *Occas. Pap. Mus. Zool. Univer. Mich.* 463:1-4.
121. Hunsaker, C.T., and D.E. Carpenter. 1990. *Environmental Monitoring and Assessment Program ecological indicators.* US EPA Report No. 600/3-90/060.

122. Hunsaker, C.T., R.L. Graham, G.W. Suter II, R.V. O'Neill, B.L. Jackson, and L.W. Barnthouse. 1990. *Regional ecological risk assessment: Theory and demonstration.* ORNL/TM-11128. Oak Ridge, TN: Oak Ridge National Laboratory.
123. Hutton, M. 1982. The role of wildlife species in the assessment of the biological impact of chronic exposure to persistent chemicals. *Ecotoxicol. Environ. Safety* 6:471-78.
124. Interagency Task Force on Acid Precipitation. 1982. *National acid precipitation assessment plan.* Washington, DC: Interagency Task Force on Acid Precipitation.
125. Jackson, D.R., and A.P. Watson. 1977. Disruption of nutrient pools and transport of heavy metals in a forested watershed near a lead smelter. *J. Environ. Qual.* 6:331-38.
126. Jenkins, D.W. 1980. *Biological monitoring and surveillance.* Biological monitoring of toxic metals, vol. 1. US EPA Office of Research and Development: EPA Report No. 600/3-80-089.
127. Johnson, A.H., T.G. Siccama, R.S. Turner, and D.G. Lord. 1984. Assessing the possibility of a link between acid precipitation and decreased growth rates of trees in the northeastern United States. In *Direct and indirect effects of acidic deposition on vegetation*, ed. R. Linthurst. Acid Precipitation Series vol. 5. Ann Arbor, MI: Butterworth Publishers.
128. Johnson, D.H. 1951. The water shrews of the Labrador Peninsula. *Proc. Biol. Soc. Wash.* 64:109-16.
129. Johnson, M.G. 1987. Trace metal loadings to sediments of fourteen Ontario lakes and correlation with concentrations in fish. *Can. J. Fish. Aquat. Sci.* 44:3-13.
130. Jop, K. 1979. Quality evaluation of roe-deer antlers from an industrial region in southern Poland. *Acta Theriol.* 24:23-24.
131. Jordan, D. 1990. *Mercury contamination: Another threat to the Florida panther*. Technical Bulletin 15 (2). Washington, DC: US Fish and Wildlife Service.
132. Karstad, L. 1967. Fluorosis in deer (*Odocoileus virginianus*). *Bull. Wildl. Dis. Assoc.* 3:42-46.
133. Karstad, L. 1970. Wildlife in changing environment. In *Environmental change, focus on Ontario*, ed. D.F. Elrich. New York: Simon & Schuster, Inc.
134. Kay, C.E. 1975. Fluoride distribution in different segments of the femur, metacarpus and mandible of mule deer. *Fluoride* 8 (2): 92-97.
135. Kay, C.E., P.C. Tourangeau, and C.C. Gordon. 1975. Industrial fluorosis in wild mule and whitetail deer from Western Montana. *Fluoride* 8:182-91.
136. Kelso, J.R.M., M.A. Shaw, C.K. Minns, and K.H. Mills. 1990. An evaluation of the effects of atmospheric acidic deposition on fish and the fishery resource of Canada. *Can. J. Fish. Aquat. Sci.* 47:644-55.
137. Kerekes, J. In press. Possible correlation of summer loon population with the trophic state of a water body. *Int. Ver. Theor. Angew. Limnol. Verh.*
138. Kettlewell, B. 1973. *The evolution of melanism.* Oxford: Clarendon Press.
139. Kettlewell, H.B.D., and M.J. Heard. 1961. Accidental radioactive labelling of a migrating moth. *Nature* 189:676-77.
140. Keystone Report. 1991. *Biological diversity on federal lands: Report of a Keystone Policy Dialogue.* Colorado: Keystone Center.
141. Klausnitzer, B., and R. Schummer. 1983. Zum vorkommen der formen von *Adalia bipunctata* L. in der DDR (Insects, Coleoptera) (Occurrence of forms of *Adalia bipunctata* L. in DDR (Insects, Coleoptera)). *Entomol. Nachr.* 27:159-62.
142. Knudsen, G.J., and J.B. Hale. 1968. Food habits of otters in the Great Lakes Region. *J. Wildl. Manage.* 32:89-93.
143. Kolb, J.A., and M. White. 1975. Small mammals of the San Bernadino Mountains, California. *Southwest. Nat.* 19 (4): 112-14.

144. Kucera, E. 1983. Mink and otter as indicators of mercury in Manitoba waters. *Can. J. Zool.* 61:2250-56.
145. Larsson, P., L. Okla, and P. Woin. 1990. Atmospheric transport of persistent pollutants governs uptake by holarctic terrestrial biota. *Environ. Sci. & Tech.* 24:1599-601.
146. Lechowicz, M.J. 1981. The effects of simulated acid precipitation on photosynthesis in the caribou lichen, *Cladina stellaris*. *Water Air Soil Pollut.* 14:133-57.
147. Lees, D.R. 1974. Genetic control of the melanic forms of the moth *Phigalia pilosaria (pedaria)*. *Heredity* 33:145-50.
148. Lees, D.R., and C.S. Dent. 1983. Industrial melanism in the spittlebug *Philaenus spumarius homoptera* Aphrophoridae. *Biol. J. Linn. Soc.* 19 (2): 115-30.
149. Leivestad, H., G. Hendrey, I.P. Muniz, and E. Snekvik. 1976. Effects of acid precipitation on freshwater organisms. In *Impact of acid precipitation on forest and freshwater ecosystems in Norway*, ed. F.J. Braekke. SNSF Project Report FR 6/76. SNSF Project, As-NLH, Norway.
150. Lemberk, V. 1989. Comparison of the ornithocenoses of spruce forest in Krkonose Mountains according to the emmisions damage degree. *Opera Corcontica* 26:131-43.
151. Lepage, P., and G.H. Parker. 1988. Copper, nickel, and iron levels in pelage of red squirrels living near the ore smelters at Sudbury, Ontario, Canada. *Can. J. Zool.* 66 (7): 1631-37.
152. Leuven, R.S.E.W., C. Hartog, M.M.C. Christiaans, and W.H.C. Helligers. 1986. Effects of water acidification on the distribution pattern and the reproductive success of amphibians. *Experientia* 42:495-503.
153. Leuven, R.S.E.W., and F.G.F. Oyen. 1987. Impact of acidification and eutrophication on the distribution of fish species in shallow and lentic soft waters of the Netherlands: An historical perspective. *J. Fish Biol.* 31:753-74.
154. Liebert, T.G., and P.M. Brakefield. 1987. Behavioural studies on the peppered moth *Biston betularia* and a discussion of the role of pollution and lichens in industrial melanism. *Biol. J. Linn. Soc.* 31:129-50.
155. Light, J.T. 1973. The effects of oxidant air pollution on forest ecosystems of the San Bernadino Mountains, Section B. In *Oxidant air pollution effects on a western coniferous forest ecosystem: Task B report,* ed. O.C. Taylor. Air Pollution Research Center, Univ. of California-Riverside.
156. Linscombe, G., N. Kinler, and R.J. Aulerich. 1983. Mink. In *Wild mammals of North America: Biology, management, and economics*, ed. J.A. Chapman and G.A. Feldhamer. Baltimore: Johns Hopkins University Press.
157. Lockerbie, D.M., and Clair, T.A. 1988. Organic contaminants in isolated lakes of southern Labrador, Canada. *Bull. Environ. Contam. Toxicol.* 41:625-32.
158. Longcore, J.R., R.K. Ross, and K.L. Fischer. 1987. Wildlife resources at risk through acidification of wetlands. *Trans. N. Am. Wildl. Nat. Resour. Conf.* 52:608-18.
159. Lowe, V.P.W., and A.D. Horrill. 1988. Ecological half-life of caesium in roe deer (*Capreolus capreolus*). *Environ. Pollut.* 54:81-87.
160. MacArthur, R.H., and J.W. MacArthur. 1961. On bird species diversity. *Ecology* 42:594-98.
161. Machie, D.W. 1960. *Ostearius melanopygius* (O.P.C.). *Bull. Brit. Spider Study Group* 8:3-4.
162. Marshall-Forbes, L. 1991. *Unprecedented risks: The effects of global climate change on U.S. wildlife resources.* Arlington, VA: Izaak Walton League of America.
163. Martin, A.C., H.S. Zim, and A.L. Nelson. 1951. *American wildlife and plants: A guide to wildlife food habits.* New York: Dover.
164. Martin, H.C., ed. 1987. *Acidic precipitation.* Parts 1 and 2. Proceedings of the International Symposium on Acidic Precipitation, Muskoka, Ontario, September 15-20, 1985. Dordrecht, Netherlands: D. Reidel.

165. Martin, M.H., and P.J. Coughtrey. 1975. Preliminary observations of the levels of cadmium in a contaminated environment. *Chemosphere* 4:155-60.

166. Martin, M.H., and P.J. Coughtrey. 1976. Comparison between levels of lead, zinc, and cadmium within a contaminated environment. *Chemosphere* 5:15-20.

167. Mascanzoni, D. 1987. Chernobyl's challenge to the environment: A report from Sweden. *Sci. Total Environ.* 67:133-48.

168 . Mason, C.F., and S.M. MacDonald. 1988. Radioactivity in otter scats in Britain following the Chernobyl reactor accident. *Water Air Soil Pollut.* 37:131-37.

169. Mason, C.F., and S.M. MacDonald. 1989. Acidification and otter (*Lutra lutra*) distribution in Scotland. *Water Air Soil Pollut.* 43:365-74.

170. Mattsson, P., L. Albanu, and A. Frank. 1981. Cadmium and some other elements in liver and kidney from moose (*Alces alces*). *Var Foda* 33 (8-9): 335-49.

171. McNeely, J.A., K.R. Miller, W.V. Reid, R.A. Mittermeier, and T.B. Werner. 1990. *Conserving the world's biological diversity*. Gland, Switzerland, and Washington, DC. Prepared and published by the International Union for Conservation of Nature and Natural Resources, World Resources Institute, Conservation International, World Wildlife Fund-U.S., and World Bank.

172. McNicol, D.K., B.E. Bendell, and D.G. McAuley. 1987. Avian trophic relationships and wetland acidity. *Trans. N. Am. Wildl. Nat. Resour. Conf.* 52:619-27.

173. McNicol, D.K., B.E. Bendell, and R.K. Ross. 1987. Studies of the effects of acidification on aquatic wildlife in Canada: Waterfowl and trophic relationships in small lakes in northern Ontario. Occasional Paper No. 62. Ottawa: Canadian Wildlife Service.

174. McNicol, D.K., P.J. Blancher, and B.E. Bendell. 1987. Waterfowl as indicators of wetland acidification. In *The value of birds*, ed. A.W. Diamond and F.L. Filion. International Council for Bird Preservation (ICBP) Technical Publication No. 6. Cambridge, England.

175. Mikkola, K. 1975. Frequencies of melanic forms of Oligia moths (Lepidoptera, Noctuidae) as a measure of atmospheric pollution in Finland. *Ann. Zool. Fenn.* 12:197-204.

176. Mikkola, K. 1984. Dominance relations among the melanic forms of *Biston betularius* and *Odontopera bidentata* (Lepidoptera, Geometridae). *Heredity* 52:9-16.

177. Mikkola, K. 1989. The first case of industrial melanism in the subarctic lepidopteran fauna: *Xestia gelida f. inferna f. n.* (Noctuidae). *Not. Entomol.* 69:1-3.

178. Mikkola, K., and A. Albrecht. 1986. Radioactivity in Finnish night-flying moths (*Lepidoptera*) after the Chernobyl accident. *Not. Entomol.* 66:153-57.

179. Mikkola, K., and A. Albrecht. 1988. The melanism of *Adalia bipunctata* Coleortera Coccinellidae around the Gulf of Finland as an industrial phenomenon. *Ann. Zool. Fenn.* 25 (2): 177-86.

180. Mikova, M., and E. Novakova. 1979. Variation of corneal glycosaminoglycan values of hares in relation to environmental pollution by industrial emissions. *J. Toxicol. Environ. Health* 5:891-96.

181. Miller, P.R., ed. 1980. *Proceedings of the Symposium on Effects of Air Pollutants on Mediterranean and Temperate Forest Ecosystems, 22-27 June 1980*. General Technical Report PSW-43, Pacific Southwest Forest and Range Experimental Station, Forest Service, US Department of Agriculture, Berkeley, CA.

182. Miller, P.R., and M.J. Eldemman, eds. 1977. *Photochemical oxidant air pollutant effects on mixed conifer forest ecosystems: A progress report, 1976*. US EPA Report No. 600/3-77-104.

183. Miller, P.R., and J.R. McBride. 1975. Effects of air pollutants on forests. In *Responses of plants to air pollution*, ed. J.B. Mudd and T.T. Kozlowski. New York: Academic Press.

184. Minns, C.K., J.E. Moore, D.W. Schindler, and M.L. Jones. 1990. Assessing the potential extent of damage to inland lakes in eastern Canada due to acidic deposition, III: Predicted impacts on species richness in seven groups of aquatic biota. *Can. J. Fish. Aquat. Sci.* 47:821-30.
185. Mitchell, B.A. 1989. Acid rain and birds: How much proof is needed? *Am. Birds* 43:234-41.
186. Miyamoto, Y. 1975. Kankyo osen no shihyo to siteno dobutsu kisetsu. Kanso-kuritsu kara mita nihon no osen bunpu (Animal phenology as an indicator of environmental pollution: Distribution of environmental pollution in Japan seen from frequency of observation). *Tokyo Kanku Chiho Kisho Kenkyukai-shi* (Geophysical Notes, Tokyo District Meteorological Observations) 8:27-29.
187. Murton, R.K., N.J. Westwood, and R.J.P. Thearle. 1973. Polymorphism and the evolution of a continuous breeding season in the pigeon, *Columba livia. J. Reprod. Fertil.* (Supplement) 19:563-77.
188. Naef-Daenzer, B., and M. Blattner. 1989. Spatial distribution of birds in relation to structure and damage of woodland, I: Oak-beech woods in northwestern Switzerland. *Ornithol. Beob.* 86:307-27.
189. National Acid Precipitation Assessment Program (NAPAP). 1990. *Acidic deposition: State of \science and technology,* Vol. 1-4. Washington, DC: Government Printing Office.
190. Newman, J.R. 1977. Sensitivity of the house martin (*Delichon urbica*) to fluoride emissions. *Fluoride* 10 (2): 73-76.
191. Newman, J.R. 1979. Effects of industrial air pollution on wildlife. *Biol. Conserv.* 15:181-90.
192. Newman, J.R. 1980. *Effects of air emissions on wildlife resources.* US Fish and Wildlife Service, Biological Services Program, National Power Plant Team: FWS/OBS-80/40.1.
193. Newman, J.R. 1984. Fluoride standards and predicting wildlife effects. *Fluoride* 17:41-47.
194. Newman, J.R., and J.J. Murphy. 1979. Effects of industrial fluoride on black-tailed deer (preliminary report). *Fluoride* 12 (3): 129-35.
195. Newman, J.R., and E. Novakova. 1979. *Effects of air pollution on nesting of the house martin* (Delichon urbica). Environmental Science and Engineering, Inc., Gainesville, FL. Unpublished Report.
196. Newman, J.R., E. Novakova, M.K. Bergdoll, and M.T. Park. 1984. *Ducks as site-specific bioindicators of trace metal pollution.* Paper read at 5th Annual Meeting of the Society for Environmental Toxicology and Chemistry, 4-7 November, Arlington, Virginia.
197. Newman, J.R., E. Novakova, and J.T. McClave. 1985. The influence of industrial air emissions on the nesting ecology of the house martin, *Delichon urbica*, in Czechoslovakia. *Biol. Conserv.* 31:229-48.
198. Newman, J.R., and R.K. Schreiber. 1984. Animals as indicators of ecosystem responses to air emissions. *Environ. Manage.* 8:309-24.
199. Newman, J.R., and R.K. Schreiber. 1988. Air pollution and wildlife toxicology: An overlooked problem. *Environ. Toxicol. Chem.* 7:381-90.
200. Newman, J.R., and M. Yu. 1976. Fluorosis in black-tailed deer. *J. Wildl. Dis.* 12:39-41.
201. Nishino, O., M. Arari, I. Senda, and K. Kuboto. 1973. Kankyo osen no suzume ni oyobosu eikyo (Influence of environmental pollution on the sparrow). *Jpn. J. Public Health* 20 (10): 1.
202. Noss, R.F., and L.D. Harris. 1986. Nodes, networks, and MUMs: Preserving diversity at all scales. *Environ. Manage.* 10:299-309.
203. Novak, I., and K. Spitzer. 1986. Industrial melanism in *Biston betularia* Lepidoptera Geometridae in Czechoslovakia. *Acta Entomol. Bohemoslov.* 83 (3): 185-91.
204. Novakova, E. 1969. Influence des pollution industriellès sur les communautes animales et l'utilization des animaux comme bioindicateurs (Influence of industrial pollution on common animals and the utilization of these common animals as bioindicators). In *Proceedings of the*

First European Congress on the Influence of Air Pollution. Wageningen, Netherlands: Centre for Agricultural Publishing.

205. Novakova, E. 1970. Influence of industrial air pollution on urine reaction in hares. In *Proceedings of the Eighth International Congress on Nutrition.* 1969, Prague.

206. Novakova, E., A. Finkova, and Z. Sova. 1973. Etude preliminaire des proteines sanguines chez le lievre commun expose aux pollutions industrielles (Preliminary study of blood proteins of the common hare exposed to industrial pollution). In *Nemzetkozi Vadaszati Tudomanyos Konferencia Eloadasai II.* Sekio, Aprovadaszda Ikodas Sopron, Budapest, 1971.

207. Novakova, E., and R. Hanzl. 1974. Prispeved k urceni potencialu krajiny pro chov nekterych druhu zvere (Contribution on the determination of the potential landscape for management of certain game species.) *Quaest. Geobiol.* 13:7-81.

208. Novakova, E., and Z. Roubal. 1971. Taux de calcium et de phosphore dans le serum sanguin des lievres exposes aux pollutions de l'air (Calcium and phosphorus ratios in the blood serum of hares subject to air pollution). In *Actes du X Congress Union Internationale Biologistes du Gibier.* 3-7 May 1971, Paris.

209. Novakova, E., and B. Temmlova. 1973. Influence de la pollution de l'air sur la syngamosis du faison commun (Influence of air pollution on syngamosis in the common pheasant). In *Actes du X Congress Union Internationale Biologistes du Gibier.* 3-7 May 1971, Paris.

210. Nyholm, N.E.I. 1981. Evidence of involvement of aluminum in causation of defective formation of eggshells and of impaired breeding in wild passerine birds. *Environ. Res.* 26:363-71.

211. Nyholm, N.E.I. 1987. Bio-indication of industrial emissions of heavy metals by means of insectivorous birds. In *Proceedings of the International Conference on Heavy Metals in the Environment*, vol. 2, ed. S.E. Lindberg, and T.C. Hutchinson. Edinburgh, Scotland: CEP Consultants.

212. Nyholm, N.E.I., and H.E. Myhrberg. 1977. Severe eggshell defects and impaired reproductive capacity in small passerines in Swedish Lapland. *Oikos* 29:336-41.

213. Office of Technology Assessment. 1987. *Technologies to maintain biological diversity.* Washington, DC: US Government Printing Office.

214. O'Gara, G. 1982. Riley Ridge: Gas sours wildlife in Wyoming. *High County News* 14:10-11.

215. Oldfield, J.E. 1972. Selenium deficiency in soils and its effect on animal health. *Geol. Soc. Am. Bull.* 83:173-80.

216. Organ, J.F. 1989. *Mercury and PCB residues in Massachusetts river otters: Comparisons on a watershed basis.* Ph.D. diss., Univ. of Massachusetts, Amherst.

217. Ormerod, S.J., N. Allinson, D. Hudson, and S.J. Tyler. 1986. The distribution of breeding dippers (*Cinclus cinclus* [L.]; Aves) in relation to stream acidity in upland Wales. *Freshwater Biol.* 16:501-7.

218. Ormerod, S.J., K.R. Bull, C.P. Cummins, S.J. Tyler, and J.A. Vickery. 1988. Egg mass and shell thickness in dippers *Cinclus cinclus* in relation to stream acidity in Wales and Scotland. *Environ. Pollut.* 55:107-21.

219. Ormerod, S.J., and S.J. Tyler. 1987. Dippers (*Cinclus cinclus*) and grey wagtails (*Motacilla cinerea*) as indicators of stream acidity in upland Wales. In *The value of birds*, ed. A.W. Diamond and F.L. Filion. International Council for Bird Preservation (ICBP) Technical Publication No. 6. Cambridge, England.

220. Overrein, L.N., H.M. Seip, and A. Tollan. 1980. *Acid precipitation—Effects on forest and fish.* Final Report SNSF Project, 1972-1980. SNSF Project, As-NLH, Norway.

221. Peterle, T.J. 1991. *Wildlife toxicology.* New York: Van Nostrand Reinhold.

222. Petersen R.C., Jr., Landner, L., and Blanck, H. 1986. Assessment of the impact of the Chernobyl reactor accident on the biota of Swedish streams and lakes. *Ambio* 15:327-31.
223. Phillips, K. 1990. Where have all the frogs and toads gone? *Bioscience* 40 (6): 422-24.
224. Pierce, B.A. 1985. Acid tolerance in amphibians. *Bioscience* 35:239-43.
225. Pierce, B.A., and J.M. Harvey. 1987. Geographic variation in acid tolerance of Connecticut wood frogs. *Copeia* 1987:94-103.
226. Popescu, C. 1979. Natural selection in the industrial melanic psocid *Mesopsocus unipunctatus* (Mull.) (Insecta: Psocoptera) in northern England. *Heredity* 42:133-42.
227. Popescu, C., E. Broadhead, and B. Shorrocks. 1978. Industrial melanism in *Mesopsocus unipunctatus* (Mull.) (Psocoptera) in northern England. *Ecol. Entomol.* 3:209-19.
228. Pough, F.H., and R.E. Wilson. 1977. Acid precipitation and reproductive success of Ambystoma salamanders. *Water Air Soil Pollut.* 7:307-16.
229. Prell, H. 1936. Die schadigung der tierwelt durch die fernwirkungen von industrieal gasen (Injury to the animal world through the distant effects of industrial waste gases). *Arch. Gewerbepathol. Gewerbehyg.* 7:656-70.
230. Raddum, G.G., and A. Fjellheim. 1984. Acidification and early warning organisms in freshwater in western Norway. *Int. Ver. Theor. Angew. Limnol. Verh.* 22:1973-80.
231. Ranta, W.B., F.D. Tomassini, and E. Nieboer. 1978. Elevation of copper and nickel levels in primaries from black and mallard ducks collected in the Sudbury District, Ontario. *Can. J. Zool.* 56:581-86.
232. Rattner, B.A., G.M. Haramis, D.S. Chu, and C.M. Bunck. 1987. Growth and physiological condition of black ducks reared on acidified wetlands. *Can. J. Zool.* 65:2953-58.
233. Richardson, D.H.S., and C.M. Young. 1977. Lichens and vertebrates. In *Lichen ecology*, ed. M.R.D. Seaward. New York: Academic Press.
234. Richkind, K.E. 1979. Genetic responses to air pollution in mammalian populations. Ph.D. diss., University of California, Los Angeles.
235. Robinette, W.L., D.A. Jones, G. Rogers, and J.S. Gashwiler. 1957. Notes on tooth development and wear for Rocky Mountain mule deer. *J. Wildl. Manage.* 21:135-52.
236. Rodhe, H., and R. Herrera. 1988. *Acidification in tropical countries.* New York: John Wiley & Sons.
237. Rose, G.A., and G.H. Parker. 1982. Effects of smelter emissions on metal levels in the plumage of ruffed grouse near Sudbury, Ontario, Canada. *Can. J. Zool.* 60:2659-67.
238. Sawicka-Kapusta, K. 1978. Estimation of the contents of heavy metals in antlers of roe deer from Silesian woods. *Arch. Ochr. Srodowiska* 1:107-21.
239. Sawicka-Kapusta, K. 1979. Roe deer antlers as bioindicators of environmental pollution in southern Poland. *Environ. Pollut.* 19 (4): 283-94.
240. Scanlon, P. 1979. Lead contamination of mammals and invertebrates near highways with different traffic volumes. In *Animals as monitors of environmental pollutants.* Symposium on Pathobiology of Environmental Pollutants: Animal Models and Wildlife as Monitors, University of Connecticut, 1977. Washington, DC: National Academy of Sciences.
241. Scanlon, P. 1990. Effects of acidification on wild mammals and waterfowl. In *Biological effects of changes in surface water acid-base chemistry.* State-of-Science/Technology Report 13. Washington, DC: National Acid Precipitation Assessment Program.
242. Scheider, W.A., D.S. Jeffries, and P.J. Dillon. 1979. Effects of acidic precipitation on precambrian freshwaters in southern Ontario. *J. Gt. Lakes Res.* 5 (1): 45-51.
243. Scheuhammer, A.M. 1991. Effects of acidification on the availability of toxic metals and calcium to wild birds and mammals. *Environ. Pollut.* 71:329-75.

244. Schindler, D.W., S.E.M. Kaslan, and R.H. Hesslein. 1989. Biological impoverishment in lakes of the midwestern and northeastern United States from acid rain. *Environ. Sci. Tech.* 23:573-80.
245. Schindler, D.W., K.H. Mills, D.F. Malley, D.L. Findlay, J.A. Shearer, I.J. Davies, M.A. Turner, G.A. Linsey, and D.R. Cruikshank. 1985. Long-term ecosystem stress: The effects of experimental acidification on a small lake. *Science* 228:1395-1401.
246. Schlesinger, W.H., and G.L. Potter. 1974. Lead, copper, and cadmium concentrations in small mammals in the Hubbard Brook Experimental Forest. *Oikos* 25:148-52.
247. Schofield, C.L. 1978. Toxicity of metals. In *Limnological aspects of acid precipitation*, ed. G. Hendrey. Upton, NY: Brookhaven National Laboratory.
248. Schreiber, R.K., and J.R. Newman. 1988. Acid precipitation effects on forest habitats: Implications for wildlife. *Conserv. Biol.* 2:249-59.
249. Schummer, R. 1976. On the problem of melanism in *Biston betularia* and *Biston strataria* Lepidoptera Geometridae in the area of East Germany. *Dtsch. Entomol. Z.* 23 (4-5): 281-94.
250. Scott, M.G., and T.C. Hutchinson. 1987. Effects of simulated acid rain episode on photosynthesis and recovery in caribou-forage lichens, *Cladina stellaris* (OPIZ) BRODO and *Cladina rangiferina* (L.) WIG-G. *New Phytol.* 107:567-75.
251. Shaw, G.G. 1981. The potential impact of sour gas plant aerial emissions on wildlife. In *Effects of sour gas on wildlife*, ed. V. Geist and A.H. Legge. Calgary, Alberta, Canada: Univ. of Calgary Faculty of Environmental Design.
252. Sheldon, W.G., and W.G. Toll. 1964. Feeding habits of the river otter in a reservoir in central Massachusetts. *J. Mammal.* 45:449-55.
253. Siccama, T.G., M. Bliss, and H.W. Vogelmann. 1982. Decline of red spruce in the Green Mountains of Vermont. *Bull. Torrey Bot. Club* 109:162-68.
254. Sigal, L.L., and J.W. Johnston, Jr. 1986a. Effects of acidic rain and ozone on nitrogen fixation and photosynthesis in the lichen *Lobaria pulmonaria (L.)* Hoffman. *Environ. Exp. Bot.* 26:59-64.
255. Sigal, L.L., and J.W. Johnston, Jr. 1986b. Effects of simulated acidic rain on one species each of Pseudoparmelia, Usnea, and Umbilicaria. *Water Air Soil Pollut.* 27:315-22.
256. Singer, R., and K.L. Fischer. 1984. Other related biota. In *The acidic deposition phenomena and its effects: Critical assessment review papers*, ed. A.P. Altshuller and R.A. Linthurst. *Vol. 2, Effects Sciences*. NCSU Acid Precipitation Program, Raleigh, NC. US EPA Office of Research and Development: EPA Report No. 600/8-83-0168BF.
257. Smith, J.B., and D. Tirpak. 1989. *The potential effects of global climate change on the United States*. Report to Congress. US EPA Office of Policy, Planning, and Evaluation, Office of Research and Development.
258. Smith, M.W., B.J. Wyskowski, C.M. Brooks, C.T. Driscoll, and C.C. Cosentini. 1990. Relationships between acidity and benthic invertebrates of low-order woodland streams in the Adirondack Mountains, New York. *Can. J. Fish. Aquat. Sci.* 47:1318-29.
259. Smith, W.H. 1974. Air pollution—Effect on the structure and function of the temperate forest ecosystem. *Environ. Pollut.* 6:111-29.
260. Smith, W.H. 1990. *Air pollution and forests: Interaction between air contaminants and forest ecosystems*. New York: Springer-Verlag.
261. Sochasky, L., ed. 1981. *Acid rain and the Atlantic salmon*. IASF Special Publication Series, No. 10. St. Andrews, New Brunswick: International Atlantic Salmon Foundation.
262. Soholt, L.F., and S. Wiedenbaum. 1981. *Oil shale: Its development and potential for air quality effects*. US Fish and Wildlife Service, Biological Services Program, Eastern Energy and Land Use Team: FWS/OBS-81/34.

263. Spry, D.J., and J.G. Wiener. 1991. Metal bioavailability and toxicity to fish in low-alkalinity lakes: A critical review. *Environ. Pollut.* 71:243-304.
264. Stastny, K., and V. Bejcek. 1983. Bird communities of spruce forests affected by industrial emissions in the Krusne Hory (Ore Mountains). In *Proceedings of the VIII International Conference. Bird census work and atlas work*, ed. K. Taylor, R.J. Fuller, and P.C. Lake, pp. 243-53. B.T.O.
265. Stastny, K., V. Bejcek, and Z. Barta. 1987. Use of bird communities as the biodiagnostical indicator of the degree of affection of spruce forests in the Krusne hory (Ore Mountains) (in Czech with an English summary). *Sbornik Ochrana Muzea* 6/84:79-103.
266. Steele, C.W., S. Strickler-Shaw, and D.H. Taylor. 1989. Behavior of tadpoles of the bullfrog, *Rana catesbelana*, in response to sublethal lead exposure. *Aquat. Toxicol.* 14:331-44.
267. Stern, A.C., ed. 1977. *Air pollution*. 3rd ed. 2 vols. New York: Academic Press.
268. Steward, R.C. 1976. Experiments on resting site selection by the typical and melanic forms of the moth *Allophyes oxyacanthae* (Caradrinidae). *J. Zool. (Lond.)* 178:107-15.
269. Steward, R.C. 1977. Melanism and selective predation in three species of moth. *J. Anim. Ecol.* 46:483-96.
270. Stickel, W.H. 1975. Some effects of pollutants in terrestrial ecosystems. In *Ecological toxicology research*, ed. A.D. McIntyre and C.F. Mills. New York: Plenum Press.
271. St. Louis, V.L., L. Breebaart, and J.C. Barlow. 1990. Foraging behavior of tree swallows over acidified and nonacidified lakes. *Can. J. Zool.* 68:2385-92.
272. Stokes, P.M., E.T. Howell, and G. Krantzberg. 1989. Effects of acidic precipitation on the biota of freshwater lakes. In *Acidic precipitation: Biological and ecological effects*, ed. D.C. Adrlano and A.H. Johnson. New York: Springer-Verlag.
273. Stoszek, M.J., W.B. Kessler, and H. Willmes. 1978. Trace mineral content of antelope tissues. *Proceedings of the Eighth Antelope States Workshop*. Jasper, Alberta, May 1-4.
274. Stripp, R.A., M. Heit, D.C. Bogen, J. Bidanset, and L. Trombetta. 1990. Trace element accumulation in the tissues of fish from lakes with different pH values. *Water Air Soil Pollut.* 51:75-87.
275. Sullivan, T.J. 1991. *Historical changes in surface water acid-base chemistry in response to acidic deposition*. State of Science and Technology Report 11. Washington, DC: National Acid Precipitation Assessment Program.
276. Suns, K., C. Curry, and D. Russell. 1980. *The effects of water quality and morphometric parameters on mercury uptake by yearling yellow perch*. Technical Report LTS 80-1. Rexdale, Ontario: Ontario Ministry of the Environment.
277. Swackhamer, D.L., and R.A. Hites. 1988. Occurrence and bioaccumulation of organochlorine compounds in fishes from Siskiwit Lake, Isle Royale, and Lake Superior. *Environ. Sci Technol.* 22 (5): 543-48.
278. Talmage, S.S., and B.T. Walton. 1991. Small mammals as monitors of environmental contaminants. *Rev. Environ. Contam. Toxicol.* 119:47-145.
279. Tansey, M.F., and R.P. Roth. 1970. Pigeons, a new role in air pollution. *J. Air Pollut. Control Assoc.* 20:307-9.
280. Taylor, O.C., ed. 1973. *Oxidant air pollution effects on a western coniferous forest ecosystem*. Task B Report. Air Pollution Research Center, Univ. of California, Riverside.
281. Tendron, G. 1964. Effects of air pollution on animals and plants. In *European Conference on Air Pollution*. Council of Europe Report No. A87-389. Strasbourg, France: Council of Europe.
282. Thompson, D.C., and K.H. McCourt. 1981. Seasonal diets of the porcupine caribou herd. *Am. Midl. Nat.* 105:70-76.

283. Tjell, J.C., T.H. Christensen, and R. Bro-Rasmussen. 1983. Cadmium in soil and terrestrial biota, with emphasis on the Danish situation. *Ecotoxicol. Environ. Saf.* 7:122-40.

284. Tome, M.A., and F.H. Pough. 1982. Responses of amphibians to acid precipitation. In *Acid rain fisheries*, ed. R.E. Johnson. Proceedings of the International Symposium on Acidic Precipitation and Fishery Impacts, Cornell Univ., Aug. 5, 1981. Bethesda, MD: American Fisheries Society.

285. Ulrich, B. 1982. Gefahren fur das waldokosystem durch saure niederchlage. In *Immissionbelastungen von Waldokosystemen* (a special 1982 number of) *Landanstalt fur Okologie Landschaftenwicklung und Forstplanung Nordrhein-Westfalen.* 4350 Recklinghausen, Federal Republic of Germany.

286. Urbanek, B. 1986. Mercury, lead and cadmium content in the plumage of pheasants and fur of hares in the Czech Socialist Republic. *Prace Volhm* 69:277-96.

287. U.S. Environmental Protection Agency (EPA). 1986. *Second addendum to air quality criteria for particulate matter and sulfur oxides (1982): Assessment of newly available health effects information.* US EPA Report No. 600/18-86/020F.

288. U.S. Environmental Protection Agency (EPA). 1990. *Environmental Monitoring and Assessment Program overview.* US EPA Report No. 600/9-90/001.

289. U.S. Environmental Protection Agency (EPA). 1990. *New source review workshop manual: Prevention of significant deterioration and nonattainment area permitting.* Office of Air Quality Planning and Standards.

290. U.S. Environmental Protection Agency (EPA). 1991. *National air quality and emissions trends report, 1989.* US EPA Report No. 450/4-91-003.

291. Villella, R.F. 1989. *Acid rain publications by the U.S. Fish and Wildlife Service, 1979-1989.* US Fish and Wildlife Service: Biological Report 80(40.28).

292. Vogelmann, H.W. 1982. Catastrophe on Camel's Hump. *Natural History* 91:8-14.

293. Walton, K.C. 1984. Fluoride in fox bone near an aluminum reduction plant in Anglesey, Wales, and elsewhere in the United Kingdom. *Environ. Pollut.* (*Series B*) 7:273-80.

294. Watt, W.D. 1987. A summary of the impact of acid rain on Atlantic salmon (*Salmon salar*) in Canada. *Water Air Soil Pollut.* 35:27-35.

295. Way, C.A., and G.D. Schroder. 1982. Accumulation of lead and cadmium in wild populations of the commensal rat, *Rattus norvegicus. Arch. Environ. Contam. Toxicol.* 11:407-17.

296. Webster, I. 1963. Asbestos in non-experimental animals in South Africa. *Nature (Lond.)* 197:506.

297. Wellings, S.R. 1970. Respiratory damage due to atmospheric pollutants in the English sparrow, *Passer domesticus*. In *Project clean air*. Research Project S-25. Department of Pathology, Univ. of California, Davis.

298. Wiener, J.G. 1987. Metal contamination of fish in low-pH lakes and potential implications for piscivorous wildlife. *Trans. N. Am. Wildl. Nat. Resour. Conf.* 52:645-57.

299. Wilson, E.O. 1988. The current state of biological diversity: In *Biodiversity,* ed. E.O Wilson and F.M. Peter, pp. 3-18. Washington, DC: National Academy Press.

300. Witkamp, M., M.L. Frank, and J.L. Shoopman. 1966. Accumulation and biota in a pioneer ecosystem of kudzu vine at Copperhill, Tennessee. *J. Appl. Ecol.* 3:383-91.

301. Woodwell, G.M. 1970. Effects of pollution on the structure and physiology of ecosystems. *Science* 168:429-33.

302. Wren, C.D. 1986. Mammals as biological monitors of environmental metal levels. *Environ. Monit. Assess.* 6:127-44.

303. Wren, C.D., H. MacCrimmon, R. Frank, and P. Suda. 1980. Total and methylmercury levels in wild mammals from the precambrian shield area of south-central Ontario, Canada. *Bull. Environ. Contam. Toxicol.* 25:100-5.

304. Wren, C.D., H. MacCrimmon, and B. Loescher. 1983. Examination of bioaccumulation and biomagnification of metals in a precambrian shield lake. *Water Air Soil Pollut.* 19:277-91.

305. Wren, C.D., and P.M. Stokes. 1988. Depressed mercury levels in biota from acid and metal-stressed lakes near Sudbury, Ontario. *Ambio* 17:28-30.

306. Wren, C.D., P.M. Stokes, and K.L. Fischer. 1986. Mercury levels in Ontario mink and otter relative to food levels and environmental acidification. *Can. J. Zool.* 64:2852-59.

307. Wrigley, R.E. 1969. Ecological notes on the mammals of southern Quebec. *Can. Field-Nat* 83:201-11.

308. Wyman, R.L., and D.S. Hawksley-Lescault. 1987. Soil acidity affects distribution, behavior, and physiology of the salamander *Plethodon cinereus*. *Ecology* 68:1819-27.

309. Yan, N.D., W. Keller, H.J. Macisaac, and L.J. McEachern. 1991. Regulation of zooplankton community structure of an acidified lake by *Chaoborus*. *Ecol. Abst.* 1:52-65.

310. Yont, W.P., and R.R. Sayers. 1927. Hydrogen sulfide as a laboratory and industrial poison. *J. Chem. Educ.* 4:613-19.

11

Air Pollution Effects on Ecosystem Processes

William H. Smith

INTRODUCTION

Regional-scale air pollution is a significant contemporary anthropogenic stress imposed on temperate forest ecosystems. Gradual and subtle change in ecosystem function and composition over wide areas of the temperate latitudes over extended time, rather than dramatic destruction of ecosystems in the immediate vicinity of point sources over short periods, must be recognized as the primary consequence of regional-scale air pollution stress. Global-scale air pollution, with its associated potential to cause rapid climate change, can dramatically alter ecosystems, especially those in the north temperate and boreal latitudes. The integrity, productivity, and sustainability of natural ecosystems are intimately linked to air quality. This chapter focuses on the ability of air contaminants to adversely affect forest ecosystem processes.

It is important, at the outset, to give perspective on this topic by providing key definitions. Ecosystems are the smallest units of nature in which all organisms, in the presence of related inorganic and organic chemicals and in the context of a specific climate, interact with their environment to produce energy flow and complete biogeochemical cycling (Botkin and Keller 1982; Odum 1971) (Table 11-1). This chapter emphasizes adverse air pollutant impacts on ecosystem processes. Impacts on ecosystem structure, for example, biodiversity, succession, and food webs, are also emphasized in other chapters in this volume. The concept of "adverse" (reduction in benefits from ecological systems) impact as used in this chapter is consistent with the definition presented by Tingey, Hodsett, and Henderson (1990). Natural ecosystems such as forests provide multiple values to society. These values may be in the form of forest products, for example, wood (lumber and paper), wildlife, or water. In addition and equally important, however, forests also provide a variety of services of value to societies (Ehrlich and Mooney 1983, Bormann 1976). These services include, among others: recreation; biological diversity (Probst and Crow 1991); landscape diversity (Forgey 1991); several environmental amenity services (Smith 1970) including microclimatic amelioration (DeWalle and Heisler 1980; Federer 1976; Heisler 1974; Hutchison, Taylor, and Wendt

TABLE 11-1. Ecosystem Elements (all units of nature with energy flow and complete chemical cycles, or ecosystems, have common components organized in structural patterns and united by functional processes)

Components	Structural Patterns	Functional Processes
Producers	Food webs	Energy flow (energy storage)
Consumers	Diversity: species variation in space	Biogeochemical cycling
Decomposers	Succession: species variation in time	
Inorganic substances		
Organic substances		
Climate		

Source: Smith 1990a.

1982), sound attenuation (Aylor 1971; Cook and Van Haverbeke 1977; Reethoff and Heisler 1976), and visual attractiveness and screening (Brush 1971; Williamson, and Fabos 1979; Payne 1973); water quality, flood and erosion management along with soil and nutrient conservation (Lull and Reinhart 1972; Lull 1971; Sopper, Lynch, and Corbett 1976); and persistent pollutant storage and detoxification (Sawhney and Brown 1989; Smith 1990a). Some of these forest values are priced and some are unpriced (Table 11-2). All values, whether forest products or forest services, whether quantified in economic terms or not, must be examined with reference to adverse air pollutant influence.

Assessment of air pollution impact on forest ecosystem processes is extremely challenging for a variety of reasons. Three of the most important include forest system variability, deficiency of understanding of ecosystem-scale phenomena, and large variation in system exposure to atmospheric deposition. Forest ecosystems have enormous variability. Forests differ in soil type, climate, aspect, elevation, species composition, age,

TABLE 11-2. Forest Ecosystem Values to Societies, Which Involve Both Products and Services

Products	Services
Wood	Recreation
Lumber	
Paper	Biological diversity
	Genes
Wildlife	Species
	Habitats
Water	
	Landscape diversity
	Amenity functions
	Microclimatic amelioration
	Sound attenuation
	Visual attractiveness/screening
	Water quality, flood and erosion management
	Soil and plant nutrient conservation
	Persistent pollutant storage/detoxification

and health. Forests may be young, uneven-aged, even-aged, all-aged, or overmature. Forests may be reproduced by seed, by coppice, or by planting. Some forests have their structure completely shaped by natural forces, some may be influenced by human forces as well as natural forces, and others may be completely artificial in design and establishment. Forests occur in urban, suburban, rural, and remote landscapes.

Forest exposure to air contaminants is extremely variable and depends significantly on downwind distances from local point sources or regional-scale area sources. Topographic considerations (e.g., elevation and aspect) and meteorologic considerations (e.g., inversion frequencies) importantly regulate forest ecosystem exposure to air contaminants. Forest systems may be arrayed along a continuum of human management efforts ranging from no management to intensive management. Forest stress, deviation from normal function, has been studied from the perspectives of individual trees, populations of trees, stands of trees, or entire forest systems. Stress considerations at the individual tree level typically involve abnormal physiology—how stress alters metabolic processes—for example, photosynthesis, respiration, translocation, or flowering. At the population level, stress impacts on rates of reproduction, morbidity, and mortality are especially significant. From the perspective of forest stands, the concepts of competition, vegetative interaction, and spatial arrangement of trees are central to stress intensity and significance. Stress alteration of forest system functions, such as biogeochemical cycling and energy flow, and ecosystem structure such as biological diversity, succession, and food webs, are encompassed in the forest ecosystem perspective. Unfortunately, our knowledge of the relationship between air pollution and forest systems is not equivalent across these perspectives. We know most about the relationship between individual trees, less about the relationship at the population and stand levels, and least about ecosystem-level responses. This difference is due to our relatively deficient understanding of complex systems compared with individual trees and to the vast differences in cost, time, and complexity of research efforts on single plants versus systems of organisms. With the exception of a few ecosystem-scale data sets, most inferences concerning perturbations to ecosystem processes are extrapolations from individual plant, population or community-scale studies.

This chapter reviews the perturbations caused by air pollutant exposure that may reduce forest values by adversely affecting forest ecosystem processes (Fig. 11-1).

ENERGY FLOW

Solar energy captured by the photosynthetic activity of ecosystem producers is moved through systems by means of two interconnected food webs, grazing and detrital. Herbivores of the grazing food webs consume living producer material and are in turn consumed by numerous carnivores and omnivores. Detrital feeders and decomposers of the detrital food webs break down organic products and dead producers. At each stage of the flow process, some energy is stored as biomass, used in metabolism, and lost as heat. Exposure to air pollutants has the potential to perturb this flow of energy and diminish certain societal values of forest ecosystems.

ENERGY FIXATION AND USE IN TREES

A most fundamental characteristic of an ecosystem is its productivity. Forest productivity (energy storage) is high relative to other ecosystems, and net productivity of 1,200 dry gm^{-2} yr^{-1} for trees and shrubs combined is quite typical for temperate forests (Whittaker

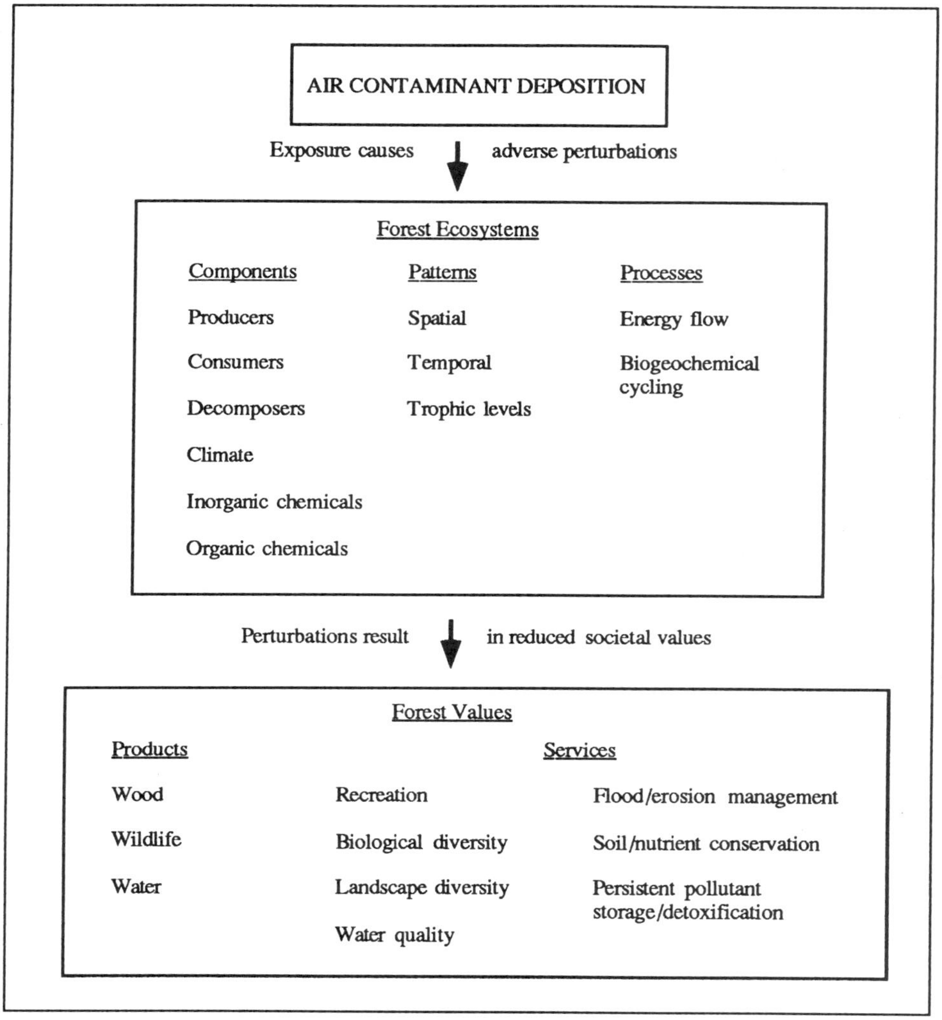

FIGURE 11-1. Linkage between air pollution exposure, forest ecosystem elements, and societal values.

1975). For forests being managed for lumber or paper production, accumulation of woody biomass is of primary concern. Exposure to air contaminants may influence biomass accumulation by adversely affecting photosynthetic rates, respiration rates, or carbon allocation in system trees.

Photosynthesis

Photosynthesis is the most fundamental metabolic process of forest ecosystems and is the primary determinant of growth and biomass accumulation. The rate of net photosynthesis of mature trees frequently is within the range of 10-200 mg of carbon dioxide taken up per gram of dry weight per day.

The rate is extremely variable, however, and is influenced by genetic, clonal, and provenance differences, season of the year, time of day, position within the crown of the tree, age of foliage, climate, and edaphic factors. Studies with a wide variety of agricultural and herbaceous species, under controlled environmental conditions, have indicated that airborne chemicals must be added to the list of environmental variables that can potentially alter the rate of photosynthesis.

Because of ease of handling and experimental design, investigators studying the relationship between air pollutants and tree photosynthesis have primarily employed tree seedlings for research material and controlled environmental facilities for growth. Evidence has been provided, under the above circumstances, for photosynthetic suppression caused by sulfur dioxide, ozone, fluoride, heavy metals, and coal dust.

The thresholds of photosynthetic toxicity for tree seedlings vary with individual species, individual pollutants, length of exposure, and other experimental conditions. For several seedlings, the threshold of sulfur dioxide photosynthetic influence may approximate 1,000 ppb (2,620 $\mu g\ m^{-3}$) for exposure of several hours or 50-100 ppb (131-262 $\mu g\ m^{-3}$) for continuous exposure for several weeks. For ozone, the threshold of photosynthetic response may approximate 100 ppb (196 $\mu g\ m^{-3}$) if the exposure occurs for several hours per day and persists for several weeks (Table 11-3).

Considerable risk is associated with extrapolation of seedling photosynthetic data accumulated in controlled environment facilities to older trees in natural forests. Excised-leaf and small-chamber techniques, therefore, have been used to assess the air pollutant influence on photosynthetic rates of trees 5 years old and older. The use of sapling-age experimental material avoids the unique characteristics of seedling metabolism. Evidence for forest tree sapling photosynthetic suppression has been presented for sulfur dioxide, ozone, and cadmium. For sulfur dioxide and ozone exposure, the sapling evidence suggests that the threshold of photosynthetic reduction may approximate 500 ppb sulfur dioxide for many hours for 2 or 3 days, and 500 ppb ozone for similar-length exposures. Chronic, many-week exposures to ozone as low as 25 ppb (49 $\mu g\ m^{-3}$) for several hours per day, however, may represent an extended term threshold (Table 11-4).

With regard to sulfur dioxide and agricultural species, Roberts (1984) has generalized as follows: 19-38 ppb (50-100 $\mu g\ m^{-3}$) for several months has produced beneficial as well as detrimental effects on yield, 38-76 ppb (100-200 $\mu g\ m^{-3}$) has produced yield losses in some studies, but not all, and 76-150 ppb (200-393 $\mu g\ m^{-3}$) generally produces significant yield losses in the selected crops studied. From the limited woody plant studies available, we find a significant impact of sulfur dioxide on forest trees may fall only in the 100-150 ppb chronic exposure range. With regard to ozone and agricultural productivity, the National Crop Loss Assessment Program has indicated yield losses in numerous crops where the growing season 7-hour mean for ozone is 40 ppb (78 $\mu g\ m^{-3}$) (Miller 1987).

Woody plant evidence available for selected species suggests forest trees may also experience important growth reductions at the ambient exposures that adversely affect agricultural species. Reich (1987) has provided a comprehensive and integrated review of ozone impact on plant photosynthesis and productivity, and has reached numerous important conclusions. At low ozone concentration (approximately 50 ppb, 98 $\mu g\ m^{-3}$), evidence from agricultural crops, hardwood trees, and conifers all suggest linear reductions in net photosynthesis and growth with respect to ozone uptake. When uptake is the same, agricultural crops are more sensitive than hardwoods, which in turn are more sensitive than conifers. These differences in sensitivity are presumed to be due to several factors, with some of the most important being that conifers generally have lower diffusive conductances than crop species and require longer exposure to ozone for comparable uptake, conifer foliage is less productive per unit of time than crop foliage, and conifer

TABLE 11-3. Threshold Dose for Photosynthetic Suppression of Selected Forest Tree Seedlings by Air Contaminants

Pollutant	Concentration	Time	Experiment Duration	Species
SO_2 6 ppm	(15.7 x 103 μg m^{-3})	4-6 hr	Single treatment	Red maple
1 ppm	(2620 μg m^{-3})	2-4 hr	Single treatment	Quaking aspen White ash
0.1 ppm	(262 μg m^{-3})	Continuous	2 weeks	White fir
0.2 ppm	(524 μg m^{-3})	Continuous	2 weeks	Norway spruce Scotch pine
O_3 0.30 ppm	(588 μg m^{-3})	9 hr day^{-1}	10 days	Ponderosa pine
0.15 ppm	(294 μg m^{-3})	Continuous	19 days	White pine
0.15 ppm	(294 μg m^{-3})	Continuous	84 days	Slash pine Pond pine Loblolly pine
0.085 ppm	(167 μg m^{-3})	5.5 hrs day^{-1}	62 days	Hybrid poplar
0.120 ppm	(235 μg m^{-3})	7 day wk^{-1} 5 day wk^{-1}	60 days	Sugar maple
F 30 μg g^{-1} d.w. basis foliar tissue				Pine (various)
Pb <10 μg g^{-1} d.w. basis foliar tissue				American sycamore
Cd <10 μg g^{-1} d.w. basis foliar tissue				American sycamore

Source: Smith 1990.

needles have a higher capacity to resist stress (low nutrient supply, herbivore feeding, microbial infection, abiotic stress) than crop foliage. Hardwoods are judged to be intermediate with regard to the above characteristics.

Ambient air over large portions of eastern North America has an average of 50-70 ppb (98-1,372 μg m^{-3}) ozone on clear summer days (natural background 20-30 ppb (39-59 μg m^{-3})), with frequent peak concentrations of 80-110 ppb (159-216 μg m^{-3}). Available evidence suggests that following 1 to 2 weeks of typical growing season pollution, mean daytime concentration of 50-70 ppb (exposure above a background of 3-7 ppm per hour), agricultural crops will exhibit significant declines in net photosynthesis and growth. Hardwood forest trees will begin to exhibit similar decreases following several additional weeks of elevated concentrations, that is, with accumulative exposure above background of approximately 10-20 ppm per hour. The threshold for conifer impact may be several months of elevated ozone concentrations resulting in exposures approximating 25-100 ppm per hour.

TABLE 11-4. Threshold Dose for Photosynthetic Suppression of Selected Forest Tree Saplings by Air Contaminants

Pollutant	Concentration	Time	Experiment Duration	Species
SO_2 1 ppm	(2620 $\mu g\ m^{-3}$)	30 minutes	Single treatment	Silver maple (excised leaves)
0.5 ppm	(1310 $\mu g\ m^{-3}$)	7-11 hr	1-2 days	Black oak
0.5 ppm	(1310 $\mu g\ m^{-3}$)	7-11 hr	1-2 days	Sugar maple
0.5 ppm	(1310 $\mu g\ m^{-3}$)	7-11 hr	1-2 days	White ash
O_3 0.5 ppm	(980 $\mu g\ m^{-3}$)	4 hr	Single treatment	White pine
0.5 ppm	(980 $\mu g\ m^{-3}$)	7-11 hr	1-2 days	Black oak
0.5 ppm	(980 $\mu g\ m^{-3}$)	7-11 hr	1-2 days	Sugar maple
0.023 ppm	(45 $\mu g\ m^{-3}$)	7 hr day^{-1} 7 day wk^{-1}	9 weeks	Red oak
Cd	$\approx$100 $\mu g\ g^{-1}$ (?)	45 hr	Single treatment	Silver maple (excised leaves)

Source: Smith 1990a.

We must recognize a variety of cautions when we consider air contaminants and photosynthesis. Most of the evidence available has been generated by seedling, sapling, or small chamber studies. Direct evidence from closed-canopy forests is very meager. Much of the seedling and sapling evidence suggests that the photosynthetic inhibition caused by sulfur dioxide and ozone is reversible if the pollutant stress is removed. Under the circumstance of variable pollutant concentration in ambient atmospheres, photosynthetic recovery might be common. Synergism, or greater stress resulting from simultaneous pollutant exposure relative to single pollutant exposure, appears frequently in the seedling and sapling literature. Evidence for synergistic photosynthetic suppression by sulfur dioxide and ozone, and fluoride and cadmium, has been presented. Studies with short-term exposure periods, have not provided evidence for an interaction with ozone and acid deposition on net photosynthesis or growth (Chappelka, Chevone, and Burk 1985; Chappelka and Chevone 1986; Reich, Schoettle, and Amundson 1986; Reich et al. 1987). Almost all the studies report photosynthetic depression in the absence of, or at least prior to, the appearance of visible foliar symptoms.

Despite the cautions, we can conclude that current levels of ambient ozone are causing declines in net photosynthesis and growth in some species in natural forests over significant portions of North America.

Respiration

Allocation of carbon resources to respiration is central to whole plant carbon dynamics. Plant growth, in effect, is the balance of photosynthetic gains and respiratory losses. Models of plant growth and productivity must describe respiration as adequately as photosynthesis (Amthor 1984). Similarly, air pollution studies must evaluate the impact of pollutant exposure on respiration as well as photosynthesis.

A useful contemporary model of dark respiratory losses partitions the losses into two functional components, a growth component and a maintenance component. The growth component is associated with the synthesis of new biomass; the maintenance component is associated with sustaining existing biomass (Amthor 1984, 1986a, b).

White pine seedlings grown in a growth chamber and exposed to 50 ppb (98 $\mu g\ m^{-3}$) ozone for 5 weeks exhibited increased respiration (Barnes 1972). Treatment of slash, pond, loblolly, and white pine with 150 ppb (294 $\mu g\ m^{-3}$) for several weeks also resulted in increased respiration by second-year needles. McLaughlin et al. (1982) recorded higher rates of respiration by branches of "sensitive" relative to "intermediate or tolerant" white pine growing in the field. This increase may have been due to ambient ozone. Reich (1983) reported an increase in respiration by leaves of hybrid poplar fumigated with low levels, 85 ppb (167 $\mu g\ m^{-3}$) and 125 ppb (245 $\mu g\ m^{-3}$), of ozone. This response was most dramatic in younger leaves.

Yang et al. (1983) exposed 2-year-old "sensitive" white pine seedlings to 100 ppb (196 $\mu g\ m^{-3}$), 200 ppb (392 $\mu g\ m^{-3}$) ozone for 50 days. These high exposures resulted in a decrease in needle respiration. Intermediate and tolerant clones did not exhibit alterations in respiration rate following ozone exposure.

Working with small chambers placed on 20-year-old Scots pine and exposed to ozone ranging from 60 ppb (120 $\mu g\ m^{-3}$) to 200 ppb (400 $\mu g\ m^{-3}$) for 1 month, Skärby, Troeng, and Boström (1987) observed that dark respiration increased throughout the treatment period, and the accumulated respiration was approximately 60 percent higher for the ozone-exposed shoots at the end of the experiment. Küpers and Klumpp (1987) exposed Norway spruce foliage to ozone, sulfur dioxide, and ozone plus sulfur dioxide under chamber conditions. They observed an elevation in dark respiration in all treatments in current-year needles. Ozone alone induced the greatest increase.

Amthor (1986b) exposed pinto bean unifoliate leaves to 90 ppb (176 $\mu g\ m^{-3}$) ozone for 6 hour per day in a growth chamber. He observed a 10 to 15 percent increase in maintenance respiration and no change in growth respiration. In open-top chamber experiments in the field, employing pinto bean, Amthor also recorded an increase in maintenance respiration and no change in growth respiration. The increase in maintenance respiration was 10 to 30 percent over three ozone exposures (zero ozone, ambient, and 2x ambient). Amthor concluded that an important mechanism for low-level ozone inhibition of plant growth and productivity may be the increase in maintenance respiration.

Carbon Allocation

In addition to fixation (photosynthesis), the utilization and allocation of carbon, along with water and mineral nutrients, within the plant is a critical regulator of plant health. Resource allocation within plants and the significance of resource allocation for yield and competitiveness are well-developed concepts (Waring and Patrick 1975; Wilson 1972; Mooney 1972; McLaughlin and Shriner 1980).

The concepts of whole-plant allocation are especially significant to the comprehension of the effects of pollutants on tree metabolism, as both sources and sinks may be impacted upon by pollutant stress (McLaughlin 1987). Almost all sustained stresses reduce canopy integrity, suppress photosynthesis, and affect storage reserves. Shade, drought, mechanical abrasion, and nutrient stress cause distinctive alterations in how photosynthate is allocated along the stem and to the roots (Waring 1987; Waring and Schlesinger 1985).

A large number of studies with nonwoody species have examined the influence of gaseous pollutants, particularly sulfur dioxide and ozone, on biomass partitioning into leaves, shoots, and roots. Generalizations are difficult, but substantial evidence suggests that exposure to these gases can reduce root biomass more than leaves or shoots, and can result in reduced root/shoot ratios (Miller 1987).

McLaughlin et al. (1982) studied the growth trends of nine 25-year-old white pine trees, in varying stages of apparent vigor, growing in a plantation in Oak Ridge, Tennessee. Tolerant, intermediate, and sensitive trees were examined for differences in patterns of photosynthate allocation that may have been due to ambient exposure to air pollutants, especially ozone. Higher retention of ^{14}C- photosynthate by foliage and branches of sensitive trees indicated that photosynthate export to boles and roots was reduced. The ratio of respiratory to photosynthetic activity was significantly higher for foliage of sensitive trees.

Reich and Lassoie (1985) exposed cuttings of hybrid poplar in chambers to daily 5.5-hour exposures to ozone at 25 ppb (49 $\mu g\ m^{-3}$), 50 ppb (98 $\mu g\ m^{-3}$), 85 ppb (167 $\mu g\ m^{-3}$), and 125 ppb (245 $\mu g\ m^{-3}$) for 10 weeks. At the end of the study, dry weights of plants in the 85 and 125 ppb treatments were 10 to 15 percent lower than in the 25 and 50 ppb treatments, but ozone did not have any influence on dry matter partitioning.

Sharpe and colleagues (NCASI 1988) have exposed loblolly pine seedlings to 120 ppb (235 $\mu g\ m^{-3}$) ozone (7 hours per day, 5 days per week) for 12 weeks in controlled-environment chambers. Treated and control seedlings were exposed to $^{11}CO_2$. Results suggest that transport of photosynthate to roots and shoots was substantially reduced by chronic exposure to ozone prior to ^{11}C treatment. Ozone exposure during ^{11}C experiments appeared to have little impact on photosynthate transport.

In addition to its influence on photosynthesis, ozone exposure may influence carbon allocation within the plant. Although evidence is not consistent, several studies have suggested reduced movement of carbon to roots following exposure. In trees, between 15 percent and 50 percent of photosynthate is allocated to produce, maintain, and replace mycorrhizal fine-root systems annually (Marx 1988). A reduction in root growth could increase the risk of drought stress. Root growth reduction could also exacerbate nutrient deficiencies. In addition, a variety of studies have indicated an increased allocation to respiration following ozone exposure. This increase may be especially significant in the maintenance component of respiratory function.

Energy Storage

The net influence of air contaminants on carbon fixation and allocation at the physiologic level can translate to adverse impact on growth (productivity = energy storage) at the forest stand level. Evidence from a variety of studies examining forest growth in the vicinity of large point sources of sulfur dioxide has indicated significantly reduced growth within a few kilometers of these facilities. Generally, the correlation of growth impact with degree of foliar injury caused by sulfur dioxide is not high. Growth retardation occurs in the absence of any visible indication of stress. Evidence for ozone suppression of forest growth has been provided by controlled-environment, open-top chamber, branch-chamber, and field studies (Smith 1990a). As in the case of sulfur dioxide, ozone suppression of growth occurs without the development of visible symptoms. Unlike sulfur dioxide, however, ozone exposure sufficient to cause growth reductions is widespread over North American forests. Oxidant-related forest growth reduction may represent the most pervasive relationship between regional-scale air contamination and forest ecosystems.

Oxidants Reduce Energy Storage

The ability of ozone to reduce the growth of agricultural ecosystems has been appreciated for approximately 30 years (Todd and Garber 1958) and has been adequately reviewed (Heck 1989). The National Crop Loss Assessment Program employed open-top field chambers in various U.S. locations to establish that ambient ozone causes yield reductions in soybean (10 percent), peanut (14 to 17 percent), turnip (7 percent), head lettuce (53 to 56 percent), and red kidney bean (2 percent). Economic analyses have indicated that the benefits to society of moderate (25 percent) ozone reductions would be approximately $1.7 billion for crops only (Adams, Hamilton, and McCarl 1985).

A large amount of evidence, generated from studies employing seedling- or sapling-size trees, also indicates that ozone can affect the growth of forest trees. Growth of plane trees in ambient greenhouse air in Washington, D.C., was observed by Santamour (1969) to be only 75 percent of the height growth in filtered air.

Jensen (1973) observed that the growth of 1-year-old sycamore seedlings was reduced by ozone doses of 300 ppb (588 $\mu g\ m^{-3}$) for 8 hours per day for 5 days per week for 5 months. Jensen (1982) also treated silver maple and eastern cottonwood seedlings with ozone at 0, 100, 200, or 300 ppb (0, 196, 392, or 588 $\mu g\ m^{-3}$) for 12 hours per day for 60 days and found that relative growth, leaf-area expansion, and leaf-weight rates declined with increasing ozone exposure.

Kress, Skelly, and Hinkelmann (1982a) found that low-dose exposure of loblolly pine seedlings to ozone, nitrogen dioxide, and sulfur dioxide in combination can result in significant height growth suppression. All three pollutant concentrations employed were below the National Ambient Air Quality Standards. Similar exposure of American sycamore to these three pollutants also resulted in significant growth suppression. Height growth was significantly suppressed by ozone alone in some cases (Kress, Skelly, and Hinkelmann 1982b). In addition, in experiments using comparable exposures to ozone and nitrogen dioxide, Kress and Skelly (1982) observed that white ash and yellow poplar exhibited significant growth stimulations when exposed to ozone at 50 ppb (98 $\mu g\ m^{-3}$) and yellow poplar and Virginia pine were the only species that failed to show any significant growth response to ozone at 150 ppb (294 $\mu g\ m^{-3}$). Fumigation with ozone or sulfur dioxide alone did not significantly affect shoot height growth or seedling dry weight of yellow poplar, but in combination with each other and with nitrogen dioxide, a greater than additive response occurred (Mahoney et al. 1984). This led John Skelly and his colleagues to emphasize that gaseous pollutants may interact synergistically to reduce plant growth.

Hogsett et al. (1985) studied the growth response of two varieties of slash pine seedlings to chronic ozone exposures. Emergent seedlings were exposed continuously to two daily peak exposure profiles of ozone having 7-hour (0900-1600) seasonal means of 104 and 76 ppb (204 and 149 $\mu g\ m^{-3}$) over a 112-day period. Destructive harvests at 7-day intervals over the exposure period were used to assess visible injury and to construct growth curves for stem diameter, plant height, top and root dry weight, and needle number and length. Visible injury was found to be slight, but all growth parameters decreased significantly with time and ozone concentration. Root growth was the most severely affected.

The use of open-top chambers in field settings has allowed growth studies to evaluate the influence of ozone on larger seedlings and sapling trees. In an early application of this exposure technique, Duchelle, Skelly, and Chevone (1982) used open-top chambers to evaluate the effect of ambient ozone in the Shenandoah National Park in Virginia on the growth of eight planted forest species native to the Virginia Appalachian Mountains. Height growth was suppressed for all species at the end of the second growing season when grown in open plots (no chamber) and ambient chambers compared with those grown in chambers with charcoal-filtered air.

Wang, Karnosky, and Bormann (1985) used open-top chambers to study ambient ozone effects on trembling aspen in Millbrook, New York (about 110 kilometers north of New York City). Over a 3-year period, four clones representing a range of pollutant sensitivities were exposed to charcoal-filtered and ambient air. Ambient ozone significantly reduced (12 to 24 percent) above-ground dry-matter production and modified tree morphology, root/shoot ratios, and rates of leaf senescence. For two clones, biomass was reduced in the absence of visible foliar symptoms. Growth reductions were not significant for eastern cottonwood or black locust tested by similar procedures (Wang, Bormann, and Karnosky 1986) (Table 11-5).

Because of the commercial importance of loblolly pine in the southeastern United States, a variety of recent studies have examined loblolly response to ambient ozone. Shafer, Heagle, and Camberato (1987) planted seedlings of four full-sib families of loblolly in a field near Raleigh, North Carolina. Open-top chambers were employed to expose the seedlings to ozone ranging from 0.5 to 1.96 times ambient ozone. Responses of stem height, stem diameter, biomass, and other characteristics were quantified by regression. All relationships were linear for three families, but one family exhibited no significant growth response. Dose-response equations suggested a maximum growth suppression of 10 percent for ambient air compared with charcoal-filtered air. In a similar experiment, Adams, Kelly, and Edwards (1988) exposed loblolly pine seedlings from five half-sib families to ambient, subambient (0.6 X ambient), and elevated [ambient + 60 ppb (120 $\mu g\ m^{-3}$)] ozone for one growing season in open-top chambers in the Tennessee Valley. Elevated ozone resulted in significantly reduced above-ground volume and secondary needle biomass relative to seedlings grown in ambient air. Subambient ozone did not result in seedling size significantly different from ambient. Evidence was presented that loblolly pine response to ozone is strongly regulated by genotype.

Kress et al. (1988) employed open-top chambers in North Carolina to examine the influence of both ozone and acid precipitation on loblolly pine growth. Exposure to elevated ozone (90 ppb, 176 $\mu g\ m^{-3}$) 12 hours per day suppressed the diameter growth and foliar biomass of seedlings in one growing season treatment. There was no apparent effect of the acidic precipitation treatments.

In light of the evidence, it is clear that ozone, at concentrations common in numerous regions in North America, can reduce the growth of some seedling and sapling trees. In his comprehensive review of this topic, however, Pye (1988) cautions that extrapolation of information from managed environment studies with young trees to predictions concerning large trees in natural environments is difficult for numerous reasons. Differences in mature tree carbon allocation, canopy structure, competition, and canopy microclimate may all mediate mature tree response to ambient ozone.

TABLE 11-5. Results from Ambient-Open, Ambient-Chambered, and Filtered-Chambered Treatments on Saplings of Three Tree Species, Millbrook, New York, 1984

Ambient- Species/Characteristics	Ambient- Open	Filtered- Chambered	Chambered
Hybrid poplar			
Leaf weight, g	34.1^{a}	41.4^{b}	50.9^{c*}
Stem weight, g	49.9^{a}	63.2^{b}	78.6^{c*}
Total weight, above grd, g	82.0^{a}	103.2^{b}	126.9^{c*}
Height, cm	189.0^{a}	208.0^{b}	231.0^{c*}
Diameter, mm	12.4^{a}	13.7^{b}	14.6^{b}
No. of leaves per tree			
on main stem	40.4^{a}	41.8^{a}	51.5^{b*}
on lateral shoots	16.6^{a}	30.0^{a}	52.0^{b}
Total no. of leaves	57.0^{a}	71.8^{a}	103.5^{b*}
No. of lateral shoots	1.1^{a}	3.7^{b}	7.9^{c*}
Cottonwood			
Leaf weight, g	107.6	93.5	113.9
Stem weight, g	86.0	76.7	88.5
Total weight, above grd, g	193.6	170.2	202.4
Height, cm	162.0^{a}	175.0^{ab}	181.0^{b}
Diameter, mm	18.6	17.1	17.8
No. of leaves per tree			
on main stem	31.8	29.7	33.6
on lateral shoots	183.0	175.0	192.0
Total no. of leaves	214.0	205.0	226.0
No. of lateral shoots	22.9^{a}	18.8^{b}	20.3^{ab}
Black locust			
Stem and leaf weight, g	106.5	76.2	91.5
Height, cm	94.1	113.0	123.0
Diameter, mm	14.6	12.0	11.9
No. of leaves per tree	51.4^{a}	36.5^{b}	41.3^{b}

Note: All weights are oven-dry weights for means of all trees per treatment. Diameters were measured approximately 5 cm above the ground. Means with different letters are significantly different at $\alpha = 0.05$ using the Duncan's option in VA-SAS; asterisk indicates differences significant at $\alpha = 0.01$. Total weight above ground may not equal leaf and stem due to missing leaf data.

Source: Wang et al. (1986).

BIOGEOCHEMICAL CYCLING

For normal forest growth, nutrients must move into, within, and out of forest ecosystems in appropriate amounts, at appropriate rates, and along established pathways. Healthy forest ecosystems conserve nutrients and continually recycle them through the system via an elaborate litterfall-decomposition-uptake intrasystem cycle.

The high productivity of forest ecosystems is achieved and maintained through efficient nutrient recycling. For most forest ecosystems essential elements required to maintain productivity cannot be sustained by annual increments from deposition and mineral substrate alone. Decomposition, mineralization, and uptake are a must. A large number of hypotheses have been proposed for air pollution interference with forest nutrient cycles. These hypotheses are concerned with direct and indirect impacts of pollutants on tree root health and on one or more soil processes or functions (Fig. 11-2).

Decomposition

The release of inorganic nutrients during decomposition of forest soil organic matter is of profound importance in the maintenance of intrasystem nutrient cycling. It has been estimated that 80 to 90 percent of net production in terrestrial ecosystems is eventually converted by decomposer organisms. Forest floors, with persistent organic matter accumulation, represent important sinks for heavy metals and persistent organic pollutants from any source including atmospheric deposition. Since this accumulation is commonly in the soil horizons with maximum root activity and maximum activity of the soil micro- and macrobiota, it is appropriate to consider the potential for heavy-metal toxicity to tree roots and other biotic components of the soil ecosystem.

The extraordinary accumulation of trace metals, particularly lead, cadmium, zinc, and copper, in the organic horizon of the forest floor (Smith 1990a), has led to the hypothesis that heavy metals depress decomposition rates. Tyler (1972) proposed that decomposition of forest litter and remobilization of nutrients will be slower or less complete as heavy-metal ions bind with colloidal organic matter and increase resistance to decomposition or exert a toxic effect directly on decomposing microbes or the enzymes they produce. The release of nutrients from organic compounds is accomplished by a large number of soil microbes and animals via a complex series of decomposition and mineralization processes. The rate of litter decomposition appears to control the rate of nutrient release. As nutrients are probably always limiting, the rate of decomposition also controls the rate of primary

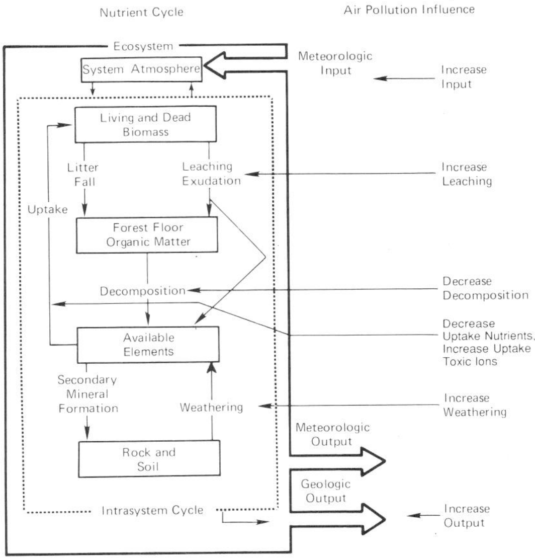

FIGURE 11-2. Potential points of interaction between nutrient cycling in forest ecosystems and air pollution.

production in forests. A significant reduction in the rate of decomposition, therefore, has the potential to importantly affect forest growth.

Over the last decade a large amount of evidence has been provided to address the hypothesis that trace metal contaminants of the soil reduce forest litter decomposition and mineralization rates (Smith 1990a). We judge that this hypothesis has been supported by the evidence provided, but only for excessively contaminated environments in the immediate vicinity of metal processing industries or other extreme sources of metal contaminants such as some urban roadside environments.

The thresholds of toxicity for numerous bacterial, fungal, insect, and other components of the soil biota are in the range of 1,000-10,000 ppm metal cation on a soil dry weight basis. These concentrations are two and three orders of magnitude greater than the concentrations of heavy metals throughout most of temperate forest ecosystems. In addition, fungi probably play a larger role in nutrient cycling in forests than other microbial groups, and they appear more resistant to trace metal influence than other microbes. In addition, there is considerable evidence for microbial adaptation to high trace metal exposure. While components of the forest floor soil biota may be affected by trace metals, decomposition rates may remain unchanged due to adaptation or shifts in species composition unreflected in gross soil process activities. There is substantial redundancy in the microbial populations capable of organic matter decomposition.

Leaching

Leaching rates regulate the flux of nutrients between the various compartments of the forest ecosystem (Fig. 11-2). A large number of hypotheses have been proposed concerning the ability of acid deposition to influence these leaching rates. Rates of particular interest include loss of nutrient cations from above-ground tree parts and migration of nutrient cations down through the soil profile. Leaching potential requires cations for nutrient displacement as well as mobile anions for nutrient migration. Acid deposition supplies both the former in the form of hydrogen and aluminum ions, and the latter in the form of sulfate and nitrate ions. In most soils, rapid biological uptake may immobilize the nitrate anion. As a result, risk of cation loss associated with this anion may be restricted to ecosystems rich in nitrogen where biological immobilization of nitrate is minimal (Johnson, Turner, and Kelly 1982). Similarly, sulfate may be immobilized in weathered soils by adsorption to free iron and aluminum oxides. In other soils, however, especially those low in free iron or aluminum, or high in organic matter, which appears to block sulfate absorption sites, sulfate may readily combine with nutrient cations and leach these elements beyond the rooting zone.

Cronon (1980) prepared microcosms using forest floor material collected from the subalpine zone (1462 m) of Mt. Moosilauke, New Hampshire. When acidity of throughfall inputs to the microcosms was increased, the forest floors exhibited increased leaching losses of calcium, magnesium, potassium, and ammonium. Based on his results, Cronon estimated that with a throughfall pH of 3.5, and in the absence of net plant uptake and recycling, approximately 0.8 percent of the total forest floor calcium pool could be lost annually. Similar calculations for magnesium, potassium, and nitrogen suggested leaching losses approximating 3, 85, and 6 to 16 percent, respectively. Mollitor and Raynal (1982) studied cation leaching in hardwood and conifer stands in Huntington Forest, Newcomb, New York. Cation leaching from the deciduous site appeared equally influenced by sulfate (pollutant) and organic (natural) anions. Sulfate and organic anion concentrations were greater in the conifer site, but organic anion leaching was judged to dominate in these stands. For selected forest soils, the risk of nutrient losses from available nutrient pools of the soil may be significant over extended time periods (decade to century scales) in regions subject to continuous acid deposition. Soils at special risk are restricted as numerous characteristics must be present; low cation exchange capacity, low base saturation, low clay or organic matter content, low sulfate and nitrate adsorption capacity, and a current pH in the 5.5 to 6.5 range.

Root Necrosis

The direct influence of heavy-metal ions on tree metabolism, especially root physiology, has the potential to reduce nutrient uptake. All of the heavy metals, biologically essential or nonessential, can be toxic to forest trees at some threshold level of exposure. Only a few metals, however, have been documented to cause direct phytotoxicity in actual field situations. Copper, nickel, and zinc toxicities have occurred frequently. Cadmium, cobalt, and lead toxicities have occurred less frequently and under more unusual conditions. Chromium, silver, and tin—in solution culture under experimental conditions—have not been demonstrated to be phytotoxic in field situations even at high doses. Direct heavy-metal phytotoxicity will result only if the metal can move from the soil to the root or from the plant surface to the plant interior. Movement via diffusion or mass flow of heavy metals from soils to roots, from roots to shoots, or from plant exteriors (leaf surfaces) to plant interiors remains poorly characterized under natural conditions. Root exposure is a function both of deposition from the atmosphere and of chemical availability. Heavy metals not available for ready exchange from binding sites or in solution are not available for root or microbial uptake. More than 90 percent of certain heavy metals deposited from the atmosphere may be biologically not available.

Heavy metals may be adsorbed or chelated by organic matter (humic, fulvic acids), clays, and/or hydrous oxides of aluminum, iron, or manganese. Heavy metals also may be complexed with soluble low-molecular-weight compounds. Soluble cadmium, copper, and zinc may be chelated in excess of 99 percent. Adsorbed heavy metals remain in equilibrium with chelated metals. Heavy metals may also be precipitated in inorganic compounds of low solubility such as oxides, phosphates, or sulfates. Miller and McFee (1983), for example, have suggested that lead may be present in the soil profile in the following forms: bound to organic matter, 43 percent; bound to ferro-manganese hydrous oxides, 39 percent; as insoluble precipitates, 10 percent; and biologically available (exchangeable), 8 percent.

Adsorption, chelation, and precipitation are strongly regulated by soil pH. As pH decreases and soils become more acid, heavy metals generally become more available for biological uptake. Natural forest soils generally become more acid as they mature. Acidification in excess of natural processes is possible, especially in soils with a pH greater than 5. Under this circumstance, soil acidification associated with acid deposition may result in increased biological availability of heavy metals in the forest floor. Rhizosphere processes also may transform heavy metals from an unavailable pool to an available pool (Smith 1990b).

Atmospheric additions of strong mineral acids to forest soils may mobilize aluminum by dissolution of aluminum-containing minerals or may remobilize aluminum previously precipitated within the soil during podzolization or held on soil exchange sites. It is clear that only dissolved forms of aluminum will be readily moved in the soil system and will be accessible for root uptake. The chemical species of greatest risk due to biological availability include free aluminum ions (Al^{3+}) and monomeric hydroxides [including $Al OH^{2+}$, $Al(OH)^{+}_{2}$, and $Al(OH)^{-}_{4}$].

Substantial evidence has been presented indicating that aluminum, once it enters forest tree roots, accumulates in root tissue (Vogt et al. 1987a, b). Direct toxic root effects may include reduced cell division associated with aluminum binding of DNA, reduced root growth caused by inhibition of cell elongation, and destruction of epidermal and cortical cells. Indirect effects of aluminum stress, especially in forest trees, may involve interference with the uptake, translocation, and/or utilization of required nutrients such as calcium, magnesium, phosphorus, potassium, or other essential elements. Shortle and Smith (1988) stressed that calcium is incorporated at a constant rate per unit volume of wood produced and is not recovered from sapwood as it matures into heartwood. As a result, when aluminum and calcium are present in approximately equimolar concentrations within the soil solution, or when the aluminum/calcium ratio exceeds one, aluminum will reduce calcium uptake by competition for binding sites in the cortical apoplast of fine roots. Reduced calcium uptake will suppress cambial growth.

Root Symbionts

The research attention given to the interaction between mycorrhizal fungi, the specialized roots they form, and air pollution is grossly out of scale with the significance of this symbiotic relationship in forest ecosystems. The potential for heavy-metal adverse impact on mycorrhizal associations appears particularly great due to the physical juxtaposition of the two entities in the forest floor. With current information, however, we cannot approximate the threshold levels of any toxic heavy metal that might exert an adverse influence on a mycorrhizal fungus in natural soil. Some data suggest aluminum may exert a toxic influence on selected mycorrhizal fungi at lower concentrations than those required to directly injure roots. Soil acidity is probably not directly restrictive to mycorrhizal development in natural forest soils. Gaseous pollutants, such as ozone, may have an adverse impact on fungal and bacterial symbionts by altering the quantity or quality of nutrient minerals supplied to the roots by the foliage. In acid soils, aluminum toxicity may play an important role in restricting symbioses involving *Rhizobium* endophytes. Evidence from forest systems is unfortunately not available.

ECOSYSTEM PROCESS PERTURBATION AND ADVERSE FOREST EFFECTS

Abnormal ecosystem energy flow resulting from altered photosynthesis, respiration, or carbon allocation by the primary producers, along with abnormal biogeochemical nutrient cycling resulting from altered soil decomposition or leaching processes, fine-root biomass, or root symbiont activity may result in adverse forest effects that stem from a reduction in forest products or a reduction in forest services. Especially significant adverse effects are associated with forest growth declines, alterations of consumer populations, and loss of biodiversity.

Growth Decline

Three regionally diverse, contemporary forest growth declines that have been linked to air quality will illustrate this risk. These are (1) growth reduction of the montane spruce-fir forest system along the Appalachian Mountain chain, (2) the southern pine systems growing on nonindustrial private land in the Southeast, and (3) components of the mixed conifer forests of California.

Evidence has been presented indicating increased mortality and morbidity or decreased radial or basal area growth increment for red spruce at low- and high-elevation sites in the Northeast. While there is reasonable consensus that growth rates have declined over the past several decades across all size classes, age classes, and elevations, there is no agreement on cause. Numerous investigations emphasize changes in climate or stand dynamics (aging, competition) as the dominant regulators of growth. Basal area data from 3,000 dominant or codominant red spruce from varied sites throughout the Northeast revealed a consistent increase in annual growth from 1910 to 1920 to approximately 1960, with general decrease during the period 1960 to 1980, with early 1980 rates 13 to 40 percent below the 1960 peak (Hornbeck and Smith 1985; Hornbeck, Smith, and Federer 1986). For low-elevation spruce this decline was judged to be primarily due to normal forest aging (Federer and Hornbeck 1987; Smith et al. 1990). Examination of high-elevation (900-1,100 m) red spruce growth on Whiteface Mountain, New York, indicated reduced growth beginning in the mid-1960's and continuing in the 1970's and 1980's (LeBlanc, Raynal, and White 1987; Raynal et al. 1988). These authors observed that the growth decreases documented were coincident with known climatic anomalies: the mid-1960's drought, which was the most severe (intensity and duration) over the past two centuries, and abnormally warm (1975-1976, 1980-1981, 1982-1983) and abnormally cold (1976-1977, 1977-1978, 1978-1979) winter temperatures. McLaughlin et al. (1987) analyzed increment cores from 1,000 red spruce distributed in high-elevation Appalachian forests in the eastern United States. They concluded that the observed growth decreases over the past 20 to 25 years were anomalous with respect to both climate change and forest aging. Federer, Hornbeck, and Smith (1987) emphasize that red spruce growth decline appears unique relative to other species and that high-elevation spruce response may be distinct in cause from low-elevation factors. The high exposure of eastern montane forests to acid deposition coupled with the high risk characteristics of some of the soils of these forests suggest air pollution stress as a potential contributor to growth decline.

Periodic timber inventories conducted by the U.S. Department of Agriculture (USDA) Forest Service documented that net annual growth of pine in the Atlantic Coast States from Virginia to Florida has peaked and turned downward after many years of increase (Sheffield et al. 1985). Average annual radial growth of southern pine under 40 cm in diameter has declined by 30 to 50 percent over the past three decades. The most pronounced declines have been measured in pines growing on nonindustrial private forest land. As in the case of red spruce, numerous factors may be responsible for the decreased radial growth in southern pine. Important factors identified by Sheffield et al. (1985) include stand aging, increased stand densities, increased hardwood competition, drought, lowered water tables, land use history (i.e., loss of old-field sites), diseases, atmospheric deposition, and interactive effects of the preceding. In his critiques of the southern pine decline, Lucier (1986) correctly indicates that the Forest Inventory and Analysis data cannot indicate the cause of decline. He further indicates that the reductions in radial and basal area growth are not equivalent to reductions in forest productivity. Clearly our understanding of the decline in net annual growth of southern pine is incomplete (Sheffield and Cost 1987). It is also clear, however, that the impact of the decline on the southern forestry industry is significant (Knight 1987) and that air quality, particularly oxidants, may be potentially involved (Zahner, Saucier, and Myers 1989).

One of the few correlations of growth parameters of large trees growing under field conditions with ambient ozone levels has been provided by the comprehensive oxidant study conducted in the San Bernardino National Forest in California (Miller 1989). The San Bernardino Mountains are part of the Transverse Range in southern California. Their position east of the Los Angeles basin allows them to function as a barrier to the movement of marine air. The position and elevation of the San Bernardino Mountains facilitates the development of inversion layers in the Los Angeles basin. Air pollutants trapped beneath the inversion layer are drawn into the forest at higher elevations in the mountains by a "chimney effect," resulting from radiant heating of the south-facing slopes. As a result, a gradient in the concentration of oxidants exists across the coniferous forest in the mountains. Monitoring stations on the southwestern edge of the forest had average hourly ozone concentrations of 120 ppb (235 $\mu g\ m^{-3}$) from May to September during 1974-78. Monitoring stations to the north and east in the forest are characterized by lesser concentrations. Camp Angeles, located 34 kilometers southeast of the stations with the highest ozone concentrations, had average hourly ozone levels of 60 ppb (118 $\mu g\ m^{-3}$) (McBride, Miller, and Laven 1985). Stem growth of ponderosa pine was determined to be limited as much as 50 percent in the highest oxidant exposure plots, and accumulated mortality from 1973 to 1978 reached as high as 10 percent. Growth and mortality of white fir was also reduced in the highest exposure areas studied. Foliage injury and up to 25 percent stem growth decrease were observed on California black oak in the highest exposure plots. Evidence for oxidant stress on California conifer forests also comes from studies in the Sierra Nevada range in the Sequoia National Forest and in the Sequoia and Kings Canyon National Parks. Tree ring analyses of Jeffrey pines growing in areas of the two national parks with different ozone exposures have been compared. Reduction of ring width index was noted in recent years for large trees growing on marginal sites in the ozone-exposed areas (Miller 1989).

If growth reductions from air contaminants do occur in commercial forests managed for wood production, the direct loss of reduced wood volume is clear and the loss of dollar value can be calculated. Less clear, and impossible to quantify in dollar terms, however, are the reduced values of forest services. Forest services, such as water quality, flood and erosion management, and soil and plant nutrient conservation (Table 11-2) depend in part on soil organic matter. Solar energy is stored in forest ecosystems in the form of wood in tree stems but also as organic matter in the soil. Approximately 75 percent of the net annual production of the northern forest ecosystem may fall to the forest floor and enter the detrital energy pathway (Gosz et al. 1978). A 55-year-old northern hardwood forest in central New Hampshire may contain 7×10^8 Kcal ha^{-1} in living biomass, which accumulates at a net rate of 1.2×10^7 Kcal yr^{-1}. In addition, approximately 1.4×10^9 Kcal ha^{-1} is stored in dead biomass, which accumulates at a net rate of about 1.2×10^6 Kcal yr^{-1} in or on the soil (Bormann and Likens 1981). Soil organic matter critically regulates ecosystem services by influencing hydrologic and biogeochemical characteristics of the area. Organic matter reduces bulk density, and thereby increases infiltration and water-holding capacity of soil, which results in reductions in peak stream flows and soil erosion. In addition, soil organic matter increases cation exchange capacity, which results in cation (nutrient) conservation along with increased potential for persistent pollutant storage and detoxification (Bormann 1976).

Forest systems further influence regional hydrologic cycles by reducing water stored in soils during the growing season via transpiration and evaporation processes. This reduction dampens summer flows, reduces storm peaks, and minimizes erosive forces. The use of solar energy in transpiration may represent approximately 10 to 15 percent of all incident solar energy falling on a forest ecosystem during 1 year. Again, in the northern hardwood forest of central New Hampshire about 3 million liters of water ha^{-1} are transpired through the leaves each year (Bormann 1976). This loss of water vapor represents the removal of actual or potential heat energy available to warm the surrounding air. As a result, the forest and its immediate environment are cooler and more conducive to human comfort than are nonforested locations. This service directly enhances the recreational, energy, and amenity values of forest systems for people. The loss of these unpriced values must be considered in comprehensive inventories of the adverse consequences of reduced growth of forest systems potentially linked to air pollutant exposure.

Alteration of Consumer Populations

Energy flows through ecosystems from primary producers to diversified consumers through food webs. If primary production energy storage is reduced, or if tree vulnerability to consumption is altered, perturbations to consuming populations may have adverse consequences for forest ecosystem products and services.

In forest ecosystems, living plant tissue available to consumers is of three primary forms; seeds and fruits, foliage, and woody tissue. These tissues are consumed by organisms in both the grazing and the detrital food webs. The major consumers in the grazing food web are leaf-eating insects (Gosz et al. 1978).

Direct or indirect stimulation of insect defoliation in forests by air pollutant exposure could have adverse effects on both forest products and forest services. In temperate zone forests, foliage-feeding insects generally consume 3 to 8 percent of the annual leaf production. Up to 40 percent defoliation appears to have little impact on productivity. Above approximately 40 percent, however, losses in wood production are proportional to foliage losses but highly influenced by other stresses on the trees, by defoliation history, and by climate (Mattson and Addy 1975). Moderate (above 40 percent) and heavy (above 75 percent) defoliation cause decreases in productivity largely due to reallocation of carbon resources within trees from wood production to refoliation.

In addition to forest product impact resulting from enhanced insect defoliation, forest services (for example, water quality protection) could be compromised. Swank et al. (1981) have provided ecosystem-level evidence that insect defoliation of forests can be a major factor in regulating the quality of stream water draining defoliated systems. These investigators documented that chronic defoliation (approximately 33 percent maximum defoliation) by the fall cankerworm (*Alsophila pometaria*) was accompanied by substantial increases in the stream export of nitrate nitrogen from three mixed hardwood forests in the southern Appalachians. Their results suggested that defoliation caused a temporary shift from wood to leaf production, increased rates of nutrient uptake, increased litter turnover rates, and accelerated rates of recycling, turnover, and loss of labile elements in forest floor horizons.

The complex interaction between forest trees, insect consumers, and air quality is also indicated by bark beetle interactions. In the latter 1960's, California investigators added ozone to the list of biotic and environmental stresses that predispose ponderosa pine to bark beetle infestation. This is perhaps one of the most completely documented examples of insect damage enhancement caused by air pollution in North America (Smith 1990a). Dahlsten and Rowney (1980) concluded that ozone-damaged trees, compared with healthier trees, produced approximately the same total brood of new beetles with fewer initial attacks. It was judged, therefore, that a given population of western pine beetles killed more trees and increased at a greater rate in stands with a high proportion of ozone-injured trees. The implications of this for forest wood production are clear. In addition, however, enhanced bark beetle activity can cause reduction in forest services such as hydrologic regulation. For two large watersheds in Colorado where bark beetles caused significant tree mortality, substantially greater water yields (with associated undesired consequences) were still evident 25 years after the bark beetle epidemic (Bethlahmy 1975).

Reduction in Biodiversity

Forest ecosystems supply an extraordinary service to human societies by providing landscape diversity and biological diversity. Comprehensively defined, biological diversity includes the total inventory of genotypes, species, and habitats contained within forest ecosystems. The multiple values provided by this diversity are extremely wide-ranging. This topic is extensively explored in other chapters in this volume, but its importance requires some consideration here.

Biological diversity is a dynamic characteristic of forest ecosystems that changes as forest systems mature through successional stages. Two major types of processes influence ecosystem succession. Autogenic processes are those resulting from biological factors within the system. In forest ecosystems, autogenic processes include site alterations caused by vegetation, the influence of one plant species on another, and the impact of native insect or disease microorganisms. Allogenic processes, on the other hand, are abiotic factors that influence succession from outside the system. Geochemical and climatic forces are especially important examples of allogenic forces that influence forest ecosystems. Stresses imposed on forest ecosystems by atmospheric pollutant deposition must be considered a 20th-century allogenic process of potential importance to forest ecosystem development and biological diversity (Smith 1980, 1990a).

The stress imposed on Californian wildland ecosystems by atmospheric pollution again provides an appropriate example. In southern California the predominant native shrubland vegetation consists of chaparral and coastal sage scrub. The former occupies upper elevations of the coastal mountains, extending into the North Coast ranges, east to central Arizona, and south to Baja California; the latter occupies lower elevations on the coastal and interior sides of the coast ranges from San Francisco to Baja California. Westman (1979) applied standard plant ordination techniques to these shrub communities to examine the influence of air pollution. The reduced cover of native species of coastal sage scrub documented on some sites was statistically indicated to be caused by elevated atmospheric oxidants. Sites of high ambient oxidants were also characterized by declining species richness.

Ponderosa pine is one of five major species of the "mixed conifer type" that covers wide areas of the western Sierra Nevada and the mountain ranges, including the San Bernardino Mountains, in southern California from 1,000 to 2,000 m elevation. Other species represented include sugar pine, white fir, incense cedar, and California black oak. The response of these five major tree species to oxidant air contaminants in the San Bernardino National Forest is variable as shown by field surveys and seedling exposures (Table 11-6). Ponderosa pine exhibits the most severe foliar response to elevated ambient ozone. A 1969 aerial survey conducted by the USDA Forest Service indicated 1.3 million ponderosa (or Jeffrey) pines on more than 405 km^2 (100,000 acres) were stressed to some degree. Mortality of ponderosa pine has been extensive. White fir has suffered slight damage, but

TABLE 11-6. Sensitivity of Conifer Tree Seedlings to Ozone Exposure of 0.36 ppm, 12 hr day^{-1}, over 25 Days in Field Chambers, Where Sensitivity is Rated in Terms of the Mean of the Log of Visible Foliage Injury

Tree Species	Mean Log Injury Score
Sensitive	
Jeffrey pine x Coulter pine hybrid	1.24
Western white pine	1.24
Ponderosa pine	1.00
Jeffrey pine	0.97
White fir	0.91
Coulter pine	0.87
Intermediate	
Red fir	0.69
Monterey pine x knobcone pine hybrid	0.69
Knobcone pine	0.51
Incense cedar	0.51
Resistant	
Big cone Douglas fir	0.41
Sugar pine	0.38
Inland ponderosa pine	0.28

Source: Miller 1983.

scattered trees have exhibited severe symptoms. Sugar pine, incense cedar, and black oak have exhibited only slight foliar damage from oxidant exposure. A 233 ha study block was delineated in the northwest section of the San Bernardino National Forest in order to conduct an intensive inventory of vegetation present in various size classes and to evaluate the healthfulness of the forest. Ponderosa pines in the 30 cm (12 inch) diameter class or larger were more numerous than other species of comparable size in the study area. These pines were most abundant on the more exposed ridge crest sites of the sample area. Mortality of ponderosa pine ranged from 8 to 10 percent during 1968-1972. The loss of a dominant species in a forest ecosystem clearly exerts profound change in that system. Miller (1989) concluded from his investigations that the lower two-thirds of the study area will probably shift to a greater proportion of white fir. It was judged that incense cedar will probably remain secondary to white fir. Sugar pine was presumed to be restricted by lesser competitive ability and dwarf mistletoe infection. The rate of composition change was deemed dependent on the rate of ponderosa pine mortality. The upper one-third of the study area, characterized as more environmentally severe due to climatic and edaphic stress, supports less vigorous white fir growth. Following the loss of ponderosa pine in this area, sugar and incense cedar may assume greater importance. Miller judged, however, that natural regeneration of the latter species may be restricted in the more barren, dry sites characteristic of the upper ridge area. California black oak and shrub species may become more abundant in these disturbed areas. Tree population dynamics were examined on 18 permanent plots established in 1972 and 1973 and on 83 temporary plots established in 1974 to investigate forest development as a function of time since the most recent fire. Generally, the data support the hypothesis that forest succession toward more tolerant species such as white fir and incense cedar occurs in the absence of fire. In the presence of fire, pine may be favored by seedbed preparation and elimination of competing species.

These more recent studies suggest a larger number of forest subtypes may exist within the forest ecosystem than initially realized.

In 1982 and 1983, plots dominated by ponderosa pine were resurveyed. Plots were grouped on the basis of foliar symptoms into severe or slight injury groups. No significant difference in seedling establishment was observed in ponderosa pine on either severe- or slight-injury plots. Differences in regeneration of other species could not be correlated with foliar symptomology. Projection of stand age structures suggested the eventual dominance of incense cedar on severe injury plots in the mixed conifer forest (McBride, Miller, and Laven 1985).

The changes in forest composition caused by oxidants in this southern California forest have created a management concern, as well as ecological change, because the forest is intensively used as a recreational resource and the loss of ponderosa pine is judged to reduce the esthetic qualities of the forest.

CONCLUSION

The fundamental processes of forest ecosystems are energy flow and biogeochemical cycling. Understanding the interaction of air pollution with these processes is essential in order to comprehensively evaluate all the adverse consequences imposed on forest products and services as a result of atmospheric deposition. In addition, future efforts to monitor forest ecosystem response to stresses, including those imposed by air pollutants, must involve assessments of forest processes.

The evidence provided in this chapter supports the conclusion that regional-scale air pollutants can influence energy flow and/or biogeochemical cycling in selected temperate zone forest ecosystems. In numerous cases, the intensity of exposure levels (dose) and variability of risk factors (soil characteristics, other biotic and abiotic stress factors, and species sensitivities) combine to cause subtle ecosystem process perturbation. This subtlety will require extended monitoring to quantify and to partition air pollution effects from other factors influencing ecosystem processes. In view of the linkage, however, between chronic air pollution stress and the values of forest ecosystem products and services, it is essential to continue to refine our understanding of this major human stress imposed on this vast and important ecosystem type.

REFERENCES

1. Adams, M.B., J.M. Kelly, and N.T. Edwards. 1988. Growth of *Pinus taeda* L. seedlings varies with family and ozone exposure level. *Water Air Soil Pollut.* 38:137-50.
2. Adams, R.M., S.A. Hamilton, and B.A. McCarl. 1985. Assessment of the economic effects of ozone on U.S. agriculture. *J. Air Pollut. Control Assoc.* 35:938-43.
3. Amthor, J.S. 1984. The role of maintenance respiration in plant growth. *Plant Cell Environ.* 7:561-69.
4. Amthor, J.S. 1986a. Evolution and applicability of a whole plant respiration model. *Theor. Biol.* 122:473-90.
5. Amthor, J.S. 1986b. Ozone-induced increase in bean leaf maintenance respiration. Ph.D.Thesis. School of For. and Environ. Studies. New Haven, CT: Yale Univ.
6. Aylor, D. 1971. Noise reduction by vegetation and ground. *J. Acoust. Soc. Am.* 51:197-205.

7. Barnes, R.L. 1972. Effects of chronic exposure to ozone on photosynthesis and respiration of pines. *Environ. Pollut.* 3:133-38.
8. Bethlahmy, N. 1975. A Colorado episode: Beetle epidemic, ghost forests, more streamflow. *Northwest Sci.* 49:95-105.
9. Bormann, F.H. 1976. An inseparable linkage: Conservation of natural ecosystems and the conservation of fossil energy. *BioScience* 26:754-60.
10. Bormann, F.H., and G.E. Likens. 1981. *Pattern and process in a forested ecosystem.* New York: Springer-Verlag.
11. Botkin, D.B., and E.A. Keller. 1982. *Environmental studies.* Columbus, OH: Merril Publishing Company.
12. Brush, R.O. 1971. The attractiveness of woodlands: Perceptions of forest landowners in Massachusetts. *For. Sci.* 25:495-506.
13. Brush, R.O., D.N. Williamson, and J. Fabos. 1979. Visual screening potential of forest vegetation. *Urban Ecol.* 4:207-16.
14. Chappelka, A.H., and B.I. Chevone. 1986. White ash seedling growth response to ozone and simulated acid rain. *Can. J. For. Res.* 16:786-90.
15. Chappelka, A.H., B.I. Chevone, and T.E. Burk. 1985. Growth response of yellow poplar (*Liriodendron tulipifera* L.) seedlings to ozone, sulfur dioxide, and simulated acidic precipitation, alone and in combination. *Environ. Exp. Bot.* 25:233-44.
16. Cook, D.I., and F. Van Haverbeke. 1977. Suburban noise control with plant materials and solid barriers. USDA Forest Service, Rocky Mt. For. Range Exp. Sta. Res. Bull. No. EM100.
17. Cronon, C.S. 1980. Controls on leaching from coniferous forest floor microcosms. *Plant Soil* 56:301-22.
18. Dahlsten, D.L., and D.L. Rowney. 1980. *Symposium on effects of air pollutants on Mediterranean and temperate forest ecosystems*, pp. 125-30. Gen. Tech. Bull. No. PSW-43. Riverside, CA: USDA Forest Service.
19. DeWalle, D.R., and G.M. Heisler. 1980. Landscaping to reduce year-round energy bills. In *1980 yearbook of agriculture. Cutting energy costs*, pp. 227-37. Washington, DC: USDA.
20. Duchelle, S.F., J.M. Skelly, and B.I. Chevone. 1982. Oxidant effects on forest tree seedling growth in the Appalachian Mountains. *Water Air Soil Pollut.* 18:363-73.
21. Ehrlich, P.R., and H.A. Mooney. 1983. Extinction, substitution, and ecosystem services. *BioScience* 33:248-54.
22. Federer, C.A. 1976. Trees modify the urban microclimate. *J. Arboric.* 2:121-27.
23. Federer, C.A., and J.W. Hornbeck. 1987. Expected decrease in diameter growth of even-aged red spruce. *Can. J. For. Res.* 17:266-69.
24. Federer, C.A., J.W. Hornbeck, and R.B. Smith. 1987. Regional dendrochronologies of red spruce and other species in New England. Gen. Tech. Rept. No. 253, Northeastern For. Exp. Sta., Broomall, PA: USDA Forest Service.
25. Forgey, B. 1991. One hundred years of serenity. *J. For.* 89:22-29.
26. Gosz, J.R., R.T. Holmes, G.E. Likens, and F.H. Bormann. 1978. The flow of energy in a forest ecosystem. *Sci. Am.* 238:92-102.
27. Heck, W.W. 1989. Assessment of crop losses from air pollutants in the United States. In *Air pollution's toll on forests and crops*, ed. J.J. MacKenzie and M.T. El-Ashry, pp. 235-315. New Haven, CT: Yale Univ. Press.
28. Heisler, G.M. 1974. Trees and human comfort in urban areas. *J. For.* 72:466-69.
29. Hogsett, W.E., M. Plocher, V. Wildman, D.T. Tingey, and J.P. Bennett. 1985. Growth response of two varieties of slash pine seedlings to chrome ozone exposures. *Can. J. Bot.* 63:2369-76.

30. Hornbeck, J.W., and R.B. Smith. 1985. Documentation of red spruce growth decline. *Can. J. For. Res.* 15:1199-1201.
31. Hornbeck, J.W., R.B. Smith, and C.A. Federer. 1986. Growth decline in red spruce and balsam fir relative to natural processes. *Water Air Soil Pollut.* 31:425-30.
32. Hutchison, B.A., F.G. Taylor, and R.L. Wendt. 1982. Use of vegetation to ameliorate building microclimates: An assessment of energy conservation potentials. Environ. Sci. Div. Publ. No. 1913, Oak Ridge, TN: Oak Ridge National Laboratory.
33. Jensen, K.F. 1973. Response of nine forest tree species to chronic ozone fumigation. *Plant Disease Rep.* 57:914-17.
34. Jensen, K.F. 1982. An analysis of the growth of silver maple and eastern cottonwood seedlings exposed to ozone. *Can. J. For. Res.* 12:420-24.
35. Johnson, D.W., J. Turner, and J.M. Kelley. 1982. The effects of acid rain on forest nutrient status. *Water Res.* 18:449-61.
36. Knight, H.A. 1987. The pine decline. *J. For.* 85:25-28.
37. Kress, L.W., and J.M. Skelly. 1982. Response of several eastern forest tree species to chronic doses of ozone and nitrogen dioxide. *Plant Dis.* 66:1149-52.
38. Kress, L.W., J.M. Skelly, and K.H. Hinkelmann. 1982a. Growth impact of O_3-hr, NO_2-hr, and/or SO_2-hr on *Pinus taeda*. *Environ. Monit. Assess.* 1:229-39.
39. Kress, L.W., J.M. Skelly, and K.H. Hinkelmann. 1982b. Growth impact of O_3-hr, NO_2-hr and/or SO_2-hr on *Plantanus occidentalis*. *Agric. Environ.* 7:265-74.
40. Kress, L.W., H.L. Allen, J.E. Mudano, and W.W. Heck. 1988. Response of loblolly pine to acid precipitation and ozone. Paper No. 88-705, 81st Annual Meeting, Air Pollution Control Association, Dallas, TX, June 19-24, 1988.
41. Küppers, K., and G. Klumpp. 1987. Effects of ozone, sulfur dioxide, and nitrogen dioxide on gas exchange and starch economy in Norway spruce (*Picea abies* L. Karsten). In *Proceedings XIV Intern. Bot. Cong.*, Berlin, FRG.
42. LeBlanc, D.C., D.J. Raynal, and E.H. White. 1987. Acidic deposition and tree growth. I. The use of stem analysis to study historical growth patterns. *J. Environ. Qual.* 16:325-33.
43. Lucier, A.A. 1986. Summary and interpretation of USDA Forest Service report on "Pine Growth Reductions in the Southeast." Tech. Bull. No. 508. New York: National Council of the Paper Industry for Air and Stream Improvement.
44. Lull, H.W. 1971. Effects of trees and forests on water relations. In *Trees and forests in an urbanizing environment*. Cooperative Extension Service. Amherst, MA: Univ. of Massachusetts.
45. Lull, H.W., and K.G. Reinhart. 1972. Forests and floods in the eastern United States. Res. Paper No. NE-226. Northeastern For. Exp. Sta., Upper Darby, PA: USDA Forest Service.
46. Mahoney, M.J., J.M. Skelly, B.I. Chevane, and L.D. Moore. 1984. Response of yellow poplar (*Liriodendron tulipifera*) seedling shoot growth to low concentration of O_3–hr, SO_2–hr, and NO_2–hr. *Can. J. For. Res.* 14:150-53.
47. Marx, D.H. 1988. Personal Communication. Southeastern For. Exp. Sta., Athens, GA: USDA Forest Service.
48. Mattson, W.J., and N.D. Addy. 1975. Phytophagous insects as regulators of forest primary production. *Science* 190:515-21.
49. McBride, J.R., P.R. Miller, and R.D. Laven. 1985. Effects of oxidant air pollutants on forest succession in the mixed conifer forest of southern California. In *Air pollutants effects on forest ecosystems*, pp. 157-67. Symposium proceedings, May 8-9, 1985. St. Paul, MN: Acid Rain Foundation.

50. McLaughlin, S.B. 1987. Carbon allocation as an indicator of pollutant impacts on forest trees. In *Woody plant growth in a changing chemical and physical environment*, ed. M. Cammell and D.L. Lavenders. Proceedings IUFRO Symposium, July 1987, Vancouver, Canada. Vienna: International Union of Forestry Research Organizations.
51. McLaughlin, S.B., and D.S. Shriner. 1980. Allocation of resources to defense and repair. In *Plant disease: An advanced treatise*, Vol. V, Ch. 20, ed. J.G. Horsfall and E.B. Cowling, pp. 407-31. New York: Academic Press.
52. McLaughlin, S.B., R.K. McConathy, D. Duvick, and L.K. Mann. 1982. Effects of chronic air pollution stress on photosynthesis, carbon allocation, and growth of white pine trees. *For. Sci.* 28:60-70.
53. McLaughlin, S., D.J. Downing, T.J. Blasing, E.R. Cook, and H.S. Adams. 1987. An analysis of climate and competition as contributors to decline of red spruce in high elevation Appalachian forests of the eastern United States. *Oecologia* 72:487-501.
54. Miller, J.E. 1987. Effects of ozone and sulfur dioxide stress on growth and carbon allocation in plants. Paper No. 10807. Raleigh, NC: North Carolina Agric. Res. Serv.
55. Miller, P.R. 1983. Ozone effects in the San Bernardino National Forest. In *Symposium on air pollution and productivity of the forest*, pp. 161-97. Washington, DC: Izaak Walton League and Pennsylvania State Univ.
56. Miller, P.R. 1989. Concept of forest decline in relation to western U.S. forests. In *Air pollution's toll on forests and crops*, ed. J.J. MacKenzie and M.T. El-Ashry, pp. 75-112. New Haven, CT: Yale Univ. Press.
57. Miller, W.P., and W.W. McFee. 1983. Distribution of cadmium, zinc, copper and lead in soils of industrialized northwestern Indiana. *J. Environ. Qual.* 12:29-33.
58. Mollitor, A.V., and D.J. Raynal. 1982. Acid precipitation and ionic movements in Adirondack forest soils. *Soil Sci. Soc. Am. J.* 46:137-41.
59. Mooney, H.A. 1972. The carbon balance of plants. *Ann. Rev. Ecol. Syst.* 3:315-46.
60. National Council of the Paper Industry for Air and Stream Improvement. 1988. NCASI-sponsored carbon tracer study nears completion at Texas A & M. *Air Qual./For. Health Program News* 3:1.
61. Odum, E.P. 1971. Ecosystem structure and function. In *Proceedings of the 31st Annual Biol. Colloquium*, ed. J.A. Wiens. Corvallis, OR: Oregon State Univ. Press.
62. Payne, B.R. 1973. The twenty-nine tree home improvement plan. *Nat. Hist.* 82:74-75.
63. Probst, J.R., and T.R. Crow. 1991. Integrating biological diversity and resource management. *J. For.* 89:12-17.
64. Pye, J.M. 1988. Impact of ozone on the growth and yield of trees: A review. *J. Environ. Qual.* 17:347-60.
65. Raynal, D.J., D.C. LeBlanc, B.T. Fitzgerald, E.H. Ketchledge, and E.H. White. 1988. Historical growth patterns of red spruce and balsam fir at Whiteface Mountain, New York. In *Proceedings of US/FRG research symposium*, pp. 291-97. Gen. Tech. Rept. No. NE-120. Broomall, PA: USDA Forest Service.
66. Reethoff, G., and G.M. Heisler. 1976. Trees and forests for noise abatement and visual screening. In *Better trees for metropolitan landscapes symposium proceedings*, pp. 39-48. Gen. Tech. Rept. No. NE-22, USDA Forest Service.
67. Reich, P.B. 1983. Effects of low concentrations of O_3 on net photosynthesis, dark respiration, and chlorophyll contents in aging hybrid poplar leaves. *Plant Physiol.* 73:291-96.
68. Reich, P.B. 1987. Quantifying plant response to ozone: A unifying theory. *Tree Physiol.* 3:63-91.

69. Reich, P.B., and J.P. Lassoie. 1985. Influence of low concentrations of ozone on growth, biomass partitioning and leaf senescence in young hybrid poplar plants. *Environ. Pollut.* 39:39-51.
70. Reich, P.B., A.W. Schoettle, and R.G. Amundson. 1986. Effects of ozone and acidic rain on photosynthesis and growth in sugar maple and northern red oak seedlings. *Environ. Pollut.* 40:1-15.
71. Reich, P.B., A.W. Schoettle, H.F. Stroo, J. Troiano, and R.G. Amundson. 1987. Effects of ozone and acid rain on white pine (*Pinus strobus*) seedlings grown in five soils. I. Net photosynthesis and growth. *Can. J. Bot.* 65:977-87.
72. Roberts, T.M. 1984. Long-term effects of sulfur dioxide on crops: An analysis of dose-response relations. *Philos. Trans. R. Soc. Lond. B* 305:299-316.
73. Santamour, F.S., Jr. 1969. Air pollution studies on *Plantanus* and American elm seedlings. *Plant Dis. Rep.* 53:482-85.
74. Sawhney, B.L., and K. Brown, eds. *Reactions and movement of organic chemicals in soils*. Spec. Publ. No. 22. Madison, WI: Soil Science Society of America.
75. Shafer, S.R., A.S. Heagle, and D.M. Camberato. 1987. Effects of chronic doses of ozone on field-grown loblolly pine: Seedling responses in the first year. *J. Air Pollut. Control Assoc.* 37:1179-84.
76. Sheffield, R.M., and N.D. Cost. 1987. Behind the decline. *J. For.* 85:29-33.
77. Sheffield, R.M., N.D. Cost, W.A. Bechtold, and J.P. McClure. 1985. Pine growth reductions in the southeast. Res. Bull. No. SE-83. Southeastern For. Exp. Sta., Asheville, NC: USDA Forest Service.
78. Shortle, W.C., and K.T. Smith. 1988. Aluminum-induced calcium deficiency syndrome in declining red spruce. *Science* 240:1017-18.
79. Skärby, L., E. Troeng, and C.A. Boström. 1987. Ozone uptake and effects on transpiration, net photosynthesis, and dark respiration in Scots pine. *For. Sci.* 33:801-8.
80. Smith, R.B., J.W. Hornbeck, C.A. Federer, and P.J. Krusic. 1990. Regionally averaged diameter growth in New England forests. Res. Paper No. NE-637. Northeastern For. Exp. Sta., Radnor, PA: USDA Forest Service.
81. Smith, W.H. 1970. Technical review: Trees in the city. *J. Am. Inst. Planners* 36:429-35.
82. Smith, W.H. 1980. Air pollution—a 20th century allogenic influence on forest ecosystems. In *Effects of air pollutants on Mediterranean and temperate forest ecosystems*, ed. P.R. Miller, pp. 79-87. Gen. Tech. Rept. No. PSW-43. Pac. Southwest For. and Range Exp. Sta., Berkeley, CA: USDA Forest Service.
83. Smith, W.H. 1990a. *Air pollution and forests*. New York: Springer-Verlag.
84. Smith, W.H. 1990b. The atmosphere and the rhizosphere: Linkages with potential significance for forest tree health. In *Mechanisms of forest response to acidic deposition*, ed. A.A. Lucier and S.G. Haines, pp. 188-241. New York: Springer-Verlag.
85. Sopper, W.E., J.A. Lynch, and E.S. Corbett. 1976. Water resources at the forest-urban interface. A problem analysis for environmental forestry research. In *Consortium for Environ. For. Studies*. Publ. No. PIEFR-PA-2. Upper Darby, PA: USDA Forest Service.
86. Swank, W.T., J.B. Waide, D.A. Grossley, and R.L. Todd. 1981. Insect defoliation enhances nitrate export from forest ecosystems. *Oecologia* 51:297-99.
87. Tingey, D.T., W.E. Hogsett, and S. Henderson. 1990. Definition of adverse effects for the purpose of establishing secondary national ambient air quality standards. *J. Environ. Qual.* 19:635-39.
88. Todd, G.W., and M.J. Garber. 1958. Some effects of air pollutants on the growth and productivity of plants. *Bot. Gaz.* 120:75-80.

89. Tyler, G. 1972. Heavy metals pollute nature, may reduce productivity. *Ambio* 1:52-59.
90. Vogt, K.A., R. Dahlgren, F. Ugolini, D. Zabowski, E.E. Moore, and R. Zasoski. 1987a. Aluminum, Fe, Ca, Mg, K, Mn, Cu, Zn and P in above- and below-ground biomass. I. *Abies amabilis* and *Tsuga mertensiana*. *Biogeochemistry* 4:277-94.
91. Vogt, K.A., R. Kahlgren, F. Ugolini, D. Zabowski, E.E. Moore, and R. Zasoski. 1987b. Aluminum, Fe, Ca, Mg, K, Mn, Cu, Zn and P in above- and below-ground biomass. II. Pools and circulation in a subalpine *Abies amabilis* stand. *Biogeochemistry* 4:295-311.
92. Wang, D., D.F. Karnosky, and F.H. Bormann. 1985. Effects of ambient ozone on the productivity of *Populus tremuloides* Michx. grown under field conditions. *Can. J. For. Res.* 16:47-55.
93. Wang, D., F.H. Bormann, and D.F. Karnosky. 1986. Regional tree growth reductions due to ambient ozone: Evidence from field experiments. *Environ. Sci. & Technol.* 20:1122-25.
94. Waring, P.F., and J. Patrick. 1975. Source-sink relations and the partition of assimilates in the plant. In *Photosynthesis and productivity in different environments*, ed. H.P. Cooper, pp. 481-99. New York: Cambridge Univ. Press.
95. Waring, R.H. 1987. Characteristics of trees predisposed to die. *BioScience* 37:569-74.
96. Waring, R.H., and W.H. Schlesinger. 1985. *Forest ecosystems: Concepts and management*. Orlando, FL: Academic Press.
97. Westman, W.E. 1979. Oxidant effects on Californian coastal sage scrub. *Science* 205:1001-03.
98. Whittaker, R.H. 1975. *Communities and ecosystems*. New York: Macmillan.
99. Wilson, W.J. 1972. Control of crop processes. In *Crop processes in controlled environments*, ed. A.R. Rees, K.E. Cockshull, D.W. Hand, and R.G. Hurd, pp. 7-30. New York: Academic Press.
100. Yang, Y.S., J.M. Skelley, B.I. Chevone, and J.B. Birch. 1983. Effects of long-term ozone exposure on photosynthesis and dark respiration of eastern white pine. *Environ. Sci. & Tech.* 17:371-73.
101. Zahner, R., J.R. Saucier, and R.K. Myers. 1989. Tree-ring model interprets growth decline in natural stands of loblolly pine in southeastern US. *Can. J. For. Res.* 19:612-21.

Part IV

Policy Issues and Research Needs

12

Policy Framework Issues for Protecting Biological Diversity

Orie L. Loucks

INTRODUCTION

For the purposes of this paper, I define *policy* as the setting of a course of action, i.e., a decision in regard to choices among possible actions. We think of "policy" commonly in terms of the foreign policy considerations of many countries; we hear of it often in energy policy, and occasionally it is the subject of a public welfare concern such as the protection of biological diversity. *Framework* indicates concern about the scope of the problems to be considered in formulating policy—local, national, or international. In the protection of biological diversity, we are examining a relatively new aspect of public welfare, concern about the variety of species in nature, many of them important for the services they perform for direct or indirect human benefit. A *policy framework* for protecting biological diversity, therefore, must address the array of interests—from scientific to social, ethical, and international—that bear on decisions affecting the risk of losing or impoverishing species due to the effects of air pollution.

The preceding chapters have suggested both scientific and policy problems: inadequate scientific understanding of the magnitude of the risks, an unclear mandate as to who should care or what it is they should care about, and problems inherent in pursuing transboundary pollution transport and its consequences. The goal of this paper is to examine a possible policy framework to answer the question of who cares and why, and how much they care. This will involve four steps: a brief review of the theory underlying such policy analysis and framework; a summary of what has happened in closely related policy questions (existing U.S. legislation and related international agreements); an evaluation of the scientific data requirements needed for a more formal policy analysis or for a suite of implementing actions; and finally, a discussion and conclusions based on the intersection of these three framework components.

DECISION-THEORY APPROACHES

The problem of how to protect biological diversity is a challenge for public policy and decision makers on several accounts. Preventing extinctions of species, an irreversible loss to the living system that is Earth, seems to represent moral high ground, but there are many who would argue that societal costs for the necessary preventive measures will be too high, regardless of the morality. These people seek, instead, a systematic, rational analysis of the benefits versus the costs of preventing the impoverishment of biological systems, including possible extinctions.

The work of Herbert Simon (1947) introduced the elements of a theory of how a decision (a policy) is made. An outcome of this early work has been articulation of what is now called the rational-comprehensive approach to decisions: rational because of the implicit formal structure and logic that seems possible; comprehensive because a wide range of options are to be considered. The process is described as having four steps (Culhane, Friesema, and Beecher 1987):

1. An assumption of agreement on the goals that govern a given decision;
2. Identification of all alternative courses of action relevant to the goals;
3. Determination of all relevant consequences for each alternative; and
4. Comparison of the sets of consequences and determination of the optimum alternative.

In Simon's original contributions to decision theory, the rational-comprehensive model was set up as a "strawman" that the author (and, subsequently, many others) picked apart before describing a more realistic view of decision making . They concluded that most decision makers cannot possibly meet the information-processing demands of the rational-comprehensive model. Moreover, comprehensiveness is not cost-effective in most circumstances, and they conclude the marginal utility of the "best" decision is usually less than the marginal cost of a comprehensive search for it (Culhane, Friesema, and Beecher 1987). Recognizing human psychology, Simon advocated a "satisficing" model in which decision makers focus on selected aspects of a problem and then adopt the first satisfactory solution they find during a search through alternatives. The starting point in searches for solutions, according to this approach, is to look at the standard operating procedures of existing organizations or agencies as they sought consensus positions in the past. These may be evident in preambles to legislation that may determine decision processes, or they may have become incorporated in new legislation or in regulations proposed for adoption. Satisficing models may have operated in the past to influence our present view of a policy framework for protecting biological diversity from air pollution effects.

Culhane, Friesema, and Beecher (1987) note that the various decision models following Simon's work all involve a modified view of the classic rational-comprehensive decision model. All are *arational* decision theories, although the models do not portray decision making as irrational, unreasonable, illogical, or inefficient.

There is a special form of the satisficing model that is of interest to us in the case of biodiversity concerns, however: establishment of a preemptive consideration, that, when present, precludes any need to search further. The Endangered Species Act of 1973 (ESA), authorizing actions to protect such species without regard to cost or benefits, is an illustration of policy determined not by rational-comprehensive considerations, but by a preemptive consideration (U.S. Congress 1973; P. Culhane, pers. communication). There was no substantive rational-comprehensive analysis even during consideration of the original legislation (although much came later). The issues instead were moral, a matter of instinctive prudence with respect to those species that could meet specified criteria of "threatened or endangered." There was no complex calculus of damages or welfare effects, no cost-benefit analysis. The data required had to do only with the prospect of extinction, and the legislation presumed this determination could be made readily.

EXISTING U.S. LEGISLATION AND RELATED INTERNATIONAL AGREEMENTS

Policy Framework Implicit in the Endangered Species Act (ESA)

Because of the way a single consideration, the risk of extinction, seems to serve as a lever facilitating actions to protect biological diversity, the ESA, now amended in relatively minor ways, could be the most important existing legislative policy under which to consider air pollution effects on biodiversity. The Congress has struggled to renew this bill every few years, and the close votes indicate support is seriously divided. However, Congress has not yet retreated on the basic principles involved. The key statement is as follows:

> The purposes of this Act are to provide a means whereby the ecosystems upon which endangered species and threatened species depend may be conserved, to provide a program for the conservation of such endangered species and threatened species, and to take such steps as may be appropriate to achieve the purposes of the treaties and conventions set forth in subsection (a) of this section.

This legislation, like most, begins with a broad statement of apparent intent, here including reference to "means whereby the ecosystems upon which endangered species and threatened species depend may be conserved." The subsequent specific objectives in the Act are more mechanistic (listings of species, review, recovery programs, etc.) and have been implemented to a fair degree, but the broad goal, conservation of ecosystems for endangered species, has not generally been implemented, as no specific means for doing so is included in the legislation.

Another important consideration in this Act is that, while recognized subspecies are to be protected, nothing is said about preventing the likely loss of genetic diversity through spatial limitation on the breeding system of species. Is this an oversight? Probably not—not because of differences in what we know now compared with 1973 when the Act was written, but because of the different informational inputs that go into a policy position such as the ESA requires (preemptive considerations), compared with the marginal benefits analysis required now to accomplish a broad, subtly measurable benefit, the protection of genetic diversity. Given the recent record of close votes, the Congress seems unlikely to use the ESA as a means of moving further toward protection of varieties or of genetic diversity.

How well has the ESA worked? The implementing measures put forward by the Fish and Wildlife Service (as the principal implementing agency) are most frequently criticized for slowness of listing and review, a complex interagency process. The ESA is criticized as well for not protecting a sufficiently broad and balanced range of biological diversity. These concerns are evident in Master's (1990) comparison of the "endangerment" status listings of The Nature Conservancy with ESA listings (Table 12-1). The interesting result is in the percentage of the total number of species qualifying as extinct, possibly extinct, and critically imperiled to rare in the Conservacy ranking (11% to 73%, with the highest percentages in the two invertebrate groups shown) compared with the percentage of the

TABLE 12-1. Status of Selected Animal Groups of North America

No. of U.S. Species Ranked (not subspecies)	Mammals	Birds	Reptiles	Amphibians	Fishes	Cray-fishes	Unionid Mussels
GX (extinct)	1	20	0	3	18	1	12
GH (historical; possibly extinct)	0	2	0	1	1	2	17
G1 (critically imperiled)	8	25	6	23	78	62	88
G2 (imperiled)	23	9	10	17	72	49	49
G3 (rare, not imperiled)	19	23	25	26	110	84	35
G4-G5 (widespread and abundant)	330	628	251	153	549	106	73
G? (not yet ranked)	62	55	9	3	24	9	26
Total	443	762	301	226	852	313	300
% of Total that is GX to G3	13	11	14	28	34	65	73
% that is E or T*	6	5	6	3	7	1	11

*The final line represents the percentage of species listed by the U.S. Fish and Wildlife Service or National Marine Fisheries Service as Endangered (E) or Threatened (T) as of 4/15/90.

Source: Master 1990.

total species listed under the ESA as "endangered or threatened" (1% to 11%, with invertebrates representing both of these extremes) For the species groups that are readily recognized by the public (mammals, birds and reptiles), the ESA protects about half of the species of concern to The Nature Conservancy, while for the less visible and less well-known groups, the ESA appears to protect only a small fraction of the species of special concern.

An additional dimension to the problem of limited numbers of protected species under ESA is seen by comparing the results in Table 12-1 with the number of known species described worldwide (Wilson 1988) as summarized in Table 12-2. Here, the vertebrate groups where ESA activity is more prominent comprise about 3 percent of the total described diversity. The vertebrate species protected under ESA, as a percentage based on the North American data in Table 12-1, are about six-hundredths of the total described diversity of just under 1.4 million species. It is important to note that the described diversity is thought to be less than one-tenth of the biological diversity that has existed on earth until recently (Raven 1988). The species-by-species approach of ESA, while it may be the best means presently available to protect threatened and endangered species in the United States, seems to fall far short of assuring the protection of any substantial part of biological diversity, either in the United States or globally.

Policy Framework Implicit in NEPA

Culhane, Friesema, and Beecher (1987) have described the outcomes of the National Environmental Protection Act (NEPA) as largely a product of the rationalist decision reform movement of the 1960's. In this respect, it is like "planning-programming-

TABLE 12-2. The Known Species Described Worldwide

	Taxon		No. of Described Species
1	Monera (Bacteria, Blue-green Algae)		4,760
2	Fungi		46,983
3	Algae		26,900
4	Plantae (Multicellular Plants)		248,428
5	Protozoa		30,800
6	Porifera (Sponges)		5,000
7	Coelentarata (Jellyfish, Corals, Comb Jellies)		9,000
8	Platyhelminthes (Flatworms)		12,200
9	Nematoda (Roundworms etc.)		12,000
10	Annelida (Earthworms, etc.)		12,000
11	Mollusca (Mollusks)		50,000
12	Echinodermata (Starfish etc.)		6,100
13	Insecta		751,000
14	Non-insect Arthropoda (Mites, Spiders, Crustaceans, etc.)		123,161
15	Pisces (Fish)		19,056
16	Amphibia (Amphibians)		4,184
17	Reptilia (Reptiles)		6,300
18	Aves (Birds)		9,040
19	Mammalia (Mammals)		4,000
		Total	1,380,912

Source: after Wilson 1988.

budgeting" (PPB) and systems analysis for decision making, also introduced in the 1960's to improve the reproducibility and reliability of decisions. In the years before NEPA, agencies could approve projects by conducting a "satisficing" comparison of a project's objectives with the agencies' narrow mission, and could ignore many adverse consequences. For example, an agency's assessment could be limited to a simple engineering feasibility study, perhaps accompanied by a narrow cost-benefit analysis. In the view of environmentalist critics, such behavior involved a narrow pursuit of economic development objectives and a disregard for adverse environmental consequences to be born by future generations. Members of the authorizing committees of Congress were often frustrated that satisficing decisions by agencies invariably presented them with a single administration-endorsed proposal and no alternatives.

Thus, the legislation establishing NEPA passed in 1969 and signed January 1, 1970, provided as follows:

> The purposes of this Act are to declare a national policy which will encourage productive and enjoyable harmony between man and his environment; to promote efforts which will prevent or eliminate damage to the environment and biosphere and stimulate the health and welfare of man; to enrich the understanding of the ecological systems and natural resources important to the Nation; and to establish a Council on Environmental Quality.

Culhane (pers. communication) believes these policy positions have not been forgotten, and that this legislation can and should be brought to bear on policy development relative to protection of biological diversity. Environmental Impact Statements, such as those of the USDA-Forest Service on their road and management plans in 1990, are as comprehensive as ever, and are reviewed just as carefully. The President in his FY 1992 budget sought 10 to 15 new positions for the Council on Environmental Quality. The public appears to expect that we will meet the broad goals implicit in NEPA, including elimination of damage to the biosphere.

Policy Framework Implicit in National Parks and National Monuments Legislation

The implementing legislation for the National Park Service (NPS) is the Organic Act of 1916, requiring the Service to "preserve the resources unimpaired for future generations." Thus, by definition, national parks are to be examples of unspoiled ecosystems, untainted by the progress of industrialization. Still, virtually no park in the contiguous 48 states remains "unimpaired" with respect to pollutants limiting visibility or persistent toxic substances accumulating in the food chain. Ten national parks have extensive damage to vegetation from ozone and sulfur dioxide (Bennett 1985), although this is probably a minimum estimate. Yellowstone, once the source of pure air sampled by researchers for comparison purposes, now has air that is contaminated by auto exhaust emissions and long-distance transport from smelters and other industrial sources. Death Valley National Monument experiences smog drifting from Los Angeles 170 miles away. Of 10 national parks and lakeshores in the upper Great Lakes area, Armentano and Loucks (1983) found all to be significantly at risk from air pollutant inputs in one form or another.

The damages from ozone are probably the best documented evidence of effects on the biota in the national parks, with Shenandoah National Park, Sequoia National Park, and Great Smokies National Park being the outstanding examples (Bennett 1985). Effects seem to be clear in regard to the loss of genetic diversity in a few species, including eastern white pine. However, enforcement of abatement measures is quite difficult for the NPS, because it must work through the Justice Department and the courts to obtain a reduction in emissions even when the sources are known. More often, experience has shown that ore bodies are exhausted and smelters are closed or become obsolete before steps to control existing sources can be accomplished.

Most important for the National Park Service and other Federal land management agencies are the new programs and policies for "Prevention of Significant Deterioration" (PSD) in areas designated as Class I under the Clean Air Act of 1977 (a designation that seeks to assure continuation of historically good air quality in many national parks and larger wilderness areas). Guidelines have been developed for review of permit applications for proposed new emissions sources with potentially negative air quality impacts in Class I areas (US EPA 1990). These guidelines define Air Quality Related Values (AQRV's) for flora and fauna, water, visibility, and other amenities, and among flora and fauna are included considerations of biological diversity along with growth, injury, and reproduction of critical species. The opportunity for park managers to find potential damage to AQRV's on these Federal lands seems likely to control many risks from identifiable future pollution sources. However, the procedures available do not provide a means of preventing deterioration due to diffuse sources of pollutants in the region, e.g., ozone and many persistent toxic substances. Concern about this question will be explored further in the section on the Clean Air Act.

National Forest Management Act of 1976 (NFMA)

The National Forest Management Act is another legislative initiative, part of which was designed to protect biological diversity on at least the Forest Service's portion of Federal land. NFMA requires national forest managers to provide detailed plans for the development and allocation of forest, grazing, water, and timber resources and to make these plans public. Included is language on measures to protect biological diversity on Forest Service lands, requiring that it be maintained at or near its condition at the time of the legislation. Over the 16 years since the passage of the legislation, the Forest Service has gradually developed policies and procedures that make the protection of biodiversity workable in relation to its on-land management activities, but air pollution effects problems are another matter.

Two other initiatives are important in considering protection of biodiversity on Forest Service lands: The Forest Health Monitoring Program (based upon the Cooperative Forestry Assistance Act of 1978) and the Forest Ecosystems and Atmospheric Pollution Research Act of 1988. This law states that

> . . . the Secretary, acting through the US Forest Service, shall . . . increase the frequency of forest inventory in matters that relate to atmospheric pollution and conduct such surveys as are necessary to monitor long-term trends in the health and productivity of domestic forest ecosystems . . .

Monitoring for these long-term trends began in parts of the United States in 1990, particularly in the northeastern states. Related work is planned soon in other areas of the country.

Finally, the implications of rules regarding prevention of significant deterioration in Class I air quality areas (particularly designations under the Wilderness Act) are important for lands managed by the Forest Service. Here, the Forest Service operates under the same guidelines for PSD as have been described above for the National Park Service (US EPA 1990), and must make a finding of potential negative impact on an "air quality related value" to assure protection from pollution of known and proposed sources.

Policy Framework Implicit in the Clean Air Act (CAA)

Federal legislation to control air pollutants in the United States was passed by the U.S. Congress in increasingly strict steps in 1955, 1963, 1967, 1970, 1977, and 1990. The amendments of 1970 and 1977 set important precedents for protection from acid gases and oxidants, many of which focus on protection of resources and biological productivity, potentially including biological diversity (Avery and Schreiber 1979).

In the earliest legislation, the main goal of pollution control was to protect human health, and measures were to be promulgated primarily by the States. A stronger Federal role emerged in the 1970 amendments where air quality criteria were required to reflect accurately "the latest scientific knowledge useful in indicating the kind and extent of all identifiable effects on public health or welfare which may be expected from the presence of such pollutant in the ambient air, in varying quantities."

In this legislation, the United States departed from the approaches of other countries by stipulating a distinction between primary and secondary standards. Primary standards were defined as those necessary for the protection of public health, while secondary standards ". . . specify a level of air quality . . . requisite to protect the public welfare from any known or anticipated adverse effects." The legislation goes on to define public welfare as follows: "includes, but is not limited to, effects on soils, water, crops, vegetation, manmade materials, animals, wildlife, weather, visibility, and climate. . . ."

Thus, prevention of some level of effects on plants and animals is recognized in the provisions of the Act, although biological diversity per se is not. Although the Clean Air Act of 1967 had authorized air quality standards, the amendments of 1970 gave the authority to the Federal Government to promulgate uniform national primary and secondary standards through the Environmental Protection Agency (Avery and Shreiber 1979). However, in the case of ozone, the primary and secondary standards are the same; for sulfur dioxide the secondary (welfare) standard is less stringent, while for total suspended particulates the secondary standard is more stringent.

To assure the prevention of significant deterioration (PSD), EPA proposed in 1974 a classification whereby each State assigned its Federal lands to one of three area classifications, depending upon the extent of air quality deterioration to be permitted. In Class I, only minor air quality deterioration would be permitted; in Class II, deterioration normally accompanying moderate, well-controlled growth was acceptable; and in Class III, a rise in pollution levels up to the National Ambient Air Quality Standard (NAAQS) was acceptable.

In the Clean Air Act Amendments of 1977, the section on PSD was expanded, increasing the potential for protection of flora and fauna. However, the specifics applied only to Class I areas, meaning that PSD could be used to protect only a limited area of the public lands. It provides for the protection of ". . . public health and welfare from any actual or potential adverse effect which . . . may reasonably be anticipated to occur from air pollution . . . notwithstanding attainment and maintenance of all national ambient air quality standards," together with the preservation, and enhancement of air quality in national parks, wilderness areas, monuments, seashores, and other areas of special national or regional recreational, scenic, or historic value. As noted above under national parks and national forest legislation, PSD is relevant principally in relation to proposed new sources of pollutants requiring permits and is relatively ineffective against diffuse sources, either current or projected.

To date, the secondary standards also have proved to be essentially meaningless in protecting "welfare" in rural and remote areas. Only two secondary standards differ from the primary standards, and these have proven to be unenforceable. Furthermore, studies (e.g., Smith 1981) are showing that many native species have a lower tolerance of pollutants than humans tolerate (even for satisfactory human health), and are negatively impacted at pollutant concentrations below the secondary standard. Indeed, the recent evidence of increased concern by EPA for better long-term protection of the environment, beyond human health concerns, may stem from new provisions in this regard in the CAA Amendments of 1990.

An important point, therefore, is that despite 20 years of experience under the 1970 CAA amendments, and further amendment in 1977 (with regulations then promulgated to prevent impoverishment of resources), progress toward abatement of air pollutant concentrations has been minimal in most rural and remote areas. Reduction in pollutant concentrations clearly has occurred, but in urban areas or in the vicinity of major point sources where the majority of the nation's monitoring has existed (for the primary concern with human health).

In summary, one can say that a general policy is in place through the CAA that speaks to prevention of the diminishment of productivity or utility ("welfare") of natural resources, rather than species or diversity itself. A policy seeking further abatement of some pollutants still could be implemented through enforcement of secondary standards if sufficient, compelling scientific data indicated benefits (in terms of either market or social values) commensurate with the costs of reaching those goals. The 1990 amendments seem to suggest effective means are available for protecting resources and related welfare interests, but this seems unlikely if an adequate scientific basis is not present. Since CAA is the only broad policy statement for the protection of biotic resources on non-Federal lands from the effects of air pollutants, it must be viewed as the strongest extant basis for broad multispecies or ecosystem protection, and the basis of a framework for developing databases on damages, benefits, and costs.

The Great Lakes Water Quality Agreement as a Means of Ecosystem Protection

A potentially relevant example of a policy directed to the protection of a whole ecosystem and the species comprising it is evident in the Great Lakes Water Quality Agreement of 1978. Article II of this Agreement (in effect a treaty between two nations, the "Parties") states

> . . . the purpose of the Parties is to restore and maintain the chemical, physical, and biological integrity of the waters of the Great Lakes Basin Ecosystem. In order to achieve this purpose, the Parties agree to make a maximum effort to develop programs, practices and technology necessary for a better understanding of the Great Lakes Basin Ecosystem and to eliminate or reduce to the maximum extent practicable the discharge of pollutants into the Great Lakes System.

This bold statement of purpose, far ahead of its time in 1978, was essentially ignored by the implementing government agencies for 6 years, until 1984 when a consensus among the States, Provinces, and Federal Governments was reached (National Research Council 1985). However, it has been embraced now by program managers in most of the eight States and in two Provinces of Canada, as well as by agencies of the two national governments. One wonders whether something similar could happen with respect to protecting the biological diversity that is similarly critical to the functioning of terrestrial systems around the Great Lakes as well as in other parts of the United States and Canada, and in Europe.

Summarizing the policy positions evident in existing U.S. legislation, we see a pattern of apparently strong intentions on the part of the Congress to minimize long-term impoverishment of biota and ecosystems, particularly for Federal lands. However, there is also a clear record of ineffectiveness in the efforts to accomplish the apparently broad goals of many of these policy positions. The preemptive determinations of the Endangered Species Act seem to have been so attractive, through their relative ease in documentation, that the complex scientific underpinnings required for a rational-comprehensive approach to protection of species and ecosystems have been passed over. Since only the CAA provides the authority to protect public welfare (possibly including the biota) of privately owned lands, new legislation may be needed (depending on how the 1990 CAA amendments are interpreted). In any case, new legislation now would be subject to the same high standards of benefit and cost analysis, and related scientific documentation, as is required under the present authority. Thus, for the present, there is every reason to accept and work within the existing legislation.

Policy Framework of International Agreements for Air Pollution Control

The mechanisms for approaching pollutant standard-setting internationally are necessarily different from the process by which domestic standards are set in any country. Sand (1991) describes these mechanisms as "transnational regimes: to recognize that the norms, roles and procedures agreed to among several nations are frequently hybrids of domestic regulation (with diverse standards), bilateral agreements between nation states, and general multi-lateral agreements, protocols or treaties." This later class ultimately requires parliamentary ratification, with an implicit, built-in time-lag, but experience shows that domestic systems of incentives lead to an informal compliance, even before enforcement measures are agreed to.

A relevant example for our discussion of policy frameworks begins with the Convention on Long-range Transboundary Air Pollution signed in Geneva, Switzerland, in 1979 (see Wetstone and Rosencranz 1983; Sand 1990). Both European and North American countries initialed the Convention, but steps toward ratification, or even informal implementation, progressed only slowly before the convention was to take force in 1983. The countries within western Europe then moved quickly, and adopted a significant protocol under the convention in 1984 to facilitate financing of the European Monitoring and Evaluation Program. The signatories decided by resolution "to contribute to financing of EMEP on a voluntary basis, in an amount equal to the mandatory contributions expected from them under the provision of the protocol if all signatories had become parties" (Sand 1990).

Furthermore, Sand (1990) notes that circumstances, and the initiative of a few countries, tended to promote "over-achievement" in emission controls within the multinational framework of Europe (and outside of the 12-member European Community structure). Within the 1979 Geneva Convention, a 10-member "club" of countries first moved ahead in 1984 by declaring a voluntary 30-percent reduction of sulfur emissions—a commitment that not all of the 31 parties to the convention were prepared to share at that time. When the 30-percent reduction was formally adopted as a protocol to the convention at Helsinki in 1985, 21 states signed it. During the negotiation of a further protocol on nitrogen oxides in 1987, a group of five like-minded states again pressed for a 30-percent reduction target; even though the target did not become part of the protocol finally signed at Sofia in 1988, 12 of the protocol's 25 signatories eventually agreed to commit themselves to a voluntary 30-percent reduction.

These examples must be understood as part of a regulatory "regime" referred to earlier, i.e., that the targets or standards may, for the present, go beyond domestic statutes, and may not yet have been ratified by the parliament of a member state. Still, the effect is to capitalize as fully as possible on the respect each nation seeks to accord concerns of neighbor nations for diminishment of welfare and resources. The process meets a reasonable standard of "rational-comprehensive" decision making, while falling short of a full-scale benefit-cost audit. Biological diversity per se has not figured prominently in the scientific underpinnings of these conventions and protocols, but risks to the productivity and distribution of a broad range of species have been considered. As the scientific understanding of these risks became more complete during the 1980's, the targets and extent of commitments to them also changed. In almost all respects the process was more dynamic, if less enforceable, than the counterpart approaches within the United States itself.

What is Biotic Impoverishment? Case Studies on the State of the Science

A third dimension of this "framework" analysis focuses on what damages have occurred, or would have to occur, to warrant a policy response, i.e., who cares and how much do they care? Measurement of damages is not a major issue in the case of preemptive decisions such as are brought to bear to achieve protection of endangered species. However, scientific data on effects are a major concern for other approaches to a policy initiative, including new legislation on air pollution controls. The U.S. legislation and international agreements nearly all require some degree of rational-comprehensive analysis, a quantitative consideration of alternatives, a determination of what societal values (benefits) are at stake, if any, and some indication of the degree of effort (cost) required to accomplish a defined standard of protection of biodiversity from air pollutant effects.

Consider what we have now come to understand as impoverishment. In its simplest form, it means only to be poorer in species, in nutrients, or in productivity (Woodwell 1990). However, the term is being extended now to include the transformation of self-maintaining ecosystems and landscapes having considerable natural resilience into systems with diminished capacity for self-regulation. The term does not need to imply whether we know the loss of regulatory capacity is due to the loss of species in the system, or to physical-chemical changes, but the loss of species (and species functions) is the clearest indication that the system is impoverished.

As an example, let us consider the well-documented record of species impoverishment at the site of an irradiation experiment in an oak-pine ecosystem at Brookhaven National Laboratory (Woodwell and Rebuck 1967; Woodwell and Houghton 1990). This experimental irradiation (expressed as rd/day) began in 1961 and exposures continued at regular intervals for 15 years. Three measures of biological diversity were used to quantify the response to ionizing radiation: diversity, the number of species per unit land acre (Figure 12-1); the coefficient of community, a comparison of the species array present in the series of plots approaching the center of the irradiation treatment in comparison to a reference area (Figure 12-2); and the percentage similarity, a comparison of the similarity of species composition along the same gradient.

In terms of the numbers of species, a small effect may be evident almost from the outer zone of the treatment, i.e., at 0.5 rd/day, 160 meters out, compared with the 0.2 rd/day at 200 meters. On the other hand, more synthetic measures such as the coefficient of community (Figure 12-2) do not indicate an apparent effect until 1 rd/day. The essential point is that the irradiation and species composition data provide a simple dose-response relationship between the dose of radiation and the surviving biodiversity. As such, they

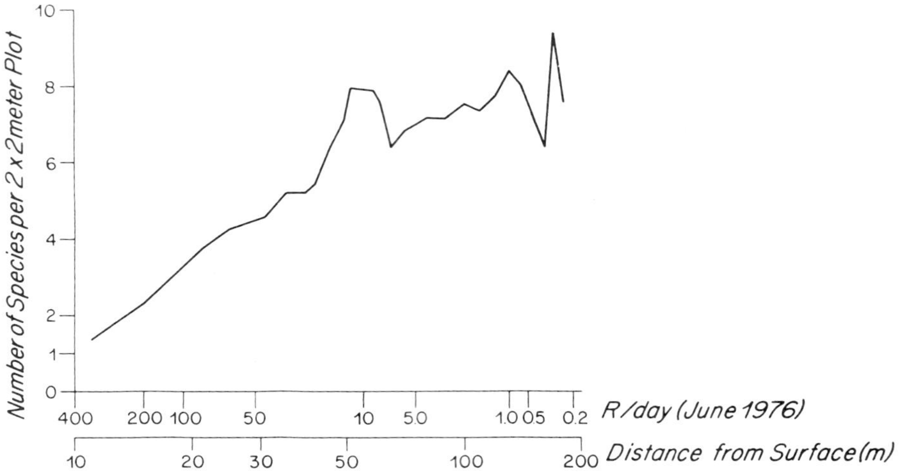

FIGURE 12-1. Diversity of higher plants along the gradient of ionizing radiation in the Brookhaven irradiated forest, June, 1976, 15 years after the start of the chronic irradiation. Diversity is expressed here as the number of species per 2 x 2-m plot.

Sources: Woodwell and Rebuck 1967; and Woodwell, G.M., and R.A. Houghton. 1990. The experimental impoverishment of natural communities: Effects of ionizing radiation on plant communities, 1961-1976. In *The Earth in transition: Patterns and processes of biotic impoverishment*, ed. G.M. Woodwell. New York: Cambridge Univ. Press (reprinted with permission).

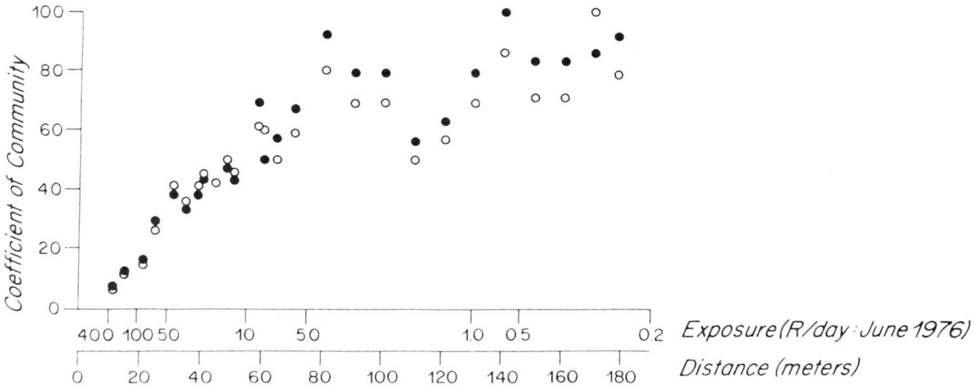

FIGURE 12-2. Coefficient of community for plant communities along the radiation gradient in 1976. Open circles represent values obtained from comparisons with the plant community at 170-178m (0-.3 R/day); closed circles represent values from comparisons with the 150-160-m community (0.4-0.5 R/day).

Sources: Woodwell and Rebuck 1967; and Woodwell, G.M., and R.A. Houghton. 1990. The experimental impoverishment of natural communities: Effects of ionizing radiation on plant communities, 1961-1976. In *The Earth in transition: Patterns and processes of biotic impoverishment*, ed. G.M. Woodwell. New York: Cambridge Univ. Press (reprinted with permission).

provide a scientific basis for a rational determination of how much "effect" might be acceptable at a given level of societal concern for these species. These results do not, however, express the loss of ecosystem amenities such as are the focus in seeking protection for the Great Lakes.

Let me use the work of McCune (1988) as a second example. Although it focuses on a relatively short-lived plant group, lichens, they have a widespread distribution and the results are potentially applicable to many horticultural and other long-lived plant species. The study was carried out by working from the city of Indianapolis out into the rural area. Air quality monitoring sites already existed in and near the city with air pollution measures as follows: mean summer season ozone, 65-77 $\mu g/m^3$, with a gradient from the southwest to the northeast, and mean annual SO_2 of 23-40 $\mu g/m^3$ (7 sites). The lichen community parameters studied by McCune included species composition (the community), species richness (the number of species), a "total cover index," and an index of atmospheric purity (IAP). Lichens were sampled on trunks of one tree species group, the ashes, to minimize the variation in substrate on which the lichens grow. The IAP was calculated for each site, based on average lichen cover classes, by species, weighted for pollution tolerance.

The differences in lichen species distribution and communities on the ash tree substrates are strongly correlated with mean annual sulfur dioxide levels across the Indianapolis urban area (Figure 12-3), much as though this segment of a biotic impoverishment measure had been taken from the main dose-response relationship of Woodwell and Houghton (1990) (Figure 12-1). Lichen species richness was correlated with SO_2, as was total cover, IAP values, and composition, regardless of whether SO_2 peaks or means were used (McCune 1988). The strongest correlations (Figure 12-3) were between mean annual SO_2

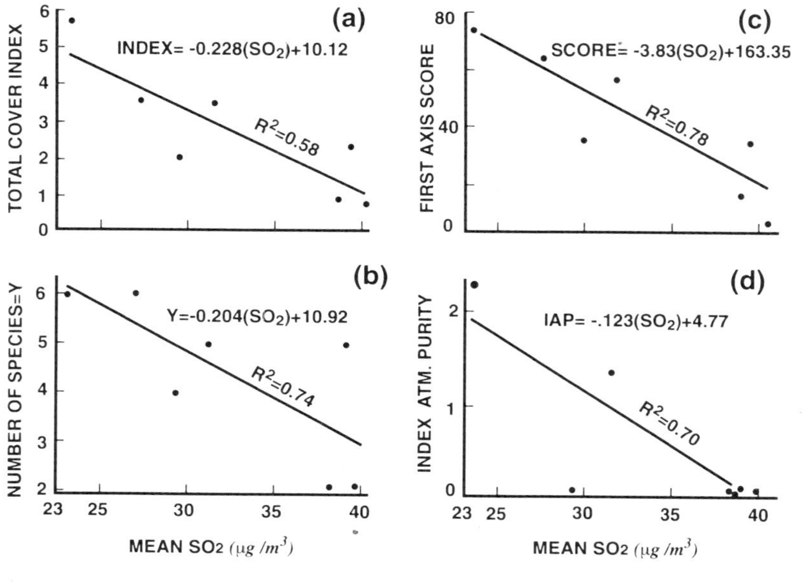

FIGURE 12-3. Regressions of lichen community parameters against mean annual SO_2 at seven air sampling sites in and around Indianapolis. -A. Lichen cover index (sum of cover class values for all species). -B. Species richness. -C First axis scores from Bray-Curtis ordination of sites in species space (data not standardized by site totals). -D. Index of atmospheric purity (IAP). All slopes are significantly different from zero at $p<0.05$.

Source: McCune 1988.

and the first axis scores of sites in lichen species space (Bray-Curtis ordination, data not relativized by site totals). There is no evidence available as to whether the effects on this group may still be spreading to the rural areas around Indianapolis at current SO_2 concentrations. Equally important, this example says nothing about the effects of ozone on many other species groups, a serious health hazard in many cities, and highly damaging to crops, horticultural material, and trees.

One interesting observation from these two examples is that, on the one hand, societal values (perhaps for reasons of health risk) precluded taking any risk of biotic impoverishment from irradiation while, on the other hand, serious effects on an "indicator" group, such as the lichens, have been accepted as inconsequential (despite evidence indicating some health effects at the mean annual SO_2 concentrations producing effects on lichens). One infers that our value system acknowledges, for the present, that biodiversity effects can be ignored under the following circumstances: for areas of priority, high-density urban use, and for nonflowering plants. Not only is this a partial indication of how much we have (or have not) cared in the past, but it also indicates there is a question of where we care (urban vs. remote).

An additional point to be made, however, is that both of the examples of effects shown in Figures 12-3 are linear; there appear to be no thresholds below which diversity decreases at a slower or zero rate. As the public welfare significance of these species' contributions comes to be better understood, the linearity of these responses is likely to be important in new standard-setting proceedings.

How do these examples comport with the information base on air pollution effects presented in this volume? I do not want to disparage the results on plant growth and diversity presented here (after all, the studies were carried out mostly for quite different purposes), but from what has been shown, there is little foundation for a broader, policy-relevant scientific synthesis on a broad array of societally important plant and animal groups. The results in our scientific information base seem to be a fragmentary record, at best. Although we may have agreed on a definition of biological diversity per se, we have not yet agreed on language linking that definition to the health and sustainability of species and ecosystem amenities and, therefore, to productivity, one of the central concepts of welfare impact measures.

If the information is inadequate for evaluating a possible course of abatement measures, is it adequate for an assessment of the risk of serious negative consequences if no action is taken? This is hard to say. Risk assessment requires a basic dose-response relationship, almost as a starting point, and that requires data on more species groups than are currently available. Also required is a clear indication of where societal values should be brought into the equation. In the case of air pollution effects on biodiversity, we may have neither the scientific data for risk assessment nor the basis for evaluating what is significant to our value system.

On the plus side, it seems likely that the information in the papers presented here will at least define the scope of the science needed to define a standard for protection of diversity useful in a benefits and cost calculus (or, at a more modest level, even in a risk assessment). Reproducible information is needed in some input/response form, distributed over most of the plant and animal world, and across a range of ecosystem/species "services." If we agree on what information is missing, we should proceed to get it and prepare for a more substantive policy analysis.

Discussion of a Policy Framework

A first impression from the legislation reviewed here is that we already may have the policy precedents needed to make the case for further protection of biological diversity from air pollution threats. We should appreciate that these statements of principles are no accident, but also recognize that, in the absence of any indication of the potential costs of accomplishing such goals, preamble statements convey no commitment to protect. Some of the apparent policy positions date from legislation passed years ago, although others are more recent. Both represent a recognition by the Congress that the public seeks protection of their natural resources from indiscriminate impoverishment as a matter of principle. The legislative statements of principle are drawn broadly, to indicate a desire to pursue them, if feasible, but in the absence of specific enforceable measures, and in the comparative absence of scientific evidence showing benefits commensurate with costs, they provide little basis for a serious policy proposition.

Is the scientific basis of our policy framework any stronger than that of the legislative basis? The answer is both yes and no. The literature here and elsewhere indicates widespread effects of air pollutants on plants and animals. Indeed, there is probably "a problem." However, as noted above, the current research does not lend itself to estimating the costs of mitigating the problem, or the benefits of meeting society's apparent intentions, as evidenced in the legislation.

Despite the limitations of authority and scientific underpinning, the material presented here probably does constitute a policy framework. For it to work, and lead to actionable measures protecting diversity from air pollution effects, measurable benefits (or at least the risks of serious damage) have to be documented. This will require a more complete synthesis of the information available (only some of it in this volume), as well as development of new information focused on the societal value concerns evident in this framework.

CONCLUSIONS

The extant policy examples reviewed here, particularly NEPA, the Clean Air Act, the Endangered Species Act, and the Great Lakes Water Quality Agreement, all seem to be in broad agreement: the Congress and the general public want protection from significant environmental impoverishment, including protection of biological diversity. However, the fact that these directives have not articulated precisely how much intrusion is considered "significant," and what level is viewed as acceptable for natural resource systems, neighboring or remote, probably indicates that in the absence of information on the "price tag" for such goals, the Congress and public are not yet ready to accept programs for implementation. There is a clear policy of commitment to the prevention of extinctions of higher plants and animals, but the vast majority of our biota are insects, other arthropods, fungi, and related groups, and the legislative initiatives reviewed here afford little protection to these groups.

From the scientific evidence, one could argue that a naive belief in the protection afforded by a species-by-species approach (under ESA) may have inhibited research on the broader and more fundamental problem of general biotic impoverishment from air pollution. This is possible, and it will need to be overcome soon if there is to be a basis for more thorough analyses on how to achieve the protection implied in the various statements of legislative intent. Only in the case of protection of the Great Lakes have there been steps toward implementing a strategy focused on restoring full biotic integrity; the costs and benefits of such a strategy are still being evaluated. That result seems to derive, in part, from the special value attached to the Great Lakes as a unique resource in this continent, and from 30 years or more of research focused on questions of system integrity (from water quality to fisheries) within the Great Lakes Basin. This example suggests the need for a national research program on the contribution of air pollution to the altering of species abundance and community composition similar to that already underway in the Great Lakes and Chesapeake Bay. Given EPA's broad mission, and its potentially supportive role for other agencies with concerns for resource protection, there may need to be effective interagency linkage in planning and implementing such research.

There is another issue of importance. The protection of ecosystems and the biodiversity they harbor is more likely to be arrived at by comprehensive analysis of long-term damages to the resources as a system, over and above effects on biodiversity. Determining whether serious system-level impoverishment has or may occur (where intergenerational economic damages can be large) will require data on system effects as well as species-level effects. Information to date is fragmented by species (leading to a lack of coherence), or is site specific, suggesting the problem does not have widespread significance. A linked assessment of multiple species functions and habitat linkages in the context of ecosystem processes is likely to be critical.

Finally, let me observe that the various strands I have tried to draw together here could leave a reader "at sea." It may be worth noting the crosscurrents in this sea, where anyone could be carried along. If this is so, we need all become strong swimmers, soon. The broader policy issue here concerns what may be the next major steps in environmental protection following the progress of the 1970's and reaffirmed in the Clean Air Act Amendments of 1990. It is clear that general impoverishment of the Earth's biota is becoming a mainstream public policy concern, and the research needed to guide its examination in detail should be given urgent, priority status.

REFERENCES

1. Armentano, T.V., and O.L. Loucks. 1983. Air pollution threats to U.S. national parks of the Great Lakes region. *Environ. Conserv.* 10:303-13.
2. Avery, M., and R.K. Schreiber. 1979. *The Clean Air Act: Its relation to fish and wildlife resources.* FWS/OBS-76/20.8. Washington, DC: USDI, Fish and Wildlife Service.
3. Bennett, J.P. 1985. Overview of air pollution in national parks in 1985. *Park Sci.* 5:8-9.
4. Culhane, P.J., H.P. Friesema, and J.A. Beecher. 1987. *Forecasts and environmental decision making: The content and predictive accuracy of Environmental Impact Statements.* Boulder, CO: Westview Press.
5. Master, L. 1990. The imperiled status of North American aquatic animals. *Biodiversity Network News* 3 (3): 1-3.
6. McCune, B. 1988. Lichen communities along O_3 and SO_2 gradients in Indianapolis. *Bryologist* 91 (3): 223-28.
7. National Research Council, National Academy of Sciences. 1985. *The Great Lakes Water Quality Agreement: An evolving instrument for ecosystem management.* National Research Council of the United States and the Royal Society of Canada. Washington, DC: National Academy Press.

8. Raven, P.H. 1988. Our diminishing tropical forests. In *Biodiversity*, ed. E.O. Wilson and F.M. Peter. Washington, DC: National Academy Press.
9. Sand, P.H. 1990. *Lessons learned in global environmental governance*. Washington, DC: World Resources Institute.
10. Sand, P.H. 1991. International cooperation: The environmental experience. In *Preserving the global environment: The challenge of shared leadership*, ed. J.T. Mathews. New York: W.W. Norton & Co.
11. Simon, H. 1947. *Administrative behavior*. New York: Macmillan Publishing Co.
12. Smith, W.H. 1981. *Air pollution and forests: Interactions between air contaminants and forest ecosystems*. New York: Springer-Verlag.
13. Tingey, D.T., W.E. Hogsett, and S. Henderson. 1990. Definition of adverse effects for the purpose of establishing secondary national ambient air quality standards. *J. Environ. Qual.* 19:635-39.
14. U.S. Congress. 1973. The Endangered Species Act of 1973. Public Law 93-205, 93rd Congress. Washington, DC.
15. U.S. Environmental Protection Agency. 1990. New Source Review Workshop Manual (Draft): Prevention of Significant Deterioration and Nonattainment Area Permitting. Research Triangle Park, NC: US EPA Office of Air Quality.
16. Wetstone, G.S., and A. Rosencranz. 1983. *Acid rain in Europe and North America: National responses to an international problem*. Washington, DC: Environmental Law Institute.
17. Wilson, E.O., and F.M. Peter, eds. 1988. *Biodiversity*. Washington, DC: National Academy Press.
18. Woodwell, G.M., ed. 1990. *The Earth in transition: Patterns and processes of biotic impoverishment*. New York: Cambridge Univ. Press.
19. Woodwell, G.M., and A.L. Rebuck. 1967. Effects of chronic gamma irradiation on the structure and diversity of an oak-pine forest. *Ecol. Monogr.* 37:53-69.
20. Woodwell, G.M., and R.A. Houghton. 1990. The experimental impoverishment of natural communities: Effects of ionizing radiation on plant communities, 1961-1976. In *The Earth in transition: Patterns and processes of biotic impoverishment*, ed. G.M. Woodwell. New York: Cambridge Univ. Press.

13

The Science-Policy Interface

Robert McKelvey and Sandra Henderson

INTRODUCTION

The subject of this chapter is the interplay between science, which spells out the effects of air pollution on biological diversity, and policy-making, which determines how society will respond. Specifically: How should scientists design their research—through the questions that they ask and the biotic quantities that they measure—so as best to inform the regulatory decision process, and so to advance the goal of conserving biological diversity?

Environmental management, whether at the local, regional, or global level, involves setting priorities and making tradeoffs. The current worldwide "biodiversity crisis" is no accident, but rather the inevitable result of unprecedented human demands on the resources of a finite world. Thus, preserving biodiversity will require giving up other things. Policy making requires, as prerequisite to rational choice, an examination of the available regulatory options and, for each, a determination of what will be gained, what will be lost, who will benefit, and who will pay. In particular, it will be necessary to find a means to quantify, and thus make visible, the multiple values ascribable to "biodiversity."

The task of "valuing" biodiversity is complicated by its multidimensional nature, by the uncertain and provisional character of expected future benefits from its maintenance, and often the irreversibility of its loss. Furthermore, many aspects of biodiversity do not fit well into the consumer-oriented, market-based framework of conventional benefit-cost analysis.

We argue here for a "multiattribute" approach to biodiversity valuation, one that forgoes the search for a single "common currency," dollars or otherwise. In this approach, the sharp boundary between science and policy-making necessarily softens, since the choice of "ecological indicator variables" and the interpretation of their significance to policy becomes a cooperative venture between scientist and policy analyst. In particular, scientists must take pains that their technical conclusions be presented in a character and format directly relevant to decision analysis.

These issues become particularly critical in the scientific evaluation of the impact of air pollution on ecological integrity and biodiversity. This is a consequence of both scientific uncertainty about the nature and extent of chronic or synergistic pollution impacts and (partially as a result) policy ambiguity about the societal significance these may entail.

BIODIVERSITY: WHAT IS IT?

Biodiversity is a many-faceted concept. In the language of the United Nations Environment Programme's Draft Convention on Biological Diversity (1991), it consists of "the sum of all plants, animals, and micro-organisms, their genes, and the ecosystems and ecological processes of which they are a part." In the Preamble to the Convention, the signatories "acknowledge that humanity shares the earth with other forms of life, and accept that these should exist independent of their benefit to humanity." Furthermore, biodiversity embodies values that are "environmental, genetic, scientific, aesthetic, recreational, cultural, educational, social, and economic," and should be preserved both "for the benefit of present and future generations and for its intrinsic value."

Thus, biodiversity is revealed to be a trans-scientific concept, embracing all of the diversity of life on the planet, its inherent qualities, and its meaning for and value to humanity. In practice, we use the word in various ways, depending very much on the context. In particular, its connotation will depend upon the geographical context that we have in mind.

At a localized site, "biodiversity" may be nearly synonymous with an ecosystem's or ecological landscape's "species richness" (or phenotypic or genotypic richness). It then becomes a quality that biologists measure (more or less satisfactorily) by a "diversity index," such as that of Shannon-Weaver.

In this localized context, biodiversity is just one of the many aspects of a natural biological system. In this interpretation, it most often is valued for its intrinsic role: providing resilience to the biological system and a buffer against future environmental perturbations.

Of course, biodiversity is not the only aspect of a local population, community, or ecosystem on which humans place value. Other inherent ecosystem properties of value include productivity and functional integrity. Also valued are "ecosystem goods and services" such as: 1) agricultural crops, 2) timber, 3) water resources, 4) fish and wildlife, 5) grazing of livestock, 6) erosion control, 7) recreation, and 8) carbon sequestration to mitigate the onset of global warming.

By contrast, in the global context, "biodiversity" embodies a complex of conceptually distinguishable qualities—aspects of the variety of life on the planet—that human society has come to value. The fundamental fact that drives our contemporary concern about biodiversity is the accelerating and irreplaceable loss worldwide of genes, species, populations, and ecosystems. Associated with this loss is an impoverishment of our choices for utilizing products obtained (or potentially obtainable) from nature, possible disruption of essential ecological processes and services, and a loss of options for biological and cultural adaptation to an uncertain future (Tingey, Hogsett, and Henderson 1990).

Some of these widely recognized values include:

1. Anticipated future utilization of genetic materials present in the biome.
2. Anticipated utilization of pharmaceuticals developed from natural sources that may be discovered in the biome.
3. The role of protected historical ecosystems as biological museums or scientific laboratories.

4. Aesthetic appreciation of the marvelously structured and intricately functioning diversity of life on the planet.
5. Ethical values, stemming from the belief that it is wrong to destroy life forms.
6. The value of preserving flexibility: that is, maintaining diversity unimpaired, so as not to forgo as-yet-unrecognized future opportunities.

This list is long; almost certainly, it is incomplete. For the most part, these points relate value at a local site to scarcity on a regional or global scale. Furthermore, the entries in the list are largely based on qualities that are difficult to quantify, and for which it seems very unnatural to assign a priori monetary values.

SETTING PRIORITIES

As we have suggested already, losses of biodiversity are the result of fundamental clashes in values and goals within our contemporary society. As a society, we are not prepared to accord absolute precedence to biological conservation over all other human aspirations. Thus, those who value biodiversity conservation must be prepared to make the case for their views, and will surely have to settle for only partial success.

Overall, the question becomes one of setting priorities: In considering the risks to biodiversity, which potential losses are most important? Which hazards are most pressing? Which tradeoffs are most acceptable? The usual way of approaching these questions has been through conventional benefit-cost assessment, which requires placing a monetary value on nonmarket "environmental goods and services."

The perceived need to express all costs and benefits in monetary values has led to extraordinary efforts by environmental economists to "value" a wide range of environmental costs, goods, and services. Conceptually, it is the most straightforward way to proceed—when it can be done. However, many biologists are profoundly uneasy about attempts to apply this approach to biodiversity valuation, and even those who favor the approach recognize the great difficulty in measuring ecological benefits.

Indeed, it is well recognized that conventional analytic techniques have consistently undervalued public goods, such as environmental benefits, as compared with conventional market values. Furthermore, environmental economists have been hobbled by the limitations of available tools, such as contingent valuation, when attempting to determine consumers' "willingness to pay" for unfamiliar and little-understood ecosystem qualities. In this decision context, the determination of who "wins" and who "loses" often is reduced to a type of political weight-lifting contest—the winner is the one who can best convince the decision maker that their analysis is more "correct."

A more fundamental objection concerns the weak theoretical background for benefit-cost analysis as commonly practiced. Benefit-cost analysis treats public welfare as a single-dimensional entity, the total net of benefits over costs *to whomever those net benefits may accrue*. Thereby, it ignores the circumstance, common in environmental affairs, that one segment of society may be reaping the benefits while a totally different segment must absorb the costs. Often, environmentalists say, it is the present generation of humanity that gains, at the expense of its descendants.

It is presumed that benefit-cost analysis captures overall societal benefits consistent with the idealized competitive market. Thus, benefit-cost analysis sets priorities in line with market power, and completely ignores questions of distributional justice. The only direct account taken of distributional effects is with respect to intergenerational tradeoffs, a matter that is dealt with by the mechanism of the discount rate. Here, too, critics raise objections, arguing that the discounting of future ecosystem values is incompatible with goals of sustainability.

Because of these characteristic oversimplifications, benefit-cost analysis (and the "utilitarianism" from which it derives) is highly suspect in economic theory (Bromley 1990). Its value is primarily as a simple technocratic tool for making pragmatic decisions.[1] Our primary objection to benefit-cost analysis is to the emphasis on measuring all values in terms of a common "currency," one based on consumer preferences and the competitive market. We believe that such an approach is particularly unnatural when applied to nonmarket ecological values, and prefer an alternative approach, one derived from the concepts of cooperative game theory and the theory of social choice (Sen 1970).

The simplest practical way to incorporate this perspective is within the framework of "multicriterion decision-making." Put simply, one backs off from attempting to find a single currency for valuation of complex environmental systems, and attempts, rather, to characterize separately the important components that will be impacted by management decisions. The tradeoffs are made clearly visible through scientific analysis, with the ultimate choices being left to regulatory institutions and the democratic political process.

Some of this conceptual thinking has spilled over into pragmatic decision making; for example, in the practice of maintaining separate accounting of regional economic impacts, and in the U.S. Forest Service's multiple-use management planning process. In both instances, there is recognition that a society's decision process consists of balancing off a number of distinct values, which are of interest to different groups within that society.

The multicriteria decision approach seems to us to be especially appropriate in approaching questions of biodiversity conservation. It is not however without its problems; and we shall return to these matters later in the chapter. First, it is necessary to inquire more closely into the rather unusual properties that biodiversity qualities display (Randall 1991).

CATEGORIES OF VALUATION

Our examination leads us to classify ecosystem valuations into four rough categories: 1) direct market values, 2) imputed monetary values, 3) random values (in the sense of probability theory), and 4) nonmonetary values, including multidimensional values. We believe that only the first two of these lend themselves to classical benefit-cost analysis.

Direct Market Values

It is straightforward to assign direct market value to the harvested goods (benefits) of an ecosystem (i.e., agricultural crops, forest products, fish, game). From this it is possible to determine a dollar amount of damage (costs) due to productivity losses resulting from pollution (Tingey, Hogsett, and Henderson 1990). For example, the current annual loss in

---

[1]We recognize that our characterization of benefit-cost analysis, while consistent with the views of many noneconomists (Constanza 1991), is at odds with the mainstream view within the economics profession. Furthermore, we need to acknowledge that a number of environmental economists have written searchingly on welfare economics as applied to environmental issues (e.g., Maler 1974; Page 1977; Lind 1982).

the United States from ozone damage to eight major agricultural crops is estimated at $2 to $3 billion (Adams, Glyer, and McCarl 1988).

Imputed Monetary Values

In attempting to capture nonmarket monetary values, economists have shown great ingenuity, devising clever means of teasing out by indirection (or "imputing") the dollar values to be attached to individuals' personal preferences. A well-known example concerns the smog and visual pollution within the Grand Canyon National Park, resulting from coal-fired electricity-generating facilities operating in the Four Corners area. Economists have used a variety of means, ranging from observing behavior to asking direct questions, in order to ascertain individuals' willingness to pay for experiencing clear air in the Canyon, or for merely knowing that the air there is clear.

Random Values

The third category, random values, consists of costs or benefits that might be naturally expressible in dollars, except that their occurrence and extent are unknown and speculative. An example concerns the value of actions taken to prevent or slow down the onset of global warming. The most direct way to do this is by cutting back the quantities of greenhouse gases (e.g., carbon dioxide, methane) that are emitted to the atmosphere. The costs of doing this are relatively easy to ascertain, but the resulting benefits are highly uncertain.

For example, it is reasonably easy to measure the cost to society of imposing lower gasoline consumption standards on new cars. It then is not too difficult to estimate the resulting reduction in overall tailpipe emissions over the course of time. But what are the benefits to society of having kept those additional tons of fossil carbon dioxide from entering the atmosphere? *If* we could predict with any confidence 1) how long the gases would remain in the atmosphere, 2) the resultant incremental climate change that would be induced, and finally, 3) the effects of this change upon human society and the biosphere, *then* we might be able to give an explicit answer in dollars. As it is, the answer can be expressed at best only as a random variable, with a very spread-out probability distribution. The plausible outcomes range from very moderate effects to the avoidance of a catastrophe.

Given the limitations in our current understanding of the global warming phenomenon, but given the possibly massive impact it may have, we are unlikely to settle for representing this random variable merely by its probabilistic expected value. Rather, it is necessary to value the entire probability distribution, and to make value comparisons among probability distributions representing alternative policies.

Humans have trouble in making such comparisons, particularly when the stakes are high. How much weight should we give to a possible catastrophic event of extremely small probability? In cases like this, it is important to spell out carefully what it is that science can know and what it cannot.

One clear principle is that valuation attaches to the *action taken*, and the degree to which its possibly adverse outcomes can be reversed. Thus, the decision to utilize a persistent pesticide, or to degrade the habitat of a rare or endangered species requires careful consideration of even low probability adverse outcomes.

Nonmonetary and Multidimensional Values

In many practical cases, setting priorities clearly does not require monetary valuation of the options. A well-known example, in the field of environmental health, is for judging the relative hazard of carcinogenic chemicals. Here it is common practice to express the hazard as the number of deaths resulting from exposure per 100,000 in the exposed population. Though imperfect, this measure permits comparing one carcinogen against another and seems adequate for regulatory purposes. Essentially, it serves as a common currency in this limited realm.

However, one may raise the objection that not all cancer deaths should be regarded as equal. Some forms of cancer are much more unpleasant than others. Also, some strike children and some strike only the elderly. Clearly it would be desirable to factor these considerations into carcinogenic hazard rankings. A natural way to take account of the last mentioned of these is to record not only cancer deaths per 100,000 but also the average age of the victims. In this way, the cancer hazard index becomes two-dimensional. One can easily imagine adding further refinements to arrive at higher dimensional indices.

A second example brings us back to the issues of global warming. We already have noted the near impossibility of assigning a monetary cost to the emission of a ton of a greenhouse gas into the atmosphere. However, it is common practice to compare the warming effects of emissions of the various gases, e.g., carbon dioxide, methane, nitrous oxide, and the various CFC's. These are measured in "carbon dioxide equivalents." For example, methane is often rated as 21 on a scale that makes carbon dioxide 1. Here again, the measure is rough, since it includes two separate effects that differ between gases: residence time of a gas in the atmosphere and heat-trapping efficiency while there. Thus, methane rates 21 on a 100-year horizon and only 9 on a 500-year horizon. Plainly, this metric too is multidimensional. (It is also random because true residence time is unknown.)

Such nonmonetary currencies clearly can be very natural and useful. The difficulty comes when the decision process requires determining an exchange rate between such a currency and dollars: e.g., how much is society willing to pay to reduce preventable cancer death rates to a specified level? The decision is a political one, but rational choice requires a clear understanding of the tradeoffs. The conversion is even more difficult when the nonmonetary currency is a vector. This conversion uncertainty represents a universal problem inherent in utilizing nonmonetary values, and we shall return to it later, especially in the context of biodiversity.

BIOLOGICAL INDICATORS

Multidimensional, nonmonetary metrics also are common in measuring the internal state of biological systems, particularly ecosystem integrity and resilience. We now briefly examine some of these, as potential tools for rational management of pollution-stressed ecosystems.

Although "ecosystem health" is in no way synonymous with "biodiversity," it is generally true that a vigorous natural ecosystem supports a greater array of plant and animal species than does one that is degraded. Thus, by characterizing the vitality of a stressed ecosystem, one is implying something about site biodiversity.

Ecologists have demonstrated that ecosystems respond in a repeatable pattern when exposed to pollutants. Bormann (1985) summarizes the responses in the following six stages, ranging from insignificant to complete ecosystem collapse:

Stage 0: Anthropogenic pollutant levels insignificant. Pristine systems.

Stage I: Anthropogenic pollution occurs at generally low levels. Ecosystems serve as a sink for some pollutants, but species and ecosystem functions are relatively unaffected.

Stage IIA: Levels of pollutants are inimical to some aspect of the life cycle of sensitive species or individuals, which therefore are subtly and adversely affected.

Stage IIB: With increased pollution stress, populations of sensitive species decline, and their effectiveness as functional members of the ecosystem diminishes.

Stage IIIA: With still more pollution stress, size becomes important to survival, and in forests, for example, large plants, trees, and shrubs of all species die.

Stage IIIB: Ecosystem collapse.

Bormann's stress-response scale might be regarded as a one-dimensional index of the ecosystem's resilience, hence of this characteristic of biodiversity.

More conventional measures of local biodiversity are the diversity indices of Hill and of Shannon-Weaver. Statistical ecologists emphasize that all such indices represent compromises, since they attempt to summarize in a single number two very distinct aspects of an ecological community: species richness and evenness (Ludwig and Reynolds 1988). Two-dimensional metrics can separate them out.

In assessing loss of ecological integrity in stressed aquatic ecosystems, regulatory ecologists have experimented with utilizing a suite of metrics, no one of which, standing alone, could adequately characterize the ecological significance of change from an idealized reference ecosystem (Plafkin et al. 1989). Some of the metrics applied to the benthic community are 1) total taxa richness, 2) taxa richness in pollution-intolerant groups, 3) indices of community similarity to the ideal standard, and 4) an index measuring extent of dominance by dominant taxa. A similar approach has been suggested for air pollution impacts by Herrman (1990).

We stress that different indicators can be expected to be useful for different environmental assessment questions. One needs to ask, "What are we assessing and why?" Many indicators will be specific to local biota and specific pollutants. For example, many lichen species are sensitive indicators of airborne pollutants and also may be important structural components of certain ecosystems.

The following is a list of characteristics that should be considered when choosing indicators. The ideal indicators would be (Cook 1976; Sheehan 1984; Munn 1988):

1. Sufficiently sensitive to provide an early warning of change (hypersensitive to stress).
2. Distributed over a broad geographical area, or otherwise widely applicable.
3. Capable of providing a continuous assessment over a wide range of stress.
4. Relatively independent of sample size.
5. Easy to measure, collect, assay, and/or calculate (in the case of an index).
6. Able to differentiate between natural cycles or trends and those induced by anthropogenic stress.
7. Relevant to ecologically significant phenomena.

No single indicator will possess all of these desirable properties; a set of complementary indicators is required. Ideally, indicators would be selected by biologists familiar with local and regional ecology (Henderson, Noss, and Ross, in press).

Of course, air pollutants differ in intensity and duration, and ecosystems differ in susceptibility to them. It is standard in the vegetation effects literature to encapsule such distinctions through the terms "injury" and "damage" (Guderian, Tingey, and Rabe 1985). Injury encompasses all measurable plant reactions that do not influence agronomic yield or reproduction. Those responses can include changes in metabolism, reduced photosynthesis, leaf necrosis, leaf drop, and altered growth or quality if yield is not impaired. In contrast, damage includes all effects that reduce the intended use or value of the plant or ecosystem. Society may be willing to accept injury to biodiversity but take action to prevent damage.

MULTICRITERIA DECISION APPROACHES

Our analysis seems now to have led us into a quandary: We know that impacts on biodiversity are multiple, differing in character, and acting on different spatial and temporal scales. We have seen, too, that values placed on impacted ecosystem themselves are multidimensional. Thus, we are forced to recognize that traditional benefit-cost approaches to environmental decision making are inadequate to resolve management and regulatory issues.

Societal decisions on the management of a natural landscape will rarely be based solely upon the impacts of anthropogenic stresses on the biota. Rather, biodiversity values will be only part of a complex of human values, many of them social, economic, or cultural in nature, which will be impacted through the decision process and effectively traded off by the management choices made.

Decision making with multiple criteria takes us beyond benefit-cost analysis, and perhaps even outside the realm of classical microeconomics. But how then are we to decide? Do we just turn over the whole undigested tangle to decision makers, and leave it to them to make sense of it? Or are there scientific principles that may be applied to organize the information and rationalize the decision process?

A modest technical literature does exist on "multicriteria optimization," which undertakes to formulate principles for the decision process (Raiffa and Schlaifer 1961). However, usually it is assumed that choices are made according to the preferences of a single point of view (that of the hypothetical "decision maker"), and thus could in principle be encapsulated in a single "utility function"—i.e., there exists a single currency.

Our perspective, rather, is that the decisions will represent compromises among affected segments of society, and that the scientists' role is to make clear to all what values will be affected and how. Societal decisions, then, are achieved through a kind of "cooperative game" (Axelrod 1984; Olson 1965).

In such a context, there are only a few general principles. But these few, we contend, are useful and, indeed, have been widely employed in environmental decision making. Here we shall describe the two that are most firmly established: the concept of a minimal acceptable security level and the concept of cost effectiveness.

It is easiest to describe these in relation to an environmental hazard: A minimum security level sets the degree of hazard that is acceptable. A least-cost analysis then determines the least expensive way in which that security level can be attained. Thus, in our cancer example, society first determines an acceptable hazard level, measured in cancer deaths per 100,000. Regulatory controls may then be devised for banning or restricting the use of particular carcinogens to achieve the security goal in the least-cost way. The Clean Air Act of 1990 contains language concerning health-based standards for airborne toxic chemicals, language that can be interpreted in a similar way.

Concerning ecosystem objectives, there are many examples in U.S. government agency practice where these two principles appear to be utilized. For example, U.S. Forest Service multiple-use management sometimes is characterized as seeking to maximize timber production values within security constraints set to protect nonmonetary environmental values.

SETTING BIOLOGICAL PRIORITIES

The cost-effectiveness criterion provides a rational guide for best attaining a specified set of biodiversity goals. However, it offers no guidance for setting priorities *among* biodiversity goals, not all of which can be optimized simultaneously. As we have seen, the biodiversity value vector is of high dimension, with individual components distributed over space and time. Some of these component values are complementary, but some are mutually antagonistic.

A principle that is dual to cost effectiveness is that of *benefit effectiveness*: given a specified funding level for supporting biodiversity objectives, which projects should be undertaken? To answer this question, it is necessary to arrive at an internal prioritization of projects, based on intrinsic biological worth and degree of risk exposure. The prioritization question arises most naturally in land-use decisions for setting aside biological reserves: Since one cannot preserve everything, which areas are most important? For example, managing an ecosystem to enhance its buffering against environmental perturbation (the bet-hedging role of biodiversity) would seem to imply ascribing higher importance to a plant or animal species exhibiting considerable genetic diversity (e.g., a mammal) than to a species able to adapt only by evolving a new species (e.g., a beetle). Managing for potentially useful pharmaceutical products might mean emphasizing herbal plants or soil fungi. Managing for genetic resources for agriculture would require an emphasis on grasses, ungulates, and certain species of fish.

Prioritization is also an issue in endangered species management. With limited resources for initiating the listing process, which species should be studied first? In designing a recovery plan, what is an acceptable security level? Further, how should limited resources be allocated between recovery efforts—say, between the Black-Footed Ferret and the California Condor?

These are all questions for the conservation ecology community to settle internally. Ideally, the allocations ought to be coordinated so that, at whatever level of funding, the overall effort is benefit-effective. The complementary question, of the overall funding level for endangered species conservation, must be settled in a broader societal context.

Benefit-effectiveness is more difficult to determine than cost-effectiveness because objectives are likely to be nonmonetary, random, and multidimensional. To grasp the difficulties caused by multidimensionality, let us return briefly to our cancer example. If one characterizes carcinogenic hazard by specifying both the number of deaths per 100,000 and the average victim age, then prioritizing requires trading off these two dimensions of the impact. Society could make the tradeoff by assigning a dollar cost to each age-specific death, but it need not, and probably will not, wish to.

A particularly instructive case, one specifically involving genetic diversity, is the salmon management philosophy adopted by the U.S. Northwest Power Planning Council. This regional intergovernmental panel provides overall policy guidance for managing the Columbia River Basin's hydropower and other electric power resources.

One goal of the Council has been energy efficiency in the Northwest. Another, however, has been to restore the diminished salmon runs on the Columbia and its tributaries. Recognizing a strong mandate for restoration, vastly exceeding any potential monetary value in the fishery, the Council set a goal (security level) early on of doubling the river's depleted salmon population. It then looked for the most cost-effective means of doing this.

Available means include direct expenditures for "habitat enhancement," for example, building fish ladders around the dams and restoring spawning beds. But it also includes modifying the hydrologic regime of the river flow, through manipulating the pattern of releases at the dams. These release decisions have profound impacts on many different river uses, not only for the timely availability of water for hydropower, but also for irrigation, navigation, flood control, safeguarding Indian treaty rights, and recreation. Management has attempted to operate cost-effectively, but the overall cost has been high: over $1 billion estimated in the past 10 years. Future costs for an expanded program could run much higher. Thus, the determination of acceptable cost must finally be resolved through the political process.

Furthermore, within the biological goal itself there are many dimensions. There is not merely a single salmon population in the river, but actually several species and a couple of hundred separate "races" or subspecies, genetically isolated through the timing and localized destination of their runs. Preserving the up-stream runs will require heroic measures, including very unnatural management, such as trucking the young pre-smolts around the dams and barging them through the slack water reservoirs.

Thus, the biodiversity goal cannot be adequately addressed by specifying an overall salmon population level. It is necessary also to specify security levels on the survival of the individual salmon runs. What these levels should be, and whether some runs are more important than others, has yet to be determined. The U.S. Endangered Species Act simply asserts that species listed as endangered should be protected, and a recovery plan devised.

No fundamental framework for establishing genetic priorities seems to exist, but a practical resolution of the issue seems imminent: Since commercial salmon harvesting and dam building began, some 200 runs have become extinct. Currently, about half of the remainder, an additional 76 runs, are thought to be in trouble. The National Marine Fisheries Service has officially listed four of these runs as endangered, on Idaho's tributary Snake River. They have thereby triggered what is predicted to be an unprecedented debate over regional natural resource management, and a severe challenge to the conservation requirements of the Endangered Species Act.

THE CAUSALITY TANGLE

Another complicating aspect of regulatory decision making results from the causality tangle of intertwining causes and effects. Thus, as discussed in earlier chapters, not only do SO_2 and ozone individually cause damage to forest tree-growth, but, acting synergistically, their combined effects are intensified. Though both require regulation, it may be most cost-effective to concentrate on one or the other.

On the other side, a single pollutant source may adversely affect both human health and ecosystem health simultaneously. Regulations promulgated to protect the former incidentally provide relief for the latter as well.

Generally, air pollution emissions and their effects are mitigated through two avenues: 1) controlling the source of emissions, and 2) treating the receptor of the pollutant. Controlling the source involves reducing emissions usually through the establishment of air quality standards, guidelines, or criteria. Protecting the receptor can involve treatment such as introducing tolerant species or adding nutrients to the system (Olem 1991).

There is not a lot of difference between the source regulatory measures for mitigating *biological effects* of air pollution and those known to be effective for mitigating *damage to human health and property*. Though the nature of the damages differ, the pollution sources and controls probably are the same. The fact that these very different kinds of impacts can be controlled simultaneously also means that much ecosystem damage is already being prevented, automatically and free of charge, by our existing regulations to protect human health. The majority of regulatory measures have been developed with the primary purpose of minimizing impacts to human health and property. If we are concerned with protecting biodiversity, we cannot afford to allow its protection to be limited to the secondary effects of human health concerns.

A recent report by the U.S. EPA's Science Advisory Board recommends that we address ecological risks with the same level of effort that we have devoted in the past to human health risks. This recommendation recognizes the intimate relationship between vital and productive natural ecosystems and the ultimate well-being of people and their habitat.

Acid rain provides an example where mitigative actions were directed at ecosystems, not at human health. This provides one of the few examples of actions taken where the ecosystem was the primary beneficiary. Although acid rain has been a key environmental issue for the past several decades, its effects were not known to cause direct injury to human health. The adverse effects of acid rain are seen most clearly in aquatic ecosystems. The public responded to visually decaying forests and "dying" lakes by demanding that policy makers "do something."

One such mitigative action was the "liming" of acidic surface waters in affected areas of North America and Europe. Liming refers to the addition of base material, most commonly, limestone, to neutralize the acidity of surface waters, sediments, and soils. The goals of liming were to increase the productivity and restore the usefulness of aquatic ecosystems for agriculture, aquaculture, industry, and habitats for wildlife (Olem 1991). Other actions taken involved addressing the source of the pollutant and fitting smokestacks with costly scrubbers. Of the two, liming is considered to be more cost-effective—it is relatively cheap and the effects are immediate. However, liming has only proved effective in the short run. If the pollutant is still being emitted, the adverse effects could reoccur (Olem 1991).

In contrast to management options involving the source or receptor, a new option being heralded as cost-effective is pollution prevention. The concept of preventing pollution recognizes that many environmental problems, such air pollution effects over broad geographic areas, are not totally amenable to traditional pollution control regulations. Pollution prevention includes the reduction, elimination, or recycling of pollutant discharges to the air, water, or land. Pollution prevention offers significant benefits to many sectors of society, benefits that are not available through traditional pollution control approaches. These benefits fall into two major categories: 1) reduction of human health and ecological risks, and 2) economic benefits.

THE BOTTOM LINE

In all of these cases, what is needed is to find some means of "boiling down" the complex biodiversity value-vector to a manageable size, somehow consolidating dimensions, so that those who must make decisions can grasp the nature of the tradeoffs being made. In effect, we need to seek a minimal set of independent currencies that might reasonably represent the broad categories of value embodied in an ecosystem or ecological landscape.

Scientists and policy analysts must accomplish this task jointly. Value is a human overlay on a natural system, but the instrumental variables that show it must be chosen on the basis of defensible scientific principles and an understanding of how natural systems work.

The bottom line is that the regulatory policy makers will be unable to deal intelligently with ecological valuation unless we as scientists help them. Scientists have an obligation to communicate their scientific knowledge to the regulatory decision makers, and in a form that is relevant to the options they face. Scientists working in the field or laboratory generally are very conscious that their knowledge is incomplete and imperfect, and they dislike speculating. Nevertheless, decisions are being made and will continue to be made, with or without their involvement.

In speaking out, scientists certainly have to distinguish carefully between knowledge that is secure and impressions that are more questionable. However, we believe that they must be willing to enter the realm of the uncertain, to express their best judgments, and to characterize the degree of assurance with which these judgments are held. In characterizing biodiversity qualities, we have to find rational, scientifically defensible principles, based on conceptions of how biological systems work. We have to avoid the multidimensionality trap of expressing scientific results in a manner so detailed and fragmented that no one can grasp the overall implications.

As scientists, we must be willing to propose ecological priorities. In so doing, we need to recognize explicitly that such prioritization rests ultimately on human value judgments, and cannot be justified on purely "scientific" grounds. But while idealized science strives for value-neutrality, ecological scientists cannot afford neutrality. Their insights and their aesthetic perceptions are essential if society is to act wisely in resolving the biodiversity crisis that urgently faces us all.

REFERENCES

1. Adams, R.M., J.D. Glyer, and B.A. McCarl. 1988. The NCLAN economic assessment: Approach, findings and implications. In *Assessment of crop loss from air pollutants*, ed. W.W. Heck et al., pp. 473-504. New York: Elsevier Science.
2. Axelrod, R. 1984. *The evolution of cooperation*. New York: Basic Books.
3. Bormann, F.H. 1985. Air pollution and forests: An ecosystem perspective. *BioScience*. 35 (7): 434-41.
4. Bromley, D.W. 1990. The ideology of efficiency: Searching for a theory of policy analysis. *J. Environ. Econ. Manage.* 19 (1): 86-106.
5. Constanza, R., ed. 1991. *Ecological economics*. New York: Columbia Univ. Press.
6. Cook, S.E.K. 1976. Quest for an index of community structure sensitive to water pollution. *Environ. Pollut.* 11:269-88.
7. Guderian, R., D.T. Tingey, and R. Rabe. 1985. Effects of photochemical oxidants on plants. In *Air pollution by photochemical oxidants: Formation, transport, control, and effects on plants*, ed. R. Guderian, Part 2, pp. 127-333. New York: Springer-Verlag.
8. Henderson, S., R.F. Noss, and P. Ross. In press. Can NEPA protect biodiversity? In *The scientific challenges of NEPA: Future direction based on 20 years of experience*. Proceedings of the Ninth Oak Ridge National Laboratory Life Sciences Symposium. Knoxville, TN: Oak Ridge National Laboratory.

9. Herrman, R. 1990. Biosphere reserve monitoring and research for understanding global pollution issues. *Parks* 1 (2): 23-28.
10. Lind, R.D. 1982. Discounting for time and risk in energy policy. In *Resources for the future*. Baltimore, MD: Johns Hopkins Univ. Press.
11. Ludwig, J.A., and J.F. Reynolds. 1988. *Statistical ecology*. New York: John Wiley & Sons.
12. Maler, K.G. 1974. Environmental economics, a theoretical inquiry. In *Resources for the future*. Baltimore, MD: Johns Hopkins Univ. Press.
13. Munn, R.E. 1988. The design of integrated monitoring systems to provide early indications of environmental/ecological changes. *Environ. Monit. Assess.* 11:203-17.
14. Olem, H. 1991. *Liming acidic surface waters*. Chelsea, MI: Lewis Publishers, Inc.
15. Olson, M. 1965. *The logic of collection action.* Cambridge, MA: Harvard Univ.
16. Page, T. 1977. Conservation and economic efficiency. In *Resources for the future*. Baltimore, MD: Johns Hopkins Univ. Press.
17. Plafkin, J.L., M.T. Barbour, K.D. Porter, S.K. Gross, and R.M. Hughes. 1989. Rapid bioassessment protocols for use in streams and rivers: Benthic macroinvertebrates and fish. U.S. EPA Report No. 444/4-89/001. Washington, DC: Office of Water Regulation and Standards.
18. Raiffa, H., and R.O. Schlaifer. 1961. *Applied statistical decision theory*. Cambridge, MA: Harvard Business School.
19. Randall, A. 1991. The value of biodiversity. *Ambio* 20:64.
20. Sen, A.K. 1970. *Collective choice and social welfare.* San Francisco: Holden Day.
21. Shannon, C.E., and W. Weaver. 1949. *The mathematical theory of communication*. Urbana, IL: Univ. of Illinois Press.
22. Sheehan, P.J. 1984. Effects on community and ecosystem structure and dynamics. In *Effects of pollutants at the ecosystem level*, ed. P.J. Sheehan, D.R. Miller, G.C. Butler, and P. Boudreau. New York: John Wiley & Sons.
23. Tingey, D.T., W.E. Hogsett, and S. Henderson. 1990. Definition of adverse effects for the purpose of establishing secondary national ambient air quality standards. *J. Environ. Qual.* 19 (4): 635-39.
24. United Nations Environment Programme. 1991. *Draft convention on biological diversity*. Ad Hoc Working Group of Legal and Technical Experts on Biological Diversity, Third Session, June 24-July 3, 1991, Madrid.

14

Air Pollution Effects on Biodiversity: Research Needs

Paul G. Risser

INTRODUCTION

Concerns about biological diversity or biodiversity have several origins, for example, interest in preserving pristine areas and the species therein, saving special species or threatened and endangered species, maintaining sustainable ecosystems for natural resource production, saving genetic material for future uses, and the reversal of the increased loss of species and habitat, especially but not only in the tropics (see White and Nekola, Chapter 2, this volume). The purpose of this book is to consider the topic of air pollution and its implications for local, regional, and global biodiversity.

Traditionally, air pollution studies have focused on measuring the effect of pollution on animals and plants, particularly crop plants and a few forest tree species, and identifying the biochemical mechanisms that are the basis for the responses. If the effects of air pollution on biodiversity are to be evaluated more thoroughly, there must be a much broader framework for considering genetic, species, and ecosystem biodiversity. Moreover, since the number of interactions between air pollutants and biodiversity is potentially very large, future research strategies must be parsimonious and focus on efficient ways of identifying the most important impacts. This chapter describes a possible approach for identifying high-priority research topics and discusses several important research questions that must be investigated to more fully understand the impacts of air pollution on biodiversity.

BIODIVERSITY

In the popular literature, "biodiversity" is frequently considered synonymous with "species diversity," that is, the number of species in an area. However, biological diversity is an important attribute at three different levels of biological organization (Noss 1990): individual organisms (genetic); populations or different kinds of organisms (species); whole ecosystems and portions of the landscape (ecosystem). For evaluating the effects of pollution on biodiversity, consideration must be given to all three of these biological organization levels. The genetic level, which is the foundation underlying all other levels, determines the genetic options available to populations or organisms and therefore underlies differences in structure and function of biological communities, ecosystems, and landscapes. Species diversity considers the number of different types of organisms in a specified area and is a measure of not only the genetic richness, but also of the number of species available to function in the ecosystem. Finally, ecosystem diversity indicates the different combinations of organisms and abiotic conditions that exist within specified spatial boundaries. Air pollutants can affect all three components of biodiversity.

Because of the complexity of the concept of biodiversity, it is important to construct an appropriate definition that includes all the components. For the purpose of this discussion, the following working definition of biological diversity will be employed (see White and Nekola, Chapter 2, this volume):

> . . . the variety of life and life processes at all organizational levels, with usual emphasis on: (1) genetic diversity within species, other taxa, or populations and/or the sum of genetic diversity within a community or geographic area; (2) species or other taxon diversity within a community or geographic area; and (3) community or ecosystem diversity across a landscape or larger region.

To consider the effects of air pollutants on biodiversity, it is necessary to have field measurements of what is meant by biodiversity. Species diversity is the most obvious component of biodiversity and it has received the most attention. Measuring the species diversity of a biological community usually consists of counting or estimating the number of species and the number of individuals of each species. In such an index the description of biodiversity involves two components, the number of classes (species) and the distribution of abundances in those classes (number of individuals of each species). The most diverse communities are those that have a large number of species (richness) with many individuals of each species (evenness).

One graphical way of describing both richness and evenness is the dominance-diversity curve. This curve consists of plotting the abundance of each species along the x-axis, ordered so the number of individuals in each species descends along the x-axis. In the resulting curves the dominant species are near the y-axis and rare species are far to the right. Steep linear curves represent low diversity, that is, communities with one or a few dominant species and a small number of intermediate and rare species. Broad curves with a lower slope represent diverse communities, usually when no one species is dominant and there are many intermediate and rare species.

The species richness and evenness that are measured in the field depend on the characteristics of the sampling design. Among the decisions to be made when developing the field sampling scheme is the spatial size or scale of the sample. Scale has two components: "grain" is the size of the actual unit of observation (e.g., sample plot size); "extent" is the distance over which the observations or sample plots are distributed (Wiens 1989). These two components of scale can independently influence the levels of diversity measured. For example, if plot sizes are relatively large, each one will contain more diversity and differences among the plots will be relatively small. Similarly, diversity indices will be different between conditions of extent where plots are distributed over a heterogeneous environmental gradient, but the heterogeneity is uniform as compared to when the heterogeneity is concentrated in certain areas of the sampled landscape. These relationships between the components of diversity indices and the consequences of various sampling procedures are well known. Peet (1974) provides a summary of measurements of biodiversity.

These general measures of biodiversity have been applied to questions about the impact of air pollution on vegetation. For example, in a North American white spruce (*Picea glauca*) association, mosses were more sensitive than vascular plants in areas heavily stressed with SO_2. As a general trend, diversity increased along a gradient of decreasing exposure, except in localized areas in which diversity increased because of the invasion by weedy angiosperms (Winner and Bewley 1978). This example demonstrates the importance of the sampling design. Sample plots must be large enough to measure these patches of invading weedy species, and the extent of the sample must be adequate to capture the species distribution along the gradient of decreasing exposure. The richness and evenness components of diversity may respond differently to air pollutants (Cribben and Sacchetti 1977). In the example case, the total number of species may be increased because the natural vegetation is eliminated or reduced, thus allowing invading species. If the dominant species are selectively harmed by the pollutants, evenness may be increased but the total number of species might be reduced.

Despite this and other examples in which various indices of biodiversity have been used to examine the response of biological systems to air pollution, a significant research question involves defining the optimum sampling approach most likely to measure the components of biodiversity that are affected by air pollution. This discussion has focused on species diversity, but it is also important to consider the genetic and ecosystem levels of organization. Can the same measures of diversity, namely richness and evenness, be applied to various levels of biological organization? Although the practical answer is affirmative, as will be discussed subsequently, other measures or indices may be required to assess the susceptibility of genetic resources and ecosystems to air pollutants.

As we begin to look beyond species diversity as the sole measure of biodiversity, it becomes more important to understand the relationship between various measures of biodiversity and the structure and function of ecosystems. Essential ecosystem functions such as productivity, sequestering of carbon and other chemicals, and rates of nutrient cycling may be affected, directly or indirectly, by air pollutants. A significant research question is whether these functions are independent of species biodiversity. If they are independent of species diversity, then these functions must be measured to determine their response to air pollution. If measures of species biodiversity are quantifiably related to these ecosystem functional characteristics, then measures of species biodiversity can be used as surrogates for ecosystem behavior.

The exact correspondence between biodiversity and ecological function, however, is not well understood and will require continued research. Karr (1990) used "biological integrity" to mean "supporting and maintaining a balanced, integrated, adaptive community of organisms having a species composition and functional organization comparable to that of a natural habitat of the region." Under this concept it is possible to measure biological integrity by recognizing the proper attributes of the community of organisms and functional organization of ecosystems. Seemingly robust measures of the Index of Biotic Integrity (IBI) have been developed for stream communities. This putative relationship of biological integrity (Karr 1990) is based on a comparison to a natural habitat, not on an underlying theory that predicts diversity from integrity or vice versa. Nevertheless, a synthetic index that considers both species diversity and ecosystem functional properties has great promise in detecting the impacts of air pollution on biodiversity at more than one level of biological organization.

AIR POLLUTION

In considering the relationship between air pollution and biodiversity, there are several components. First, it is necessary to recognize that organisms have evolved methods for counteracting toxic materials, but not all organisms have the same detoxifying capabilities. The pathways of exposure and toxicity mechanisms for plants and animals are known for several common air pollutants (see Newman, Schreiber, and Novakova, Chapter 10, and Musselman et al., Chapter 4, this volume). For example, in plants atmospheric deposition (wet, dry, and gaseous) interacts through direct surface contact and through stomatal uptake, although the leaves of some plant species have a buffering capacity against both acidic and gaseous deposition. If exposure is through the leaves, gaseous pollutants produce ionic species and free radicals in the extracellular fluid of the stomatal cavity that may damage cellular membranes and metabolic processes (Mansfield and Freer-Smith 1981). Various materials, such as carotenoids, certain phenolic compounds, peptides, enzyme systems, polyamines, and organic buffering systems (Larsson 1988), have evolved in plants to counteract toxic molecules.

A second consideration is that air pollutants frequently do not occur alone, but rather in mixtures. As an example, the toxic action of sulfur dioxide (SO_2) arises partly from the direct effects of its solution products, SO_3^{2-} and HSO^{3-}, on photosynthetic processes of photophosphorylation and ATP synthesis, but further toxic effects of SO_2 arise from free-radical-generating chain reactions. Thus, biochemical scavenging mechanisms to remove oxygen species are important in SO_2-stressed as well as O_3-stressed plants. Exposures to combinations of SO_2 and nitrogen dioxide (NO_2) are more detrimental than either gas alone. Because SO_2 inhibits the normal NO_2 detoxification pathway, toxic levels of nitrite and nitrate ions may accumulate in the choroplasts (Tingey et al. 1971; Wellburn et al. 1981). Similarly, in areas of industrial pollution, animals are also simultaneously exposed to several pollutants, which cause several symptoms (see Newman, Schreiber, and Novakova, Chapter 10, this volume). The effects can be both direct, caused by ingestion of chemicals, or indirect, caused by changes in habitat or food quality. Thus, the effects of air pollution on biodiversity will depend upon the impacts of single and multiple gases and particulates on individual plants or animals and their habitats. It will be important to understand which species have detoxification mechanisms.

A third consideration is that the effects of air pollutants on organisms also depend upon the environmental conditions. For example, ozone (O_3) toxicity from free radical oxidation of membrane sulfhydryl groups and the double bonds of unsaturated fatty acids is reduced at low O_3 concentrations (< 200 ppb) by natural antioxidants, but not at higher concentrations. Unsaturated hydrocarbons (e.g., ethylene) may add to O_3 damage. So, under drought conditions when ethylene is produced, the interactive effects of O_3 and other environmental stresses may exacerbate air pollution toxicity. However, the soil moisture-ozone interactions with individual species have been inconsistent, with some cases showing no change in O_3 sensitivity due to water stress while other experiments demonstrated reduced O_3 sensitivity under stress, or additive or synergistic effects (see Colls and Unsworth, Chapter 6, this volume). Thus, in the case of ozone, more investigations are required to determine the conditions under which each of these sensitivity cases can be expected. With respect to biodiversity, it is well known that various plant species have quite different tolerances to drought. Therefore, the impact of air pollution on biodiversity will depend on a much greater understanding about the relationships of drought tolerance, ethylene production, and O_3 toxicity. It will be a powerful diagnostic tool if these mechanisms can be categorized according to genetic or taxonomic groups of organisms.

A fourth consideration in the investigation of air pollutants and biodiversity is the way in which air pollutants affect the competitive ability of the organisms. Increased enzyme activity and the production of specific antioxidants may follow exposure to air pollutants. Since antioxidants and their effectiveness may vary with nutrient availability, seasonal status and other environmental conditions, the ultimate prediction of air pollutant effects on biodiversity depends on both the biochemical characteristics of the species and the prevailing environmental conditions. Moreover, the production of antioxidants places an extra demand on the production of chemical reduction reactions such as the production of ATP and NADPH. These higher energy demands may mean that the plant grows more slowly and is less competitive during periods of exposure to pollutants (Pitelka 1988; see Wolfenden et al., Chapter 5, this volume). So, while the regulation of alternative biochemical pathways for antioxidants may be under relatively simple genetic control, the interactions with the prevailing environmental conditions make it difficult to predict the success of species even though the genetics of detoxification mechanisms may be known. These same principles may also determine whether the exposed organism is able to withstand predation and disease.

A fifth dimension has to do with the exposure regime of air pollutants. The movement of airborne chemicals downwind and their direct deleterious effects on plants and animals have been recognized since early in the century. It is now clear that chemicals move by long-range transport mechanism over regional and even global scales (Eriksson et al. 1989). Thus, the potential effects of air pollutants on biodiversity cannot be evaluated by examining obvious localized conditions, but rather must take into account regional atmospheric conditions, prevailing weather conditions and chemical transformations that occur in the atmosphere.

Measuring the effect of air pollution is not a simple matter. As just discussed, consideration must be given to the mechanisms of exposure, the biochemical and physiological mechanisms of detoxification, the interaction between the exposure regime and the environment, the energetic costs of resisting pollutants and, finally, the local, regional, and even global patterns of pollutant dispersions. Thus, there are complexities in measuring biodiversity, and measurement of the effects of air pollutants on organisms must consider several mechanisms. The following section proposes a general approach for providing some order to this complicated task of measuring the impacts of air pollution on biodiversity.

APPROACHES TO MEASURING THE IMPACT OF AIR POLLUTANTS ON BIODIVERSITY

Investigations on the effects of air pollutants on biological systems have involved two basic approaches, namely, experimentally manipulating air pollutant exposures and using air pollution gradients at increasing distances from a source. In either case, measurements can be made on the pathway of exposure and then the responses as various measures of biodiversity, for example, at the gene, species, and/or ecosystem level (Figure 14-1). Although this strategy appears to be straightforward, there are two major challenges. First, a decision must be made as to how to measure biodiversity at the biological level of interest. Second, it is necessary to extrapolate these results to local, regional and even

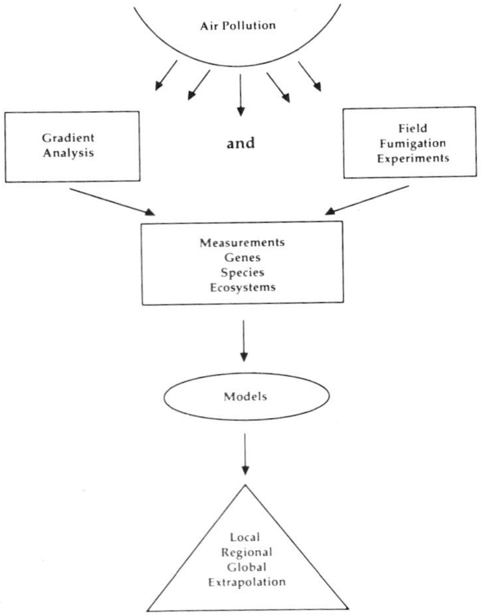

FIGURE 14-1. Approach for extrapolating localized experimental results to larger spatial scales.

global spatial scales if the relationship between air pollution and biodiversity is to be understood. Extrapolation from the individual measurements to larger spatial scales requires a model. As will be discussed below, Grime's (1979) plant strategy type represents one type of model. A future research need, however, is to develop and validate additional models.

There are two different conceptual approaches for extrapolating the effects of air pollution on genetic characteristics and individual species to much broader spatial scales. One strategy involves identifying resistance or susceptibility mechanisms, determining the distribution of these morphological or biochemical mechanisms in different types of organisms, and then predicting the consequences of air pollution on the recipient species and ecosystems (Figure 14-2). In contrast to this mechanistic approach, it should be possible to empirically identify guilds of species (or genotypes) that have similar responses to air pollution, and then use the distribution of these guilds to predict the consequences on biodiversity at different spatial scales (Nash and Wirth 1989). Part of these two approaches can be described from the current information base on air pollution and biodiversity. This approach would also seem plausible for extrapolating to the ecosystem level because a considerable amount is known about ecosystem behavior. The prospects for extrapolating genetic-level characteristics are not so clear. A large and comprehensive project must be established in the future to test the feasibility and effectiveness of these two basic alternatives for measuring the impact of air pollution on biodiversity.

MEASURING THE EFFECTS OF AIR POLLUTION

As discussed previously, measuring the effects of air pollutants on biodiversity requires the selection of one or more genetic, species, community, ecosystem and/or landscape

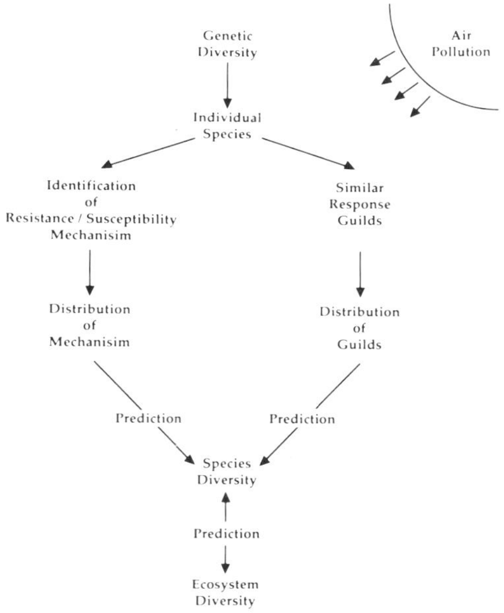

FIGURE 14-2. Two pathways for relating genetic and species diversity of ecosystem diversity.

response property. Since there is an almost infinite number of potential properties, selection criteria need to be established for choosing the most useful and information-rich responses. Among the criteria that might be invoked for measuring responses to air pollution and that measure or indicate effects on biodiversity are: parameters sensitive to air pollutants (e.g., visual symptoms), but that indicate subsequent differential survival, reproduction, or production capabilities of the organism; parameters that provide the earliest and most unequivocal indications of the health of the biological system (e.g., productivity or changes in species/ecosystem condition); and characteristics that are themselves important or that mimic these important biological characteristics (e.g., reproductive rates of individual species or nutrient cycling rates of ecosystems). These selected field or laboratory measurements may focus on the presence, abundance, location, and vigor of species or genotypes, the interactions among species or genotypes, or the fluxes and exchanges of energy, water, and nutrients within and among biological communities and portions of the landscape (see Musselman et al., Chapter 4, this volume). The difference between these measurements and those conventionally used in air pollution dose-response studies, is that these selected measurements must be related to the survival of genetic, species, or ecosystem diversity. In other words, the selected parameters must have some predictive value in anticipating the effects of air pollution on biodiversity.

Decisions about measurement parameters to indicate susceptibility to air pollution must consider the types of effects, especially in the context of biodiversity. For example, first-order or primary effects are the direct effects of air pollutants on plants or animals (e.g., phytotoxicity of O_3 and SO_2 but also positive effects such as NO_x and fertilization). Second-order or indirect effects include the onset of premature senescence from O_3 injury, delayed frost hardening caused by late-season fertilization from atmospherically deposited nitrogen, or changes in terrestrial or aquatic animal habitat caused by changes in vegetation. Third-order or tertiary effects include such processes as increased frequency or intensity of fire caused by changes in the conditions of the vegetation from air pollutants, increased insect populations and disease occurrence from altered habitats or species interactions, and changes in biogeochemical cycles or successional status caused by air pollutants (see Musselman et al., Chapter 4, this volume).

Considerable information exists from which to make the initial decisions. For example, plants with high stomatal conductances are generally more vulnerable to SO_2 or O_3 pollution than those with low conductances. From the context of biodiversity, it is also important to recognize that there are interclonal differences in this sensitivity, e.g., clones of *Populus tremuloides* that were found to be more tolerant to SO_2 had lower stomatal conductances (Kimmerer and Kozlowski 1981). Similarly, Steubing et al. (1989), studying the effects of SO_2, NO_2, and O_3 on the herb layer in a German beech (*Fagus*) forest, found interspecific variations in responses to the pollutants. These relative sensitivities could be related to stomatal responses of individual species, to life forms, and in some cases, tentatively to carbon assimilation rates. So, stomatal conductances represent a potential framework for predicting the responses of plant species, genotypes, or life forms to air pollutants—and thus, to predict the consequences to biodiversity. However, in many situations these comparative sensitivity responses are influenced by water availability and other environmental conditions. In addition, competitive relationships are not always straightforward (Marks and Strain 1989) so that simple categorizing of different populations as to tolerance and subsequent success in the community will require additional study.

It may be possible to identify other markers of sensitivity to pollutants and to relate these markers to environmental and competitive conditions. The most feasible approach for enhancing the predictive framework is to examine growth and reproduction responses to pollution over a broad array of environmental conditions. The design of these studies must also include consideration of competitive relationships among the target organisms. This ambitious empirical approach will require considerable effort, but it can be made much more efficient by using our current knowledge about indicators such as stomatal conductance and by incorporating our understanding of organism responses to environmental controlling variables and competitive interactions.

There is an enormous range of potential variables to be measured for detection of the influences of air pollutants on biodiversity. To help answer major research questions, a refined template should be developed for making these decisions about what to measure. Such a template would systematically evaluate air pollution responses against direct or indirect measures of biodiversity. Since there are too many plants, animals, and microorganisms to be tested, it will be necessary to search for common characteristics, for example, detoxification pathways, morphological and physiological characteristics that confer tolerance, and taxonomic or ecological groupings of organisms that demonstrate resistance or susceptibility to air pollutants.

As an example of a system that might contribute to a template for relating air pollution to biodiversity, consider the U.S. Forest Service Forest Health Monitoring Program, which has three levels of monitoring (see Musselman et al., Chapter 4, this volume). The three layers of observation are as follows:

a. Detection Monitoring—recording the condition of forest ecosystems, estimating baseline conditions, and detecting changes from those baselines over time. This level of observation would be useful for identifying taxonomic or ecologic groupings of species that demonstrate common responses to air pollutants. Moreover, since these are field measurements, there are opportunities to incorporate consideration of interacting environmental conditions, competitive relationships, and susceptibility to disease, herbivory, and predation. The results of these field measurements can be incorporated into measures of biodiversity.

b. Evaluation Monitoring—determining the causes of detected changes, if possible, or else hypothesizing causes that can be tested experimentally. Under this level, common physiological and morphological adaptations can be identified, and once again, related to field measures of biodiversity.

c. Ecosystem Monitoring—providing very high-quality, detailed information for rigorous assessment of cause/effect relationships on a small set of intensive sites from representative ecosystems. This would be the most intensive level where detailed studies can be made of genetic systems, biochemical pathways, and other mechanisms that confer tolerance or susceptibility to air pollution. Throughout these analyses, the purpose would be to relate these response mechanisms to behavior in the field that determines biodiversity.

This approach appropriately recognizes the multiple levels of measurement that are required for detection of the influences of air pollution on biodiversity. The ultimate power of this scheme will depend upon its use under a wide range of environmental conditions, because from these studies will come a sufficiently comprehensive set of measurements that prediction can be made for a variety of species and ecosystems.

Effects of Air Pollution on Species, Communities, and Ecosystems

At the genetic level, the evolution of resistance of any population to environmental stress depends on (a) the availability of genetically determined, plant-to-plant and animal-to-animal variation in response to stress, and (b) natural selection for resistance (see Newman, Schreiber, and Novakova, Chapter 10, this volume; Pitelka 1988; and Taylor and Pitelka, Chapter 7, this volume). Many plant and animal species have demonstrated considerable variability in resistance to air pollution, but it is not always possible to determine whether the resistance trait is a direct evolutionary response to the pollutant or is an indirect response to another environmental condition. Moreover, the costs of different mechanisms of resistance are unlikely to be identical, and these differential costs may influence the outcome of competitive interactions of the organisms in the field. The potential loss of genetic diversity due to air pollution is a serious concern (see Taylor and Pitelka, Chapter 7, this volume), but one that has not been evaluated in many situations. Genetic loss could occur from local extinctions, genetic drift, selection of resistant genotypes or loss of less resistant species in competitive conditions. There is some evidence for loss of genetic diversity from local extinctions and genetic drift as a consequence of air pollution. However, these shifts in gene pools are subtle and difficult to detect (Scholz, Gregorius, and Rudin 1989). On a broader scale, it is difficult to differentiate between the role of adaptations based on phenotypic plasticity from the genetic basis of physiological and biochemical adaptations that are subject to selection. Thus, the major issue is how to identify those populations that are most likely to experience changes in their genetic structure as a result of exposure to air pollution. The task includes not only determining the genetic characteristics of the individual populations, but assessing the conditions of environmental variability as well as fragmentation of habitats and other constraints imposed by human activities (see Taylor and Pitelka, Chapter 7, this volume).

Much of the research to date on the direct effects of air pollutants that can be applied to the species level of biodiversity has been accomplished on crop plants and a few species of forest trees (Hutchinson and Meema 1987; Schultze 1989). As a result, extrapolation to the concept of biodiversity is difficult because the experimental results are limited to a relatively narrow spectrum of plants rather than the very broad array that is found in natural ecosystems.

One useful approach to providing some order to the task has been the use of the PHYTOTOX database, which consists of approximately 8,000 chemicals, 2,000 plant species, and 50 responses. A portion of the database includes the quantitative responses of plant species to specific chemicals (Fletcher, Johnson, and McFarlane 1988; Royce et al. 1984). Nellessen and Fletcher (1991) selected native species in Illinois and evaluated the possible effects of two herbicides, trifluralin and alachlor. These two chemicals were selected because they are commonly used on corn and soybeans, and because they volatilize from the soil surface. The chemicals are released by volatilization and can drift downwind to nontarget areas. Central Illinois (U.S.A.) is dominated by croplands with interspersed oak-hickory (*Quercus-Carya*) woodlots. Therefore, it is reasonable to evaluate whether the chemicals applied to the agricultural land can inadvertently subject nontarget organisms to toxic airborne chemicals (Marrs et al. 1989). The database is relatively depauperate for native species, but it was possible to consider 18 native species that occur in this habitat. When the potential concentrations of trifluralin drift were calculated, representatives from eight genera were projected to be potentially affected based on the dose-response data in PHYTOTOX (*Acer, Cassia, Dioscorea, Ilex, Rhododendron, Solanum, Thuja, and Urtica*). Alachlor is less volatile, but still had the potential of affecting 6 of the 18 species. Even though the database contains relatively few native species, and most of the dose-response data are of herbicides, this approach has a potential as a beginning process for evaluating the potential effects of air pollutants on the biodiversity of both localized and regional areas.

Since databases such as PHYTOTOX can never be expected to include all biological species, it will be necessary to develop ways of partitioning species into those more or less likely to be affected by air pollutants. The conceptual framework of Grime (1979) provides an opportunity to organize the biochemical and cellular processes of air pollution responses into three categories of plants: ruderal, competitor, and stress-tolerant species (see Wolfenden et al., Chapter 5, this volume). These are:

a. Ruderal species, which exploit conditions of low stress and high disturbance, are adapted to regimes where disturbance is high but other stress factors are low;
b. Competitor species exploit conditions of low stress and low disturbance; and
c. Stress-tolerant species exploit conditions of high stress and low disturbance.

Many crop plants have developed from ruderal species and are annual or biennial, often with mesomorphic leaves, fast completion of their life cycle, high rates of dry matter production, abundant seed crops, and low resilience to adverse environmental conditions (see Wolfenden et al., Chapter 5, this volume). On the other hand, stress-tolerators tend to be long-lived perennials, are frequently evergreens with leaves possessing xerophytic characters, and their physiological activities display a resilience to adverse climatic factors.

As Wolfenden et al. discuss in this volume, short-term, acute exposures to SO_2 and O_3 might have a greater effect on a ruderal species than on a stress-tolerant species, because the stomatal conductance of ruderals tends to be higher, whereas the xeromorphic characteristics of stress-tolerators tend to confer low stomatal conductance, which conserves water and reduces the uptake of gaseous pollutants. However, long-term exposures may increase effects on stress-tolerators because their leaves remain metabolically functional whenever growth conditions are favorable. Ruderal species tend to respond to stress by switching from the vegetative to reproductive state. This life-cycle acceleration may reduce the number of seeds, but the longer term effect on the subsequent generation is minimized.

The responses of competitor species are more difficult to predict because they acquire resources relatively rapidly, with high stomatal conductances under favorable conditions but with the capability to switch to lower conductances when the circumstances change. These classifications of plant strategies are promising as a mechanism for extrapolating results of air pollution studies to the broader context of biodiversity. However, ruderal species are the most comprehensively studied because many crop plants belong to this category; there are fewer studies of competitors or stress-tolerator species. Moreover, there has not yet been a comprehensive test of these plant strategies as a template for relating biochemical, physiological, and morphological responses to air pollutants to broader-scale biodiversity measures.

Airborne pollutants can impact agricultural and natural communities by reducing the diversity in plant and animal communities, and by altering successional patterns (Weinstein and Birk 1989). These effects occur through individual plants and animals to the community and ecosystem levels of organization. As noted above, some species will be more susceptible or tolerant than others, depending on individual genotypes, and on the environmental conditions and physiological status of the organism. The sensitive species will no longer be able to compete with the more tolerant ones and will be replaced or reduced in importance. The effects may be more subtle; that is, decreased vigor may render the plant or animal more susceptible to insect damage or disease, or may reduce its reproductive potential.

Among the most studied plant communities are the southern California chaparral and coastal sage scrub communities that have been exposed to long-term elevated atmospheric oxidants. In this instance, the effects of air pollutants have been determined by phytosociological techniques rather than on the basis of physiological tolerances. Various woody species were found to demonstrate different levels of tolerance to the air pollutants, and as a result, the biodiversity of the region changed. Community changes in response to severe exposures to air pollutants, such as near smelters, are rather obvious (Gordon and Gorham 1963); however, measuring community responses to chronic, low-level exposures is much more difficult (see Armentano and Bennett, Chapter 9, this volume; Sigal and Suter 1987). Studies around the Sudbury, Ontario, smelter indicated that species diversity indices were less sensitive to smelter effects than the understory canopy cover or basal area of the trees; the number of species per area was also more sensitive than species diversity indices (Freedman and Hutchinson 1980). These results may be confused in part because of the differences in recovery time for the understory and overstory, but in general, the overstory consisting of longer lived organisms with larger respiration demands was a better indicator of pollution effects than the understory (Rapport, Regier, and Hutchinson 1985). Similarly, changes in species composition, species diversity, and density response patterns to air pollution are possible to describe. However, the successional status of an ecosystem also contributes to the observed responses (McClenahen 1978). In a successionally stable, mixed-grass prairie exposed to SO_2, Lauenroth and Preston (1984) showed a decrease in biomass but no effect on species diversity. The same study might have had a quite different effect on a community that was undergoing successional changes. Although it is possible to begin to describe some general trends in the relationship between air pollution and conventional measures of biodiversity (e.g., decreased diversity under the most severe conditions but increased diversity at intermediate levels of pollution), these research results make it clear that the responses are not yet entirely predictable. Our current understanding of air pollution and biodiversity is simply insufficient in many cases to differentiate between exposure responses and differences in the characteristics of the ecosystems themselves (see Armentano and Bennett, Chapter 9, this volume).

Ecosystem properties and the patterns of air pollution are variables of space and time. Because the relationships between air pollutants and ecosystem properties consist of two highly variable components, the relationships are difficult to measure (see Smith, Chapter 11, this volume). Nevertheless, considerable information is available about specific processes such as fixation of carbon (i.e., photosynthesis) and how these processes are affected by air pollution. Thresholds of photosynthetic toxicity depend on the individual species and its physiological status, the type of pollutant and exposure regime, and the prevailing environmental conditions. Air pollution during the growing season from O_3 in North America routinely reduces plant productivity (Reich 1987). Several studies have suggested that air pollutants tend to reduce the allocation of carbon to roots, which in turn may decrease the plant's ability to withstand drought conditions (McLaughlin et al. 1982; Waring and Schlesinger 1985). Although high concentrations of trace metals may suppress soil decomposition rates (Smith 1990), except for acid deposition, the effects of air pollution on nutrient cycling rates in most ecosystems is not known. Thus, little is known about the effects of air pollutants on biogeochemical cycles, despite the importance of nutrient cycles in determining the occurrence and maintenance of biodiversity (Noble, Martin, and Jensen 1989; Waring and Schlesinger 1985).

SUMMARY

This chapter has described the research needs for defining the most information-rich measures of biodiversity; addressing certain important questions concerning the biochemical and physiological processes that determine the effects of air pollution on biological systems; and examining the relationships between air pollution and genotypic, species, community, and ecosystem characteristics of biodiversity. A summary of these research priorities follows:

- Measurement of the loss of genetic diversity in relation to measured loss of species diversity, particularly under conditions of environmental variability and habitat fragmentation.
- Development of an efficient framework to test the sensitivities of organisms to air pollutants, and translating these results into the consequences for biodiversity at the genetic, species, and ecosystem levels of biological organization.
- Definition of optimum sampling strategies for measuring the effects of air pollution on the various scales of biodiversity.
- Identification of the same or different indices of biodiversity for measuring the impact of air pollution on genetic, species, and ecosystem biodiversity.
- Determination of whether losses of genetic and species diversity from air pollution affect the behavior of ecosystems, both under current conditions and as these

ecosystems might respond to changing climate or alternative land uses.

- Determination of the consequences of single and multiple pollutants on plants and animals, especially as these relationships are affected by environmental conditions as well as biotic influences such as disease, herbivory, and predation.
- Development of a classification of groups of genotypes, species, and ecosystems that demonstrate similar responses to air pollutants.
- Development of an optimum sampling design for testing long-term, subtle effects of air pollutants on biodiversity, especially over large geographical areas and over changing climatic conditions.

In addition, Moser, Barker, and Tingey (1991) have proposed useful research priorities:

- Identify and prioritize the most critical airborne contaminants and sensitive ecosystems. A comprehensive computer-based system would be useful for conducting preliminary risk assessments of the numerous airborne toxic chemicals and their effects on vegetation, and would provide research guidance.
- Quantify and model the exposure, deposition velocity, and absorption of air toxins to plants.
- Determine the biochemical and physiological responses of plants to chronic exposures to air toxic chemicals and develop exposure-response functions. The research then could be extended to quantify the response of plant populations and simulated plant communities.
- Initiate long-term studies to determine sensitive elements of plant community structure and function that would lead to significant change and degradation from air toxic exposure. This research would identify unacceptable change in plant communities and identify early warning signals.

Many other specific research questions still exist and are mentioned in the various chapters of this book. For example, a more complete understanding is needed of the relationships between topography and the air pollutant-scavenging ability of the vegetation, of how to predict air pollutant uptake from morphological characteristics of leaves, and of how to identify additional biological markers for air pollutants in various types of organisms.

Although there is a considerable amount of information on air pollution and its effects on organisms and biological systems, much of the information is fragmented. Indeed, the research needs described in this chapter are only the beginning in developing a policy framework from which the public can make decisions that involve the relationships between air pollution and biodiversity (see Loucks, Chapter 12, this volume; Winner, Mooney, and Goldstein 1985). To develop specific policy measures, it will be necessary to define clearly the cost and benefits of the interrelationships between air pollution and biodiversity, and to evaluate the recognized societal value of maintaining biological diversity.

REFERENCES

1. Cribben, L.D., and D.D. Sacchetti. 1977. Diversity in tree species in southeastern Ohio *Betula nigra* communities. *Water Air Soil Pollut.* 8:47-55.
2. Eriksson, G., S. Jensen, H. Kylin, and W. Strachan. 1989. The pine needle as a monitor of atmospheric pollution. *Nature* 341:42-44.

3. Fletcher, J.S., F.L. Johnson, and J.C. McFarlane. 1988. Database assessment of phytotoxicity data published on terrestrial vascular plants. *Environ. Toxicol. Chem.* 7:615-22.
4. Freedman, B., and T.C. Hutchinson. 1980. Long-term effects of smelter pollution at Sudbury, Ontario on forest community composition. *Can. J. Bot.* 58:2123-40.
5. Gordon, A.G., and E. Gorham. 1963. Ecological aspects of air pollution from an iron-sintering plant at Wawa, Ontario. *Can. J. Bot.* 41:1063-78.
6. Grime, P.J. 1979. *Plant strategies and vegetation processes.* Chichester, England: John Wiley & Sons.
7. Hutchinson, T.C., and K. Meema, eds. 1987. *Effects of air pollutants on forests, agriculture and wetlands.* New York: Springer-Verlag.
8. Karr, J.R. 1990. Biological integrity and the goal of environmental legislation: Lessons for conservation biology. *Conserv. Biol.* 4:244-50.
9. Kimmerer, T.W., and T.T. Kozlowski. 1981. Stomatal conductance and sulfur uptake in five clones of *Populus tremuloides* exposed to sulfur dioxide. *Plant Physiol.* 67:990-95.
10. Larsson, R.A. 1988. The antioxidants of higher plants. *Phytochem.* 27:969-78.
11. Lauenroth, W.K., and E.M. Preston. 1984. The effects of SO_2 on a grassland. A case study in the northern great plains of the United States. Ecological Series 45. New York: Springer-Verlag.
12. Mansfield, T.A., and P.H. Freer-Smith. 1981. Effects of urban air pollution on plant growth. *Biol. Rev.* 56:343-68.
13. Marks, S., and B.R. Strain. 1989. Effects of drought and CO_2 enrichment on competition between two old-field perennials. *New Phytol.* 111:181-86.
14. Marrs, R.H., C.T. Williams, A.J. Frost, and R.A. Plant. 1989. Assessment of the effects of herbicide spray drift on a range of plants species of conservation interest. *Environ. Pollut.* 59:71.
15. McClenahen, J.R. 1978. Community changes in a deciduous forest exposed to air pollution. *Can. J. For. Res.* 8:432-38.
16. McLaughlin, S.B., R.K. McConthay, D. Duvick, and L.K. Mann. 1982. Effects of chronic air pollution stress on photosynthesis, carbon allocation, and growth of white pine trees. *For. Sci.* 28:60-70.
17. Moser, T.J., J.R. Barker, and D.T. Tingey, eds. 1991. *Ecological exposure and effects of airborne toxic chemicals: An overview.* EPA Report No. 600/3-91/001.
18. Nash, T.H., III, and V. Wirth. 1989. *Lichens, bryophytes and air quality.* Berlin and Stuttgart: Cramer.
19. Nellessen, J.E., and J.S. Fletcher. 1991. Use of PHYTOTOX database to estimate the influence of herbicide drift on natural habitats in agroecosystems. In *Ecological exposure and effects of airborne toxic chemicals: An overview,* ed. T.J. Moser, J.R. Barker, and D.T. Tingey, pp. 102-6. EPA Report No. 600/3-91/001.
20. Noble, R.D., J.L. Martin, and K.F. Jensen. 1989. Air pollution effects on vegetation, including forest ecosystems. In *Proceedings of the Second U.S.-U.S.S.R. Symposium.* Broomall, PA: U.S. Department of Agriculture, Forest Service, Northeastern Forest Experiment Station.
21. Noss, R.F. 1990. Indicators for monitoring biodiversity: A hierarchical model. *Conserv. Biol.* 4:355-64.
22. Peet, R.K. 1974. The measurement of species diversity. *Annu. Rev. Syst. Ecol.* 5:285-307.
23. Pitelka, L.F. 1988. Evolutionary responses of plants to anthropogenic pollutants. *Trends Ecol. & Evol.* 3:233-36.
24. Rapport, D.J., H.A. Regier, and T.C. Hutchinson. 1985. Ecosystem behavior under stress. *Am. Nat.* 125:617-40.
25. Reich, P.B. 1987. Quantifying plant response to ozone: A unifying theory. *Tree Physiol.* 3:63-91.

26. Royce, C.L., J.S. Fletcher, P.G. Risser, J.C. McFarlane, and F.E. Benenati. 1984. PHYTOTOX: A database dealing with the effect of organic chemicals on terrestrial vascular plants. *J. Chem. Inf. Comput. Sci.* 24:7.
27. Scholz, F., H.R. Gregorius, and D. Rudin, eds. 1989. Genetic effects of air pollutants in forest tree populations. Berlin: Springer-Verlag.
28. Schultze, E.D. 1989. Air pollution and forest decline in a spruce (*Picea abies*) forest. *Science* 244:776-83.
29. Sigal, L.L., and G.W. Suter II. 1987. Evaluation of methods for determining adverse impacts of air pollution on terrestrial ecosystems. *Environ. Manage.* 11:675-94.
30. Smith, W.H. 1990. The atmosphere and the rhizosphere: Linkages with potential significance for forest tree health. In *Mechanisms of forest response to acidic deposition*, ed. A.A. Lucier and S.G. Haines, pp. 188-241. New York: Springer-Verlag.
31. Steubing, L., A. Fangmeier, R. Both, and M. Frankenfeld. 1989. Effects of SO_2, NO_2, and O_3 on population development and morphological and physiological parameters of native herb layer species in a beech forest. *Environ. Pollut.* 58:281-302.
32. Tingey, D.T., R.A. Reinart, V.A. Dunning, and W.W. Heck. 1971. Vegetation injury from the interaction of nitrogen dioxide and sulphur dioxide. *Phytopathol.* 61:1506-11.
33. Waring, R.H., and W.H. Schlesinger. 1985. *Forest ecosystems: Concepts and management.* Orlando, FL: Academic Press.
34. Weins, J.A. 1989. Spatial scaling in ecology. *Funct. Ecol.* 3:385-97.
35. Weinstein, D.A., and E.M. Birk. 1989. The effects of chemicals on the structure of terrestrial ecosystems: Mechanisms and patterns of change. In *Ecotoxicology: Problems and approaches,* ed. S.A. Levin, M.A. Harwell, J.R. Kelly, and K.D. Kimball, pp. 181-212. New York: Springer-Verlag.
36. Wellburn, A.R., C. Higginson, D. Robinson, and C. Walmsley. 1981. Biochemical explanations of more than additive inhibitory low atmospheric levels of SO_2 and NO_2 upon plants. *New Phytol.* 88:223-37.
37. Winner, W.E., and J.D. Bewley. 1978. Contrasts between bryophyte and vascular plant synecological responses in an SO_2-stressed white spruce association in Central Alberta. *Oecologia* (Berlin) 33:311-25.
38. Winner, W.E., H.A. Mooney, and R.A. Goldstein, eds. 1985. *Sulphur dioxide and vegetation: Physiology, ecology and policy issues.* Stanford, CA: Stanford Univ. Press.

Index